Second Edition

GET READY FOR ORGANIC CHEMISTRY

JOEL KARTY

Elon University

PEARSON

Boston Columbus Indianapolis New York San Francisco Upper Saddle River
Amsterdam Cape Town Dubai London Madrid Milan Munich Paris Montréal Toronto
Delhi Mexico City São Paulo Sydney Hong Kong Seoul Singapore Taipei Tokyo

Editor in Chief: Adam Jaworski
Executive Editor: Jeanne Zalesky
Associate Editor: Jessica Neumann
Marketing Manager: Scott Dustan
Editorial Assistant: Lisa Tarabokjia
Marketing Assistant: Nicola Houston
Managing Editor, Chemistry and Geosciences: Gina M. Cheselka
Project Manager, Production: Maureen Pancza
Composition/Full Service: Element Thomson North America
Project Manager, Full Service: Kelly Keeler
Senior Art Specialist: Connie Long
Art Project Manager: Ronda Whitson
Text Permissions Manager: Beth Wollar
Text Permissions Researcher: Jenny Bevington
Design Director: Jayne Conte
Cover Designer: Suzanne Behnke
Senior Manufacturing and Operations Manager: Nick Sklitsis
Operations Supervisor: Maura Zaldivar
Cover Photo Credit: Shutterstock

Credits and acknowledgments borrowed from other sources and reproduced, with permission, in this textbook appear on the appropriate page within the text.

Library of Congress Cataloging-in-Publication Data
Karty, Joel.
Get ready for organic chemistry / by Joel Karty.—2nd ed.
p. cm. Includes index.
ISBN 978-0-321-77412-5
1. Chemistry, Organic—Textbooks. I. Title.
QD253.2.K365 2012
547—dc23 2011031151

1 2 3 4 5 6 7 8 9 10—EB—15 14 13 12 11

PEARSON

www.pearsonhighered.com

ISBN-10: 0-321-77412-4
ISBN-13: 978-0-321-77412-5

Contents

Preface

Welcome to organic chemistry, the branch of chemistry that underlies all biological processes at the molecular level and guides us in the synthesis of important compounds and materials, including antibiotics, cancer therapeutics, plastics, and even materials used to make light emitting diodes (LEDs)—components that are incorporated into the screens of electronic devices such as televisions, computers, and phones. Because of these far-reaching implications, your upcoming organic chemistry course has so much to offer—and could be among the most valuable courses you ever take. But the sad truth is that many (dare I say *most*?) students in the course never come to fully realize this. Either they end up struggling to the point that their only concern is passing or, if they do find themselves doing reasonably well, they still finish the course seeing organic chemistry as nothing more than a bunch of "stuff" to know for the exams.

Why does this happen? Almost inevitably it has to do with a tendency to memorize as much as possible. But the fact of the matter is that you can't memorize it all! And you shouldn't try to. If you do, then you will quickly find yourself overwhelmed, and nothing will seem to make sense. The good news, though, is that there are relatively few basic concepts that drive organic chemistry, and, if you can *understand* them and *apply* them in various ways, organic chemistry can become crystal clear. Moreover, you will be able to complete your year of organic chemistry with an appreciation for everything the course has to offer.

The question becomes, what are those concepts? That's where this book comes in. Throughout this book, I will take you through some of the most important organic chemistry concepts and show you how they are applied in various scenarios. If you truly work hard to understand these concepts and their applications, you will be ready to tackle your yearlong course.

That brings up one more question—the question of when to read and work through this book. Without a doubt, it is best to complete it prior to or shortly after the start of your first term of organic chemistry. It is vitally important that you start your yearlong course on the right track by maintaining a focus on learning, understanding, and applying basic concepts. The reason is that, if you start on the track of memorization, you will find it very difficult to switch tracks, and the longer you go, the more difficult switching tracks will become—you will have increasing amounts of material to catch up on as your class continues to push forward on new material.

That said, it is not impossible to switch tracks from memorizing to focusing on basic concepts. I have heard numerous testimonials where students learned of this book midway through their term, read it, and were able to have incredible turnarounds. So this book truly is a case of "it's never too late."

Once you complete this book, hold on to it. Review it from time to time. The ideas that are introduced here are designed to give you a strong start to organic chemistry, but the concepts are relevant to material throughout the entire year.

Although I have spent a good deal of time talking about the importance of understanding the basic concepts that drive organic chemistry, **Chapter 1** revisits the topic of memorizing versus learning and understanding, but discusses it in greater depth. The aim of the chapter is to convince you, once and for all, that learning and understanding is the way to go and, in order to do so, goes through some examples. The chapter also touches on study habits and addresses some elements of the organization of this book compared to that of a typical organic chemistry textbook.

Chapter 2 covers Lewis dot structures, covalent bonds, formal charges, and resonance. You likely spent a good deal of time on these topics in your general chemistry course, and you might even feel quite comfortable with them. But in organic chemistry, that's not good enough—you need to *master* them. To that end, the chapter shows you how to draw Lewis dot structures quickly and accurately and also teaches you how to draw all resonance structures of a molecule.

In **Chapter 3,** we discuss the three-dimensional geometries of molecules and how they play roles in molecular polarity and intermolecular interactions. As will be seen in this chapter, these concepts are vital to aspects of physical properties such as boiling point, melting point, and solubility, as well as to the existence of *cis/trans* isomers. Later in the book, you will also see how intermolecular interactions help drive reactions.

Chapter 4 deals with isomerism. In your general chemistry class, you might have touched on the idea of constitutional isomers, but this chapter takes the topic much further. It introduces other types of isomers, including *enantiomers* and *diastereomers*, as well as the notion of *chirality*, or handedness, of molecules. Further, the chapter teaches you conclusions to reach about physical and chemical properties of molecules, depending on their specific isomeric relationship, and it also shows you how to draw all constitutional isomers and all stereoisomers of a molecule.

Chapter 5 is the first of two chapters on *reaction mechanisms*—the roadmaps that show how reactions occur in a step-by-step manner. Reaction mechanisms are among the most important things to focus on in organic chemistry—ignoring them spells disaster for most students. Although mechanisms are generally comprised of multiple steps, this chapter focuses on the 10 individual steps that are most common (and therefore most important) to organic chemistry. Beyond just teaching you to recognize certain types of steps, this chapter provides insight into *how* and *why* such steps occur.

In **Chapter 6**, we delve into the notion of *charge stability*, one of the most important driving forces for reactions. You will learn how to determine the relative stabilities of molecular structures based on their charges and the atoms and types of bonds involved with those charges. These ideas are applied to determining the relative strengths of acids, bases, and nucleophiles (types of reactants you will also learn about in this chapter).

Chapter 7 is the second chapter on reaction mechanisms. Whereas Chapter 5 focuses on the details of individual reaction steps, this chapter focuses on reaction mechanisms consisting of multiple steps. It begins by teaching you how to read and write multistep mechanisms and continues by discussing various rules of thumb that will help you distinguish reasonable mechanisms from unreasonable ones. It ends with an application dealing with relatively lengthy mechanisms.

Chapter 8 deals with one of the first classes of reactions you will encounter in your yearlong course—nucleophilic substitution and elimination. One of the reasons for including this chapter is that it requires you to apply numerous concepts introduced in previous chapters, reinforcing the importance of maintaining your focus on concepts at all times. In addition, nucleophilic substitution and elimination reactions are among the most challenging reaction classes you will face in your entire year of organic chemistry, so this is where many students hit the proverbial brick wall in their traditional course. The aim of this chapter is therefore to provide you with the tools that will allow you to break through that wall.

Finally, **Chapter 9** reviews some of the basic strategies discussed in previous chapters. It also gives some advice as to how to keep your focus on the right things—and ultimately conquer organic chemistry.

Some special features in each chapter include the following:

- **Your Starting Point:** Tests your grasp of the chapter content before you start. Answers are provided for all of these.
- **Quick Check:** Asks you to recall or apply what you have just read in order to keep your eyes from scanning the page while your brain is on holiday. The answer is provided on the same page.
- **Picture This:** Asks you to visualize scenarios and then answer questions about them in order to help you better understand the topics.
- **Time to Try:** Is a simple experiment or quick assessment in which you perform an active exercise.
- **Why Should I Care?:** Highlights the relevance of the material so you understand its importance in the big picture.
- **Reality Check:** Has you connect what you learn to the real world.
- **Look Out:** Highlights the possible pitfalls or challenges to a student. These paragraphs focus on areas where a novice practitioner may not understand the consequences of a particular action or inaction.
- **Keys:** Highlight the main themes for reinforcement and easy review. These can be spotted by the symbol.
- **What Did You Learn?:** Asks you to apply material from throughout the chapter to answer these end-of-chapter problems.

Acknowledgments

I must begin by thanking Maureen Cullins, director of the Multicultural Resource Center at the Duke University School of Medicine. Nine years ago, she hired me to teach a preparatory organic chemistry course as part of the Summer Medical and Dental Education Program (SMDEP)—a program joined by students from several dozen colleges and universities around the country. It was in my first summer teaching that course that I learned how critical it is for students to be prepared for organic chemistry—with the mindset of understanding and applying as opposed to memorizing—before the first day of their traditional year-long course. My students from that class went on to have tremendous success in organic chemistry at their home institutions. It was also in my first summer there that I learned that there were essentially no resources out there to help students prepare. So I decided to write one.

My thanks also go out to Jim Smith, acquisitions editor at Benjamin Cummings, who signed me to write *The Nuts and Bolts of Organic Chemistry: A Student's Guide to Success* (the first edition of this book). The book did not fit the traditional model of an organic chemistry supplement, so it was a bit of a risk for him. But fortunately Jim shared my vision of having a significant positive impact on education in the sciences at the college level. Many thousands of students are glad that he did.

I would also like to thank Nicole Folchetti and Jessica Neumann, both at Pearson Education. Nicole, who at the time was the editor-in-chief of the chemistry titles at Pearson, originally proposed converting *The Nuts and Bolts of Organic Chemistry* into this edition, *Get Ready for Organic Chemistry*. Jessica, current associate editor in chemistry at Pearson, has overseen this conversion, and it has not been the easiest of projects. I am grateful for her patience.

My students deserve a lot of thanks as well, particularly my SMDEP students. It is because of my SMDEP students nine years ago that I was able to design an effective curriculum that really helps students everywhere prepare for organic chemistry. And every summer since then my SMDEP students have allowed me to tweak and fine-tune that curriculum, ultimately leading to the development of this book. Truly, my experience with SMDEP has been a case where I learn more from my students than they learn from me.

Finally, I am grateful to my wife, Valerie, and my two sons, Joshua and Jacob. It is perhaps most accurate to say that they have *endured* my love of teaching organic chemistry and my passion for writing. Over the years, they have seen as much of the back of my head as they have the front. They have seen me as much with a laptop as they have without. And the notes they find around the house more often have organic structures written on them than anything else. Thank you.

1 Why Do Most Students Struggle with Organic Chemistry?

1.1 Introduction

The complete and utter impossibility of organic chemistry is a complete and utter misconception. So why is it, then, that organic chemistry is *the* class? It is *the* weeding-out class; it is *the* most difficult course required for premedical students and others preparing for the allied health profession; it is *the* class that is most heavily weighted by medical school admissions committees. From my experiences, both as a student and as a professor, I find that the answer is *not* that organic chemistry is intrinsically so difficult, but rather that organic chemistry is simply *different* from any class you have ever taken. Many students struggle because they fail to see that it is such a different course by its very nature, and they fail to see exactly how it is different. Therefore, the goals of this chapter are (1) to provide insight into how organic chemistry differs from other classes you have taken and may have excelled in; (2) to convince you that the seemingly obvious way to approach organic chemistry will, in actuality, be your downfall; (3) to provide you with the right mind-set for organic chemistry and to explain, to some extent, the right strategy for the course; and (4) to introduce some important features of this book, which will help you focus on that strategy.

1.2 How Organic Chemistry Is Different

Organic chemistry is different because it probably demands more analysis, critical thinking, and reasoning than any class you have previously taken. Although most students don't realize it, organic chemistry is much more a course in problem solving than anything else. In fact, it is for this reason that a friend of mine, who is a biology professor, believes that organic chemistry is the best, and the single most important, course that a biology major can take—not simply because of the specific applications it has in biology, but more importantly, because of the exposure it gives students to new ways of thinking. It is the ability to think critically that has application to biology—and elsewhere.

Medical schools take much the same view. The grade that you earn in organic chemistry will likely be singled out by any medical school you apply to, perhaps making this grade the single most important one on your transcript. Many students believe this is because medical schools use this grade to see how well you "perform" under pressure and stress. While there may be some truth to this, medical schools are much more interested in knowing how well you can think and how you apply what you already know to new situations. Of the course grades that typically appear on transcripts in their applicant pools, they believe that the organic chemistry grade provides them the most insight.[1] I believe they are correct.

The way the Medical College Admission Test (MCAT) is written reflects the importance of critical thinking and problem solving in organic chemistry. Just turn to the biological sciences portion of an MCAT exam book, and look at some of the organic questions—especially the ones pertaining to a passage. These questions may ask you why it is, based on the results of some experiment described in the passage, that Scientist A believes that the reaction occurs by Scheme I instead of Scheme II. Or you may find a question asking what the product of Reaction IV would be if one of the reagents was replaced with a similar reagent. These questions are not simply asking for regurgitation of information. In fact, these questions often cover material you are not expected to have seen before. They are asking you to *extend* the fundamental concepts you learn in organic chemistry to new situations and new problems. This is no accident; the American Association of Medical Colleges (AAMC)—the organization that composes and publishes the MCAT—specifically does not want the next generation of doctors to be good at only regurgitation. Its members maintain that the amount of knowledge out there is simply too great to memorize, and doctors are routinely called upon to extend the ideas that they do know to new problems and new situations.[2]

1.3 The Wrong Strategy

Before we discuss "the right strategy" for organic chemistry, we should first discuss the wrong strategy, in hopes of dispelling a common, but grave misconception. This misconception is that organic chemistry requires an inordinate amount of memorization. The reality is that it most certainly does *not*.

The more you rely on memorization in this course, the worse you will do!

Despite reality, the vast majority of organic chemistry students each year enter the course with the idea that memorization is the key—so much so that they quickly resort to flash cards, either homemade or store bought. Now, that's not to say that flash cards are necessarily evil or that they are not useful. Instead, what I have observed, time and again, is that flash cards are horribly misused. For a given reaction, students using flash cards have a tendency to memorize the reactants, the products, the necessary reaction conditions, and a few other important things specific to that reaction. As a result, their focus is removed from understanding *why* it is that the reaction forms the products it does as well as *why* it is that changing the reaction conditions only slightly may yield an entirely different product! Understanding why such things happen is far more useful.

[1]Brenda Armstrong, director of admissions, Duke University School of Medicine.

[2]Lois Colburn, assistant vice president, AAMC Division of Community and Minority Programs.

One reason it is so important to understand instead of memorize is the problem of short-term retention of the material:

Memorizing is "easy in, easy out."

If you simply memorize a reaction without understanding why the particular products are formed, or why those reagents are necessary to carry out that reaction, or what drives that reaction, then you will quickly forget whatever you memorized. There is no foundation. There is no context. Consequently, when it comes to your final exam for Organic I, which will likely be cumulative, you will have to memorize a second time. Organic II courses frequently use final exams that are cumulative over the entire year, so you may have to memorize a third time. Finally, it could be a year or longer between the end of your organic chemistry course and the MCAT (or other standardized test) exam date. You will have to memorize yet again. By contrast, if you understand a reaction to greater depth than just being able to regurgitate reactants and products, your retention of that reaction (and everything about it) will be much longer. Instead of having to memorize several times, you will need to learn only once.

A second reason it is important to understand material rather than memorize is that it *simplifies* your focus.

In organic chemistry, there are a handful of fundamental concepts that can explain a wide variety of chemical reactions and phenomena.

These reactions and phenomena are tightly connected, even though they might sometimes appear to be quite disparate. Therefore, if you rely heavily on memorization, each of the reactions and phenomena that you encounter throughout a year of organic chemistry (all told, perhaps hundreds) will remain independent, disconnected pieces of information. That represents *a lot* of effort, for little gained.

In thinking about keeping your focus on fundamental concepts when it comes to reactions, consider this. When two reactants are mixed together, do they rely on having memorized flash cards in order to determine whether and how they should react? Of course not! Any reaction that takes place is an outcome of those molecules abiding by certain laws of nature. Those laws of nature are what lead to the fundamental concepts that you must learn and apply.

A third reason to focus on understanding has to do with standardized exams.

The MCAT and similar exams seem to expect that the majority of students rely on memorization.

As I alluded to earlier, in order to make distinctions among students in a competitive and talented pool, questions on these exams are often designed to favor those who understand the few basic concepts and who can extend that knowledge. In other words, such questions are essentially designed to distinguish between those who relied on memorization and those who did not.

Memorizing is attractive. But if you want to succeed in this course, you must keep it to an absolute minimum. As a student, I found organic chemistry to be one of the most straightforward courses in college, and I never wrote or made use of any flash cards. As a professor, I observe that my top students, year in and year out, refrain from memorizing and using flash cards. By contrast, those students who tell me that they relied heavily on flash cards are the ones who found organic chemistry indescribably frustrating.

One of my favorite stories as a professor involves such a notion. One year a student taking my Organic Chemistry I course earned a C on the first hour exam, an F on the second, and a D on the third. Immediately after the third exam, she came to my office, quite frustrated and overwhelmed, proclaiming that she was going to drop the course; she couldn't stand any more frustration, even though we were only three weeks away from finishing the semester—one more exam plus the final exam. It took me only a few minutes to convince her not to drop, and to finish out the semester. To help solve the problem, I asked her how she studied. Sure enough, flash cards and memorization were among the first things she mentioned. I asked her: "Why the flash cards and memorization, even though I warned you, and even though I showed you all along how to use the right strategy?" I was quite impressed with her answer: "How do you expect us to undo what we've been trained to do for 18 years?" Although this would be a good end to the story, there's one more piece to it. I spent the next hour reviewing with her some of the basic strategies I had stressed all semester. After that meeting, I didn't see her again in my office. She earned a B+ on that fourth hour exam and, shortly thereafter, a B on the cumulative final exam. The following semester, in Organic II, this same student earned a 96 percent on one of the exams—the highest grade in the class and an incredible turnaround.

1.4 The Right Strategy

I've spoken quite a bit about the pitfalls that accompany memorization. It is clear to me, and to many whom I have taught, that memorizing your way through organic chemistry is the wrong strategy. So what is "the right strategy"?

The right strategy is to focus primarily on learning and understanding fundamental concepts that underlie organic chemistry, and to gain competency in applying them to each chemical reaction and phenomenon you encounter.

You might guess that this would yield results that are no better than memorization and would be more time-consuming. On the contrary, it will be easier, less frustrating, and a time saver in the long run. Instead of memorizing hundreds of reactions—their reactants, products, reaction conditions, and so on—you need learn and understand only the small number of fundamental concepts and be able to apply them to a large number of situations.

Let me elaborate on this using an example from geometry. (I choose geometry in part because it tends to elicit many of the same feelings that organic chemistry does.) Suppose that in class your professor covered interior angles of regular polygons (all sides and angles identical). Later, on the exam, you are faced with the following question: "What is the interior angle in a nine-sided regular polygon?" There are three ways you might imagine having studied for a question like this. First, you could have memorized the interior angles of as many regular polygons as possible—60° for a triangle, 90° for a square, 108° for a pentagon, 120° for a hexagon, and so on. This is a mistake because (a) there is a limit to the number of angles that you can memorize and you may not have been able to memorize the angle of the nine-sided polygon; and (b) even if you did memorize it before the exam, you run the risk of drawing a blank come exam time.

A second strategy is to memorize the equation that yields the interior angle of an n-sided regular polygon, as shown in Equation 1-1:

$$\theta_{\text{interior}} = 180° - (360°/n) \tag{1-1}$$

where θ_{interior} is the interior angle. The advantage is that memorizing this equation cuts down significantly on the amount of material that you must memorize. However, the problem with this strategy is that it is still memorization. Therefore, you still run the risk of drawing a blank on the exam, particularly because there are numerous other equations that you will have had to memorize while studying. Or perhaps you remember the equation's general form, but you don't remember the details, such as whether 180° is outside the parentheses and 360° is inside, or vice versa. Both such problems result from not having any context for the equation—it provides the correct answer, but you don't know why it works.

The best strategy to ensure success on a problem like this is to learn and understand simple fundamental concepts that are easy to grasp and that allow you to derive the answer quickly and in a way that makes sense. Because the concepts and the derivation make sense, the risks involved with memorization are eliminated. Furthermore, your retention of the material will be lengthened tremendously.

For our problem at hand, there are two fundamental concepts to apply: (1) There are 360° in a circle, and (2) there are 180° in a line. Simple enough? Imagine, then, driving a car along the perimeter of a regular polygon with n sides (Figure 1-1). You end up driving in a straight line until you encounter a vertex of the polygon, at which point you make your first turn. The angle by which you change direction will be an *exterior angle* of the polygon, θ_{exterior}. After completing the turn, you again drive in a straight line until you encounter the second vertex and make the second turn. Eventually, you end up at your original position and direction after having made n turns, at which point you will have turned a total of 360° (the total number of degrees in a circle). Therefore, each turn you made must have been $360°/n$. That is,

$$\theta_{\text{exterior}} = 360°/n \tag{1-2}$$

Finally, to calculate the interior angle, we notice that the exterior and interior angles sharing the same vertex define a line and therefore must sum to 180°. This is shown in Equation 1-3.

$$\theta_{\text{exterior}} + \theta_{\text{interior}} = 180° \tag{1-3}$$

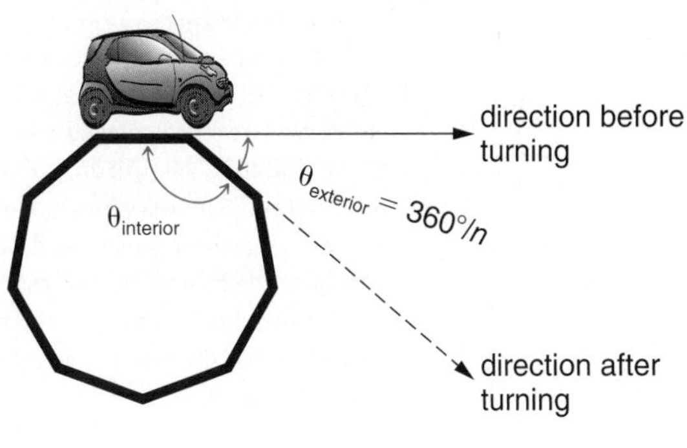

FIGURE 1-1 Derivation of the formula for the interior angle of a regular polygon. The solid arrow represents the direction you are traveling initially. The dashed arrow represents the direction you are traveling after making a turn at the first vertex of the polygon. The angle by which you have turned is an exterior angle, θ_{exterior}.

Substituting $(360°/n)$ for $\theta_{exterior}$ yields Equation 1-4.

$$(360°/n) + \theta_{interior} = 180° \tag{1-4}$$

Finally, subtracting $(360°/n)$ from both sides of the equation, we arrive at Equation 1-5.

$$(360°/n) + \theta_{interior} - (360°/n) = 180° - (360°/n) \tag{1-5}$$

After canceling the two $(360°/n)$ terms on the left side of the equation, we arrive at the general formula for the interior angle of a regular polygon, shown previously in Equation 1-1.

The approach we just used in this geometry problem is one that is widely applicable throughout organic chemistry. Just as we employed fundamental, easy-to-grasp concepts to derive the previous answer, there are a handful of fundamental concepts in organic chemistry that we will learn, understand, and apply to a variety of scenarios. The challenge therefore is for you not only to learn such fundamental concepts, but also to work hard to see how they apply in solving various problems.

1.5 Organization and Goals of the Book

The material in this book is essentially the same as that in my lectures for a six-week organic chemistry prep course that I teach in the summer at the Duke University School of Medicine. Students in that class come from around 40 different colleges and universities across the country. The following fall, they return to their respective institutions to take their full year of organic chemistry. From the organization and focus of my course, students invariably feel well prepared for, and confident about, the organic chemistry challenges they will face. The vast majority of those students end up earning an A or a B in the course!

The remaining chapters of this book are organized quite differently from the traditional textbook of 1000 pages or more. The traditional textbook is typically divided into chapters that focus primarily on specific types of compounds or specific reactions. In contrast, the chapters in this book are organized by specific fundamental concepts. The first several sections of each chapter introduce the concept(s); toward the end of each chapter, we examine specific examples of the application of that concept to a variety of organic reactions and phenomena. Several of these applications will be among the most difficult you will encounter in organic chemistry.

Organizing this book in such a fashion will (1) demonstrate the importance and the power of understanding each concept instead of memorizing, (2) make connections between seemingly different reactions much more transparent than studying organic chemistry one reaction type or molecule at a time, and (3) eliminate much of your fear about your upcoming organic chemistry course.

Of the fundamental concepts we will examine, probably the most important is the *reaction mechanism*, discussed in Chapters 5 and 7. Reaction mechanisms are at the center of organic chemistry; a reaction mechanism provides a detailed, step-by-step description of what happens behind the scenes in an overall reaction. Most importantly, reaction mechanisms allow us to understand *why* reactions occur as they do. They are analogous to our exercise in deriving Equation 1-1, where we learned *why* the equation takes the form it does.

Despite the importance of reaction mechanisms, they are introduced somewhat late in this book because understanding reaction mechanisms requires that you first be comfortable with a handful of fundamental concepts. Those concepts are introduced in Chapters 2 through 4 as well as Chapter 6.

Chapter 8 is different from the other chapters. Instead of focusing on specific concepts, it focuses on a specific set of reactions. This chapter is included because these reactions are typically among the first truly new organic reactions encountered in a full year of organic chemistry. Typically, they are introduced midway through the first semester. The good news is that individually these reactions are relatively straightforward to deal with. The not-so-good news is that there are subtle differences between them, making it difficult for many students to determine which reaction is favored under which set of conditions. As a result, these reactions alone often cause the unprepared student's grade to plummet. Chapter 8 demonstrates how several of the concepts from previous chapters can be brought together to help make quite a bit of sense out of this set of reactions. As a result, when you encounter them in your organic chemistry course, you should feel quite confident and competent.

In order to benefit you as much as possible, I've tried to make this book accessible. I include novel analogies, and I personify molecular species when talking about the energetics of chemical processes, using language like "happy," "unhappy," "wants to," "likes," and "doesn't like." In taking this approach,

Anyone who has completed a year of general chemistry should be comfortable reading this book.

In fact, I have already used this book in my organic prep course I previously mentioned. Most students in that class have completed one full year of general chemistry, a significant percentage have had only the first semester, and some of them have not taken any general chemistry in college. The feedback has been outstanding—they felt my book was quite readable, and many claimed that they *finally* understood some of the general chemistry concepts they didn't understand the first time around.

That said, the intended audience for this book is anyone who has had a full year of general chemistry in college. However, because the level of rigor of a general chemistry class is different at each college and university, Chapters 2 and 3 are designed to get everyone up to speed. You might be tempted to skip those chapters or to skim over them lightly. I recommend not doing so. Read through them carefully. Much of the material will be presented differently from what you have seen before, which will provide additional insight and a deeper understanding. Furthermore, the discussions in those chapters are more focused on the types of problems that you encounter in organic chemistry. Finally, and perhaps most importantly, in those chapters we spend time on a number of trouble spots that tend to plague the majority of students throughout organic chemistry.

1.6 Features of This Book

I have already spent a lot of time impressing upon you the importance of focusing on learning, understanding, and applying fundamental concepts. But the truth is that some of you will naturally be able to do so much more easily than others. Recall the story about my student who asked: "How do you expect us to undo what we've been trained to do for 18 years?" The question is a very good one. The answer is rather simply stated:

Stay active as you learn!

No, this doesn't mean that you should be reading this book while running on a treadmill. Rather, it means that you need to keep your brain connected to your hand—and make sure your hand is holding a pencil. You should be underlining and writing notes in the margins. Work through the problems as I present the solutions. Review concepts from earlier in the book, as they are used to explain things later in the book. And don't believe everything I say—prove it to yourself by working things out.

To help you stay active in these ways, there are a number of features of the book that you should know about and utilize as you encounter them.

- **Your Starting Point:** Tests your grasp of the chapter content before you start. Answers are provided for all of these.
- **Quick Check:** Asks you to recall or apply what you have just read in order to keep your eyes from scanning the page while your brain is on holiday. The answer is provided on the same page.
- **Picture This:** Asks you to visualize scenarios and then answer questions about them in order to help you better understand the topics.
- **Time to Try:** Is a simple experiment or quick assessment in which you perform an active exercise.
- **Why Should I Care?:** Highlights the relevance of the material so you understand its importance in the big picture.
- **Reality Check:** Has you connect what you learn to the real world.
- **Look Out:** Highlights the possible pitfalls or challenges to a student. These paragraphs focus on areas where a novice practitioner may not understand the consequences of a particular action or inaction.
- **Keys:** Highlight the main themes for reinforcement and easy review. These can be spotted by the symbol .
- **What Did You Learn?:** Asks you to apply material from throughout the chapter to answer these end-of-chapter problems.

With this, it's time to tackle some organic chemistry. Get a pencil, and let's get started!

2 Lewis Dot Structures and the Chemical Bond

When you complete this chapter, you should be able to:

- Use the systematic procedure to construct a Lewis structure for any relatively small organic molecule, given its connectivity.
- Explain why covalent bonds form between two atoms.
- Calculate the formal charge of any atom within a molecular species, given the complete Lewis structure.
- Explain what resonance is and why it is important.
- Draw the resonance hybrid of a species, given its individual resonance structures.
- Construct a complete Lewis structure for a fairly complex organic molecule relatively quickly, given its connectivity.
- Draw all resonance structures of a species, given its Lewis structure.
- Use shorthand notation to represent a relatively complex molecule and draw a complete Lewis structure from its shorthand notation.

Your Starting Point

Answer the following questions to assess your knowledge about Lewis structures.

1. The three types of particles that make up an atom are ______________________________.

2. The nucleus, which is located at the center of an atom, is comprised of what kind of particles? ______________________________

3. The size of an atom is governed primarily by the space taken up by which of its particles? ______________________________

4. What are *valence electrons*? ______________________________

5. A bond in which valence electrons are shared between two atoms is called ______________________________.

6. How can atoms benefit when they share valence electrons? ______________________________

7. What is the phenomenon called when a molecule has two or more valid Lewis structures? ______________________________

8. *Resonance structures* are related by the movement of ____________________, while ____________________ remain frozen in place.

Answers: 1. Protons, neutrons, and electrons. 2. Protons and neutrons. 3. Electrons. 4. The electrons found in an atom's outermost shell. 5. A covalent bond. 6. By sharing electrons in the form of a covalent bond, atoms in molecules can gain octets. 7. Resonance. 8. Valence electrons; atoms.

2.1 Introduction

Organic chemistry is predominantly concerned with the reactions that organic molecules undergo. A **chemical reaction** can be defined as the conversion of reactants to products through *the breaking and formation of chemical bonds between atoms*. It is therefore clear that the way in which atoms bond together to form molecules is central to the chemical behavior of those molecules. In this chapter, we discuss why molecules are constructed the way they are, focusing on Lewis structures and the theory of resonance. In doing so, we also delve into the basic idea of formal charge.

The end-of-chapter applications of the principles we cover in this chapter are heavily geared toward organic problems. Section 2.6 focuses on drawing Lewis structures quickly. In Section 2.7, we learn shorthand notation for molecules. Finally, in Section 2.8, we work through exercises in drawing resonance structures of molecules.

2.2 Lewis Dot Structures

Lewis dot structures are the most convenient way in which chemists represent atoms, molecules, and ions. Lewis structures provide information about electrons in a particular **species** (i.e., the collection of nuclei and electrons), and they also provide information about bonding—especially which atoms are bonded together and by what types of bonds. Before we talk about using Lewis structures in molecules, however, we begin with the representation of atoms.

2.2A ATOMS

Lewis structures are concerned only with **valence electrons,** which are *electrons in the outermost (or valence) shell of an atom*. Valence electrons are distinguished from **core electrons,** which are all of the electrons occupying the shells that are not the valence shell—that is, the inner shells. The reason that valence electrons are the only ones shown in a Lewis dot structure is that *the valence electrons are exposed to other atoms and molecules*. Thus,

Valence electrons are the electrons that are ultimately involved in bonding and chemical reactivity. ■

Fortunately, the periodic table (found on the inside front cover of this book) can be used to quickly determine the number of valence electrons an isolated atom has.

The group number of the group in which an atom is found is the same as the number of valence electrons it has. ■

In a Lewis structure representation of an atom, the nucleus is represented by the element corresponding to the atomic number (the number of protons). A nucleus that contains seven protons, for example, is the nucleus of a nitrogen atom and is represented by N, whereas a chlorine nucleus, which has 17 protons, is represented by Cl.

The Lewis structure representation of an atom is completed by drawing the valence electrons as dots around the nucleus. For example, representations of a nitrogen atom and a chlorine

FIGURE 2-1 **Lewis Structures of Atoms.** (*left*) Lewis structure representation of the nitrogen atom, showing its five valence electrons. (*right*) Lewis structure representation of the chlorine atom, showing its seven valence electrons.

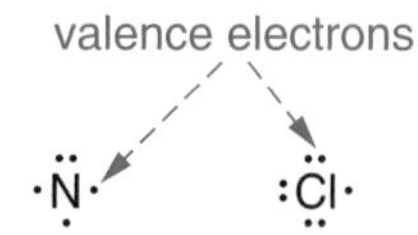

atom are shown in Figure 2-1. Notice that the nitrogen atom is shown to have five valence electrons, as it is located in Group 5 of the periodic table, and the chlorine atom has seven valence electrons, as it is located in Group 7. Notice also that valence electrons typically appear as four groups about the nucleus. Nitrogen has three individual electrons, and one pair, called a **lone pair of electrons.** Chlorine, on the other hand, has one individual electron and three lone pairs. Such electron groupings reflect the fact that these valence electrons reside in four *orbitals* and each orbital can hold up to two electrons.

Lewis structures can also represent a species that carries a charge. To do so, we must simply realize that a negative charge is produced when an atom has an excess electron, wheras a positive charge is produced when an atom loses an electron. For example, N^+ has four valence electrons (one fewer than N's group number), whereas Cl^- has eight (one greater than Cl's group number).

✔ **QUICK CHECK**

For each species, specify the number of valence electrons.

1. C^+ __________
2. O __________
3. N^- __________

Answers: 1. Three. 2. Six. 3. Six.

2.2B MOLECULES

The primary difference between Lewis structures for atoms and those for molecules is that some electrons in molecules are involved in bonding, and the Lewis structures must show this. In particular, the type of bonding shown explicitly by Lewis structures is **covalent bonding,** which is *the sharing of a pair of electrons between two separate atoms.* To begin to understand covalent bonding, we make note of a peculiarity of quantum mechanics:

Atoms are especially stable ("happy") when they have completely filled valence shells.

This means two electrons in the first shell (the so-called **duet**), and eight electrons in the second shell (the so-called **octet**). As evidence of this stability, note that the element with a completely filled first shell (two total electrons) is helium and the element with a completely filled second shell (10 total electrons) is neon. Both of these elements are part of the family of elements called the **noble gases,** which are highly unreactive.

Let's now examine the Lewis structures of several molecules. The first is the simplest—the H_2 molecule. Each hydrogen atom (H), when isolated, has only one electron, which is one short of a complete valence shell (what it desires). If two hydrogen atoms are in close proximity, then the two total electrons can be *shared* between the two hydrogen nuclei, giving rise to a single covalent bond (the sharing of a pair of electrons between two nuclei). A covalent bond can be

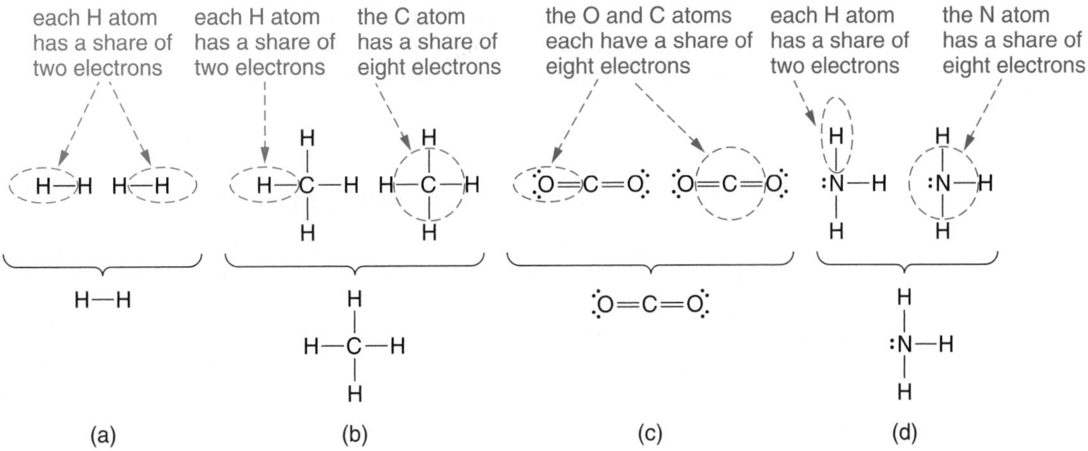

FIGURE 2-2 **Complete Valence Shells in Molecules.** (*bottom*) Lewis structures of various molecules. (*top*) Two different ways of assigning the valence electrons to atoms for each Lewis structure. The dashed ovals/circles represent the different electron assignments. In every Lewis structure, each atom has a share of enough electrons to complete the valence shell.

represented by a dash (X–Y) or a pair of adjacent dots (X:Y); in this book, we will use dashes. With the sharing of that pair of electrons, each hydrogen nucleus is surrounded by two electrons and therefore "feels" that it has a complete valence shell. Figure 2-2a illustrates this point.

In the case of carbon, an isolated atom has four valence electrons, which is four fewer than an octet. In order to feel that it has an octet, the carbon atom needs to be surrounded by another four electrons, which can be accomplished in various ways. One way is to form four single bonds with hydrogen atoms (Figure 2-2b), making a molecule of CH_4 (methane). Each hydrogen atom with which the carbon atom forms a bond provides a share of one additional electron. Another way for the carbon atom to feel that it is surrounded by eight total electrons in its valence shell is to form two double bonds with oxygen atoms, yielding a molecule of CO_2 (Figure 2-2c). Each oxygen atom provides a share of an additional two electrons.

The isolated nitrogen atom has five valence electrons and therefore needs three more electrons to fulfill its octet. Ammonia (NH_3) is one molecule that allows nitrogen to be surrounded by a total of eight electrons (Figure 2-2d). Notice in this case that nitrogen has three bonds to hydrogen, leaving two electrons as a lone pair.

TIME TO TRY

The Lewis structure of a molecule is repeated three times below. In the first structure, circle the electrons that constitute each H atom's duet. In the second structure, circle the electrons that represent each C atom's octet. In the third structure, circle the electrons that represent the N atom's octet and the O atom's octet.

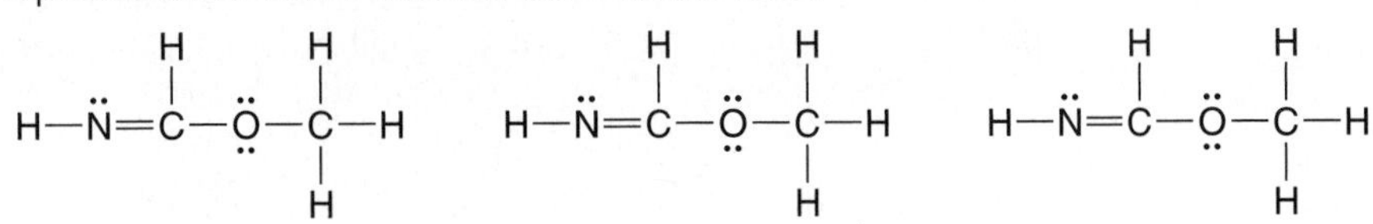

With the understanding that atoms are especially stable when they have complete valence shells, we can use the following systematic procedure to construct a Lewis structure of a molecule if we are given the **connectivity** (i.e., which atoms are covalently bonded together) and the molecule's total charge.

SYSTEMATIC PROCEDURE FOR CONSTRUCTING LEWIS STRUCTURES OF MOLECULES

1. **Count the total number of valence electrons.**
   a. The number of valence electrons contributed by each neutral atom is the same as its group number (e.g., C = 4, N = 5, O = 6, F = 7).
   b. If the total charge is −1, −2, or −3, ADD 1, 2, or 3 valence electrons, respectively.
   c. If the total charge is +1, +2, or +3, SUBTRACT 1, 2, or 3 valence electrons, respectively.
2. **Write the skeleton of the molecule, showing only the atoms and a single covalent bond connecting each pair of atoms that you know must be bonded together.**
3. **Subtract two electrons for each single covalent bond drawn in step 2.**
4. **Distribute the remaining electrons as lone pairs around the atoms.**
   a. Start from the outside atoms and work inward.
   b. Attempt to achieve a complete valence shell for each atom.
5. **If an atom is lacking an octet, convert lone pairs from *neighboring* atoms into bonding pairs of electrons, thereby creating double and/or triple bonds.**

Note that in step 5 the atom whose lone pair is being converted into bonding electrons does not actually lose its share of those electrons. Therefore, if that atom initially has an octet, it keeps its octet. The octet-deficient atom gains a share of an additional two electrons.

Applying the above five steps, let's construct a Lewis structure of NO_2^-, where N is the central atom. To satisfy step 1, we count five valence electrons for N and six valence electrons for each O, for a total of 17. We add one electron for the overall −1 charge, for a total of 18 valence electrons. This is shown in Figure 2-3a. Step 2 has us write down the skeleton of the species. Because N is the central atom, each O must be bonded to the N, giving us O—N—O as the skeleton. The two necessary covalent bonds account for four valence electrons, leaving another 14 that must still be shown. This takes care of step 3, as shown in Figure 2-3b. In step 4, we distribute those 14 electrons around the molecule, beginning from the outer atoms and working inward, with the intent of fulfilling each atom's octet as we go. Therefore, we can place six of those 14 electrons as three lone pairs on the O atom on the left. Those three lone pairs, plus the single covalent bond, give that O atom its octet. That leaves eight electrons yet to be accounted for. We do the same with the O atom on the right, which takes care of another six valence electrons. That leaves two valence electrons not yet shown. The only place remaining for them is on the N atom. We have now completed step 4, and the resulting structure is shown in Figure 2-3c.

After completing step 4, there remains one atom that does not have its octet—the N atom. It has a single covalent bond to each of the two O atoms (for a share of four electrons) and a lone pair (for another two electrons), such that the N atom is surrounded by a total of six

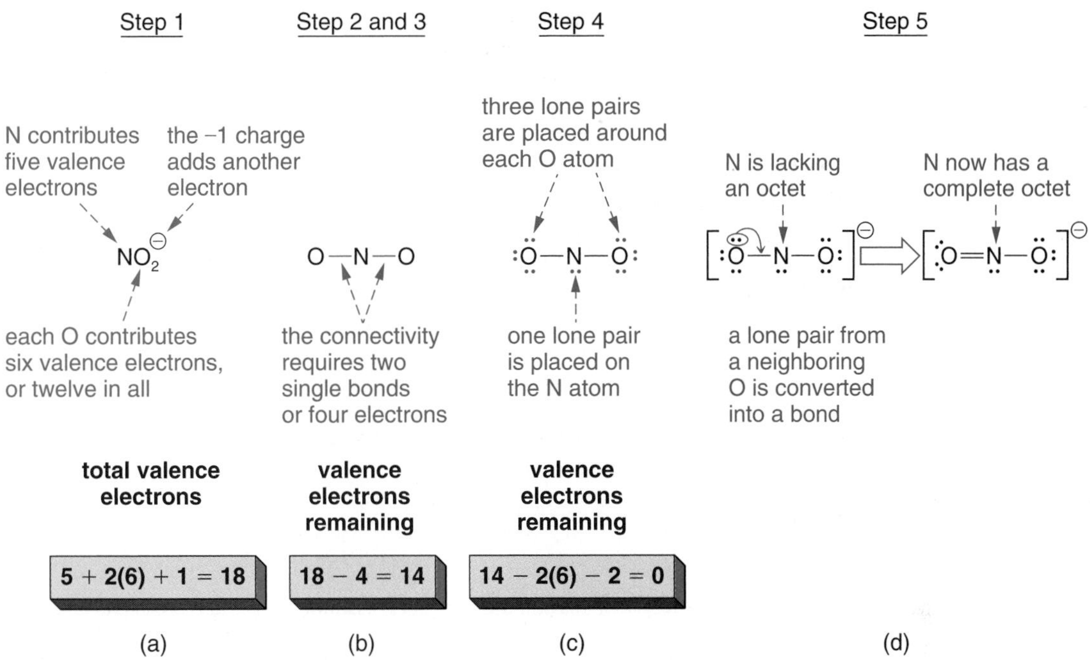
Step 1
Step 2 and 3
Step 4
Step 5
N contributes five valence electrons
the −1 charge adds another electron
each O contributes six valence electrons, or twelve in all
the connectivity requires two single bonds or four electrons
three lone pairs are placed around each O atom
one lone pair is placed on the N atom
N is lacking an octet
N now has a complete octet
a lone pair from a neighboring O is converted into a bond
total valence electrons
valence electrons remaining
valence electrons remaining
5 + 2(6) + 1 = 18
18 − 4 = 14
14 − 2(6) − 2 = 0
(a)
(b)
(c)
(d)

FIGURE 2-3 **Steps in Constructing the Lewis Structure of NO_2^-.** (a) We count 18 total valence electrons in NO_2^-: five from N, six from each O, and one for the −1 net charge. (b) Two electrons are used for each covalent bond connecting N to an O, leaving 14 valence electrons yet to be accounted for. (c) Each O atom receives three lone pairs to complete its octet, leaving two electrons to place as a lone pair on the central N atom. (d) The N atom fulfills its octet when a lone pair of electrons from a neighboring O atom is converted into an additional covalent bond.

valence electrons. Step 5 suggests that we convert a lone pair of electrons from one of the neighboring O atoms into an additional covalent bond between that O atom and the N atom. The resulting structure (Figure 2-3d) shows that the N atom has a double bond (for a share of four electrons), a single bond (for a share of two electrons), and a lone pair (for another two electrons). Eight total electrons now surround the N atom, giving it an octet. Additionally, the double bond on one O gives that O a share of four electrons, and the two lone pairs give it an additional four electrons, so that it has an octet. And the O atom with three lone pairs does not change during the conversion—its three lone pairs give it six valence electrons, and the single covalent bond gives it a share of another two electrons for a total of eight.

TIME TO TRY

In the space provided here, repeat step 5 in Figure 2-3d using the lone pair of electrons on the right-hand O atom.

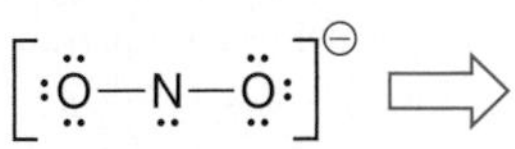

One last thing pertaining to Lewis structures:

For convenience, lone pairs of electrons are often not shown explicitly. Unless you are provided enough information to suggest otherwise, you can assume that all atoms will have their octets.

Therefore, if an O atom has only one bond to it and no lone pairs are shown, you can assume that the remaining six electrons needed to fulfill its octet are present in the form of three lone pairs. Similarly, if you see an N atom that has only a triple bond, you can assume that the remaining two electrons are present as a single lone pair.

As you will see throughout the next few sections, it is often important to keep track of *all* valence electrons. When you encounter Lewis structures in which lone pairs have been omitted, it is best to explicitly draw in those lone pairs, at least in the beginning. As you become more accustomed to working with Lewis structures, this will be less and less crucial. For now, don't cheat yourself—draw in those lone pairs explicitly!

TIME TO TRY

In each species shown below, draw in all lone pairs, assuming that each atom has its octet.

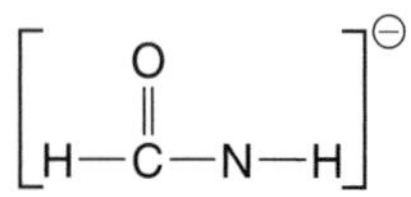

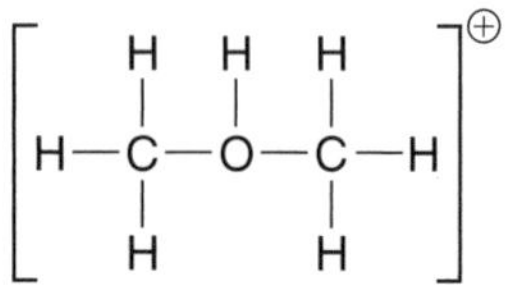

2.3 Formal Charge

Although electrons that are part of a covalent bond do not truly belong to a single nucleus (because they are shared), it is convenient to account for such valence electrons by *assigning* them to specific nuclei. One way to do so is a method associated with what is called **formal charge.**

WHY SHOULD I CARE?

Formal charge is important in understanding the reactivity of molecules, as you will see throughout the rest of this book. Specifically, knowing how to calculate formal charge is important in being able to apply one of the concepts introduced in a later chapter: "Charge is bad."

Most traditional textbooks suggest that you use an equation to calculate the formal charge. I recommend *not* using an equation because it is simply one more thing to memorize; as I stressed in Chapter 1, *keep memorization to a minimum.* Instead, all we must do is assign valence electrons to atoms in the following manner.

FORMAL CHARGE RULES FOR ASSIGNING VALENCE ELECTRONS TO ATOMS

- A lone pair of electrons on an atom is assigned to that atom.
- For each bonding pair of electrons between two atoms, one electron is assigned to one atom, and the other electron is assigned to the other atom.

PICTURE THIS

When assigning valence electrons using the formal charge rules, envision splitting each covalent bond in half, as shown here.

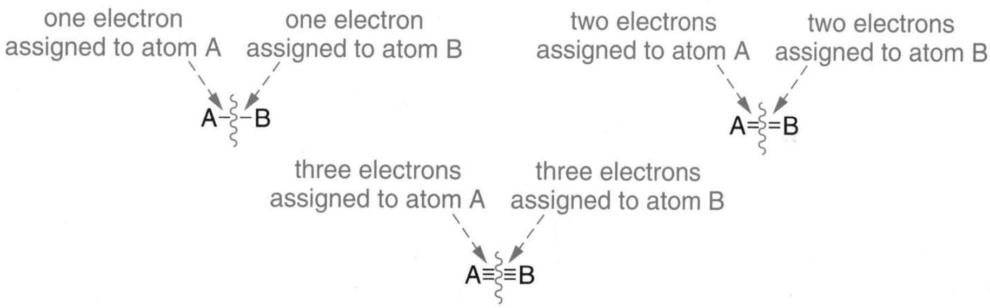

QUICK CHECK

Using the formal charge rules, determine how many *total* valence electrons are assigned to each atom in the following species.

1. 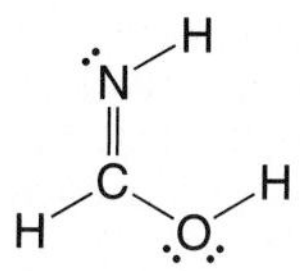

2. HC≡O:

Answers: 1. Each H is assigned one valence electron; C is assigned four; N is assigned five; O is assigned six.
2. H is assigned one; C is assigned four; O is assigned five.

With valence electrons assigned in this way, it is then simply a matter of comparing the number of valence electrons assigned to an atom to the number of valence electrons the atom would have as an isolated neutral atom. Recalling that the number of valence electrons an isolated neutral atom has is the same as its group number, we arrive at the following generalizations:

- An atom in a molecule has a formal charge of 0 if the number of valence electrons it is assigned is the same as its group number (C = 4, N = 5, O = 6, F = 7).
- Each extra electron contributes an additional −1 formal charge to the atom.
- Each missing electron contributes an additional +1 formal charge to the atom.

Figure 2-4 shows how these rules are applied to the atoms in CH_4 and $[CN]^-$. In Figure 2-4a, we can see that the electrons in each of the four C—H bonds are split evenly between the C and H atoms. We therefore assign C a total four valence electrons, the same as its group number. In other words, C is assigned the same number of valence electrons as is found in an isolated, neutral C atom, so in CH_4, C receives a formal charge of 0. In addition, each H atom is assigned a single valence electron, the same as its group number, so each H atom receives a formal charge of 0 as well.

Figure 2-4b shows how formal charges are assigned in $[CN]^-$, whose Lewis structure is :C≡N: (you should be able to come up with this from the five steps provided earlier). In this species, N has a lone pair and a triple bond. Both electrons of the lone pair are assigned to the N atom. In the triple bond, there are six electrons total, three of which are assigned to the N atom.

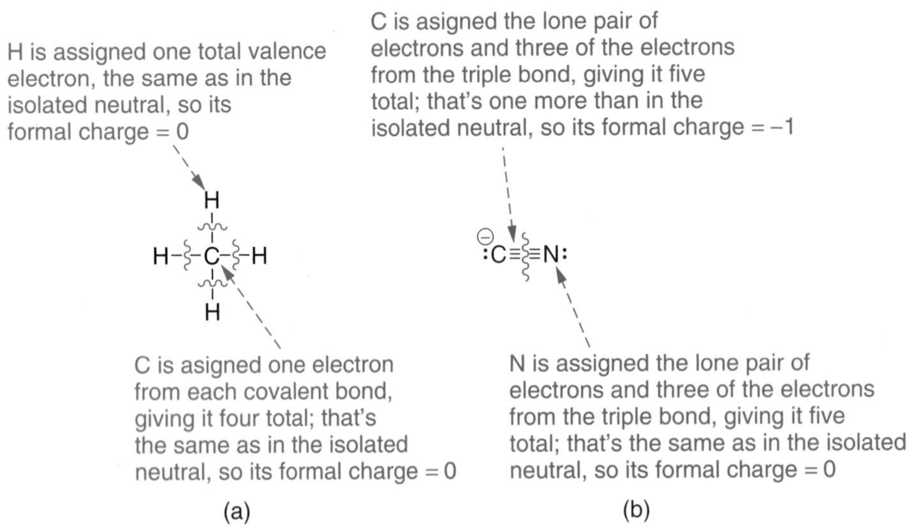
H is assigned one total valence electron, the same as in the isolated neutral, so its formal charge = 0
C is assigned the lone pair of electrons and three of the electrons from the triple bond, giving it five total; that's one more than in the isolated neutral, so its formal charge = −1
H
H
H
H
C
:C≡N:
C is asigned one electron from each covalent bond, giving it four total; that's the same as in the isolated neutral, so its formal charge = 0
N is assigned the lone pair of electrons and three of the electrons from the triple bond, giving it five total; that's the same as in the isolated neutral, so its formal charge = 0
(a)
(b)

FIGURE 2-4 Determining Formal Charges of Atoms in Molecules. (a) The CH_4 molecule. In each C—H bond, one electron is assigned to H, and the other is assigned to C. Therefore, each H atom is assigned a total of one valence electron, the same as in an isolated neutral H atom, so its formal charge is 0. The C atom is assigned a total of four valence electrons, the same as in an isolated neutral C atom, so its formal charge is 0. (b) The $[CN]^-$ anion. Three electrons from the triple bond are assigned to C, and the remaining three are assigned to N. The lone pair on C is assigned to C, and the lone pair on N is assigned to N. Therefore, the C atom is assigned a total of five valence electrons, one more than in an isolated neutral C atom, so its formal charge is −1. The N atom is assigned a total of five valence electrons, the same as in an isolated neutral N atom, so its formal charge is 0.

In all, the N atom is assigned five valence electrons, which is the same as N's group number. Thus, the formal charge on N is 0.

Now let's turn our attention to C, which is assigned both of the electrons from the lone pair and three electrons from the triple bond. This gives C a total of five valence electrons, which is one more than its group number. The C atom is therefore assigned one electron more than in the isolated neutral C atom, giving it one extra negative charge, or a formal charge of −1. Notice in Figure 2-4b that this is indicated by placing the negative charge near the carbon atom.

QUICK CHECK

What is the formal charge on each atom in the following species? (All valence electrons are shown.)

1. 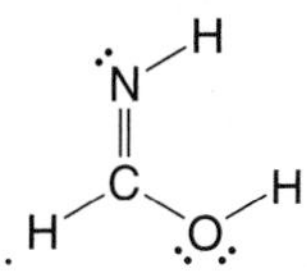 ____________________

2. $HC{\equiv}O\!:$ ____________________

Answers: 1. All atoms have a formal charge of 0. 2. The H and C atoms have a formal charge of 0, and the O atom has a formal charge of +1.

A second way we could have deduced that −1 is the formal charge on C in $[CN]^-$ is to calculate by difference using the following rule:

The sum of the formal charges on all the atoms in a species must equal the total charge of the species.

In this case, the total charge is −1, which should equal the sum of the formal charges on C (i.e., FC_C) and on N (i.e., FC_N). We have already determined that $FC_N = 0$, such that $-1 = FC_C + 0$. Solving this equation, we get $FC_C = -1$.

This rule applies to our previous example of CH_4 as well. Recall that for CH_4, $FC_C = 0$ and $FC_H = 0$. Thus, the total charge of the molecule is $FC_H + 4(FC_H) = 0 + 4(0) = 0$.

QUICK CHECK

Compute the sum of the formal charges on each of the species provided in the previous Quick Check.

Answers: 1. Total charge $= 3(FC_H) + FC_N + FC_C + FC_O = 3(0) + (0) + (0) + (0) = 0$.
2. Total charge $= FC_H + FC_C + FC_O = (0) + (0) + (+1) = +1$.

2.4 Resonance

Resonance is a phenomenon that occurs within a species that has *two or more valid Lewis structures.* Each valid Lewis structure is called a **resonance structure.** A good example is NO_2^-. Earlier we derived a valid Lewis structure by going through the five steps outlined in Section 2.2B. Note the situation we have after the completion of the first four steps (Figure 2-3c), shown again at the left of Figure 2-5. The species is symmetric, with three lone pairs and a single covalent bond on each O atom and with one lone pair and two single bonds on the N atom. Previously, the Lewis structure was completed with step 5 by converting the lone pair of electrons from the O atom on the left into a bonding pair between the O and N atoms—this is shown again in Figure 2-5a. However, we could have used a lone pair from either O atom—both are equally good. If, instead, a lone pair is used from the O atom on the right, a second Lewis structure is obtained—this is shown in Figure 2-5b. It therefore appears that there are two equally good Lewis structures for NO_2^-. In other words, there are two resonance contributors of NO_2^-.

At this stage, it is common for students to say: "But if I flip the first structure over, I get the second one! So they can't be resonance structures of each other!" The first part of that statement is true, but the second is not. Indeed, because the two structures appear to differ only by their orientation in space, they are called **equivalent resonance structures.** However, they are still resonance structures of each other because the completion of each Lewis structure (step 5) involves a different O atom. Specifically, notice that the double bond in Figure 2-5a involves the O atom labeled "1," whereas the one in Figure 2-5b involves the O atom labeled "2."

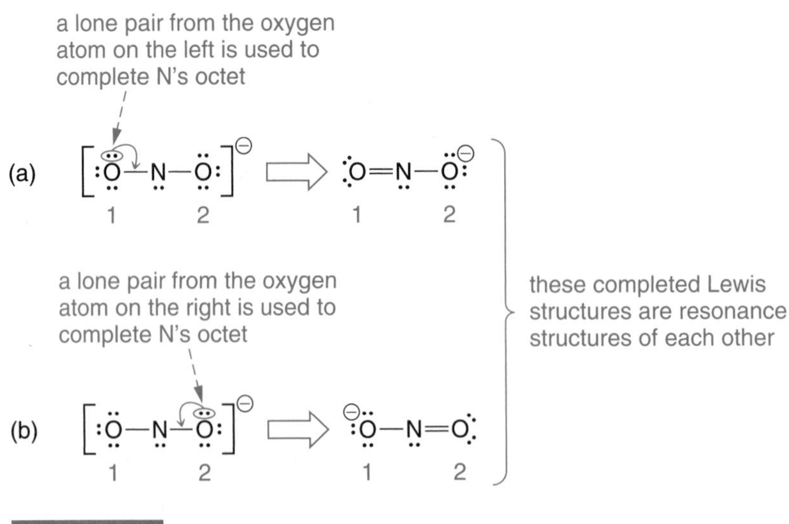

FIGURE 2-5 **Resonance Structures of NO_2^-.** (a) Conversion of a lone pair of electrons from the O atom on the left into a bonding pair leads to one resonance structure. (b) Conversion of a lone pair from the O atom on the right into a bonding pair leads to the other. The O atoms are labeled "1" and "2" in order to emphasize that they are separate atoms.

It is important to know that, although there are two resonance structures that can be drawn for NO_2^-, the NO_2^- species has *only one* definite structure. Experimentally, we know that NO_2^- has two of the exact same nitrogen–oxygen bonds and that each one behaves as something intermediate between a single and a double bond. We also know from experiment that each oxygen atom is identical and carries a partial negative charge—that is, a charge that is somewhat less than a full −1 charge.

Strangely, each of the two resonance structures of NO_2^- (Figure 2-5) appears to disagree substantially with what we know to be true about the real species. For example, in each resonance structure, there appear to be two different bonds—one single bond and one double bond. There also appear to be two different oxygen atoms—one that has no charge and one that has a charge of −1. In other words, NO_2^- appears to be a species that cannot be described accurately by the rules of Lewis structures we learned previously—Lewis structures have limitations! As it turns out, any species that has two or more valid Lewis structures—that is, has resonance—suffers from the same problem.

How do we reconcile the difference between the Lewis structure of a species like NO_2^- and its actual structure? The answer is quite simple:

We take each resonance structure to be an *imaginary* species.

The actual structure behaves as something like the *average* of all of its resonance structures.

This average is called a **resonance hybrid,** and the resonance structures that are involved are called **resonance contributors.** In Figure 2-5, one resonance contributor shows that there is a double bond between N and O(1), whereas the other resonance contributor shows a single bond between those same two atoms. The resonance hybrid should resemble something of an average between a single bond and a double bond, or about 1.5 bonds. Likewise, in one resonance contributor, there is a double bond between N and O(2), whereas the other resonance structure shows that there is a single bond there. Again, in the resonance hybrid, there is something like 1.5 bonds between N and O(2). This is indicated in Figure 2-6 by a solid bond plus a dashed bond between N and each O.

The same analysis can be done with the formal charge. In one resonance contributor, O(1) has a formal charge of −1, and O(2) has a formal charge of 0. In the other resonance contributor, it is the reverse—O(1) has a formal charge of 0, but O(2) has a formal charge of −1. In the resonance hybrid, the charge on O(1) is approximately the average of the two, or about −½. By the same token, the charge on O(2) should be about −½. In the hybrid in Figure 2-6, this is indicated by a "δ^-" on each O, which stands for "partial negative charge" and is read as "delta minus."

WHY SHOULD I CARE?

Resonance is a very simple, yet powerful tool that helps us get a good "feel" for molecular behavior–its structure, stability, and reactivity. In Section 2.6, we begin to see how resonance contributes to stability. And in Chapter 6, we see how resonance plays a specific role in chemical reactivity.

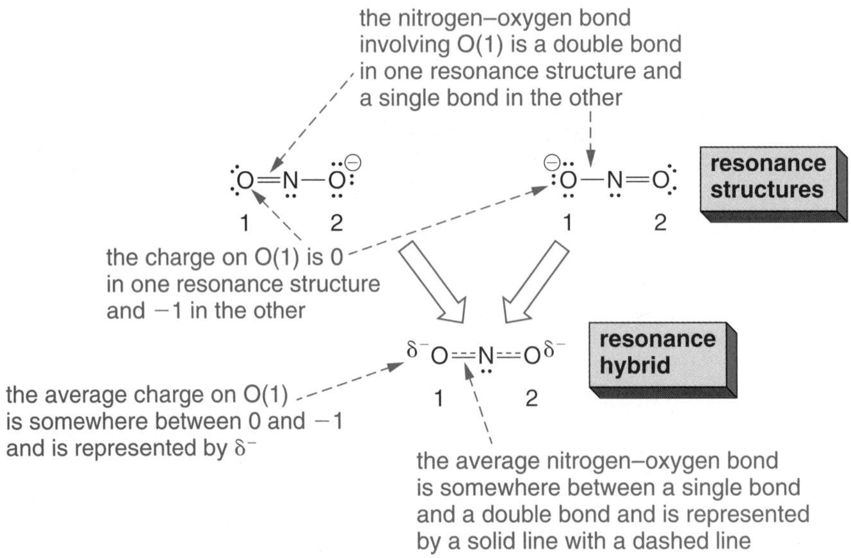

the nitrogen–oxygen bond involving O(1) is a double bond in one resonance structure and a single bond in the other
1
2
1
2
resonance structures
the charge on O(1) is 0 in one resonance structure and −1 in the other
resonance hybrid
the average charge on O(1) is somewhere between 0 and −1 and is represented by δ⁻
the average nitrogen–oxygen bond is somewhere between a single bond and a double bond and is represented by a solid line with a dashed line

FIGURE 2-6 **The Resonance Hybrid of NO_2^-.** The individual resonance structures, or resonance contributors, of NO_2^- are shown at the top. The resonance hybrid, which is a more accurate depiction of the true species, is shown at the bottom. Each nitrogen-oxygen bond in the hybrid is represented as a single bond plus a partial bond (dashed line) because it appears as a single bond in one resonance structure and a double bond in the other. Each charge on oxygen is a partial negative charge (represented by δ^-) because the formal charge on oxygen is 0 in one resonance structure and −1 in the other.

REALITY CHECK

It is useful to pause for a moment and consider an analogy to the idea of a resonance hybrid—a duck-billed platypus. Just as the actual NO_2^- is most accurately described as a hybrid of its two resonance structures, a platypus can be thought of as a hybrid of a duck and a river otter, as illustrated in Figure 2-7.

Realize that the platypus is the actual animal that exists—it does not spend time interconverting between a duck and an otter. The same is true with molecules. A resonance hybrid of a molecular species is the one, true structure that exists—the species does *not* spend time interconverting between the individual resonance contributors.

In order to arrive at any resonance hybrid, we must be able to draw all of its resonance contributors and then take their average. To draw all resonance contributors (as we learn to do in Section 2.6), we need to understand the relationship between any pair of them.

Resonance contributors are related by the rearrangement of electrons, while the atoms themselves remain frozen in place.

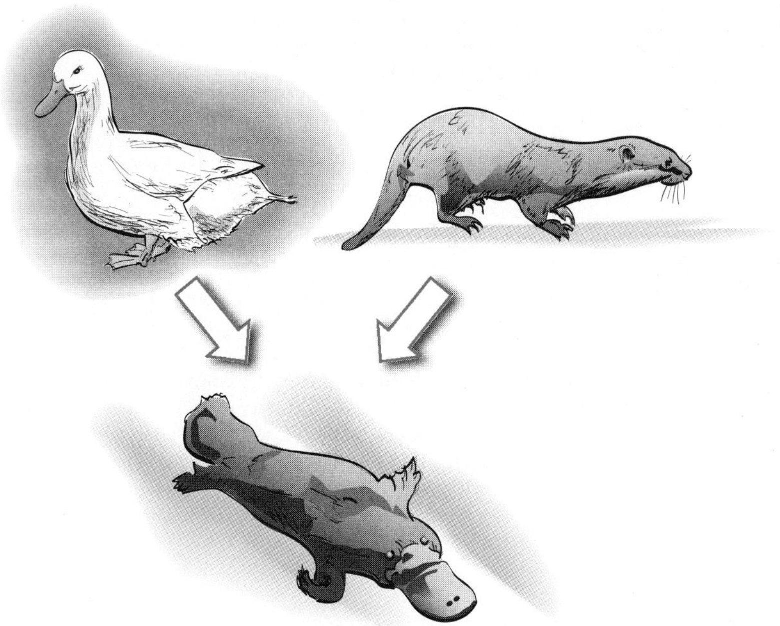

FIGURE 2-7 **Analogy of a Resonance Hybrid.** A resonance hybrid is analogous to a duck-billed platypus (*bottom*). The platypus can be viewed as a hybrid between a duck (*top left*) and a river otter (*top right*). The platypus is the actual animal—it does not spend time interconverting between a duck and an otter.

This can be seen for the resonance contributors of NO_2^-, given that each one is the result of having moved a different pair of electrons in step 5 of the Lewis structure rules.

As a consequence of their relationship, resonance contributors can be interconverted by the movement of electrons only. To convert one resonance form of NO_2^- into the other (Figure 2-8), a lone pair of electrons on O^- is converted into a covalent bond to form an N═O double bond. As a result, the formal charge on the O atom goes from −1 to 0. Simultaneously, a pair of electrons from the N═O double bond on the other side of the N atom is kicked over as a lone pair on the O atom that is initially uncharged. The result is the conversion of the N═O double bond to an N—O single bond, with the formal charge on that O atom going from 0 to −1. In addition, the original N—O single bond has become a double bond, with that O atom's formal charge going from −1 to 0.

Notice in Figure 2-8 that there are two types of arrows used when working with resonance contributors. One is the double-headed straight arrow (↔), which, when placed between two structures, indicates that they are resonance contributors of the same overall species. The other type of arrow is the curved, single-headed arrow (⤻), which shows the movement of a pair of electrons. In Figure 2-8, one curved arrow originates from a lone pair of electrons and points to the region between the N and O atoms, which signifies the conversion of that lone pair of electrons into a covalent bond. The other curved arrow originates from the middle of the double bond and points to an O atom, which signifies the conversion of a pair of electrons from the double bond into a lone pair of electrons.

When drawing resonance contributors, it is important to remember that a resonance contributor, by definition, is a valid Lewis structure; we must therefore adhere to the octet rule.

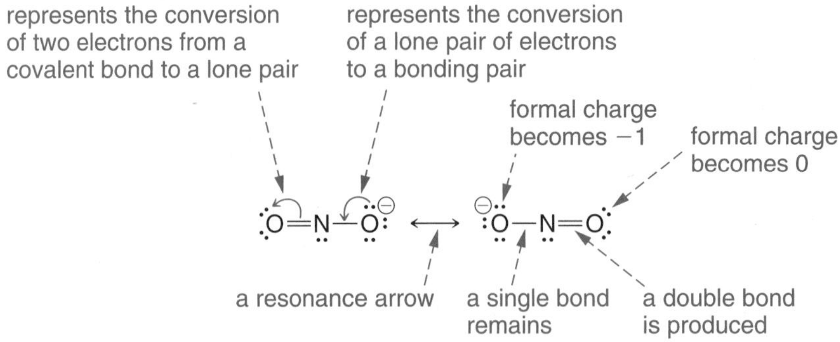

FIGURE 2-8 **Interconversion of Resonance Structures of NO_2^-.** To convert the resonance structure of NO_2^- on the left into the one on the right, two pairs of electrons must move, represented by the two curved arrows. The curved arrow on the left represents a pair of bonding electrons converting to a lone pair on O, leaving behind a single O—N bond and introducing a −1 formal charge on that O. The curved arrow on the right represents the conversion of a lone pair of electrons into a bonding pair, producing an N═O double bond and leaving the O atom uncharged. The two resonance structures are connected by a resonance arrow.

Common atoms encountered in organic chemistry—such as carbon, nitrogen, and oxygen—are not allowed to exceed their share of eight electrons (for hydrogen, it is two electrons). At the same time, atoms want to attain their octets. Therefore,

A valid resonance form should have as many atoms with octets as possible.

In our example with NO_2^-, it is not appropriate to draw only one of the two curved arrows in Figure 2-8. If the curved arrow on the right was drawn without the curved arrow on the left, this would result in four bonds and one lone pair on the N atom, for a total share of 10 electrons; that would therefore exceed the octet (Figure 2-9a). If, however, the curved arrow on the left was drawn without the one on the right, the result would be an N atom with a share of only six electrons (Figure 2-9b). This is not considered a viable resonance structure, since we

FIGURE 2-9 **Invalid Resonance Structures of NO_2^-.** (a) Shifting the electrons according to the curved arrow in the structure on the left produces an invalid resonance structure because the N atom has exceeded its octet (*right*). (b) Shifting the electrons according to the curved arrow on the left produces an invalid resonance structure because the N atom has lost its octet (*right*).

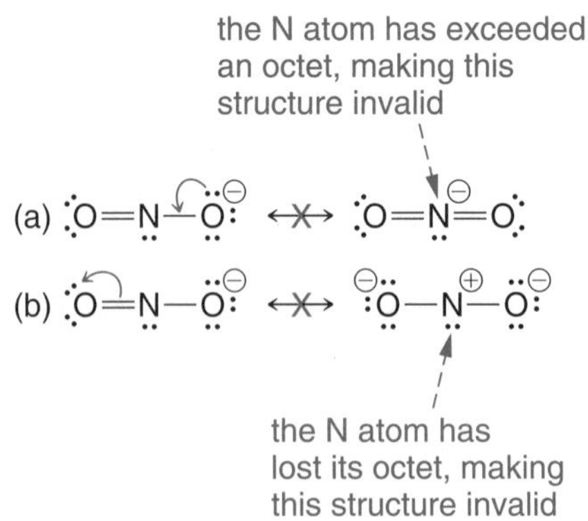

know that it doesn't have the maximum number of atoms with their octet; there are two other resonance forms in which all three atoms have their octets (Figure 2-8).

2.5 Application: Drawing Lewis Structures of Complex Molecules Quickly

In Section 2.2, we reviewed the systematic five-step procedure used to construct Lewis structures. That procedure is extremely useful, but realize that, unless the species has relatively few atoms, carrying out the complete procedure can become very cumbersome. For example, even a moderately small organic molecule with the formula $C_7H_6O_2$ requires us to manage 46 valence electrons!

In situations like this, our job is simplified dramatically by recognizing that atoms tend to prefer a certain number of bonds and a certain number of lone pairs to achieve their complete valence shell. These preferences are summarized in Table 2-1.

Careful examination of these preferences reveals a very important result.

An atom prefers the combination of bonds and lone pairs that gives it a formal charge of 0.

For example, Table 2-1 shows that C prefers to have four bonds and no lone pairs. As we learned earlier, when computing formal charge, one electron from each of those bonds is assigned to C, for a total of four valence electrons. That's the same number as in the isolated neutral atom, giving C a formal charge of 0.

With this in mind, we can complete Lewis structures very quickly if we know exactly which atoms are bonded together (i.e., the connectivity) and if the formal charges of all atoms are 0. Typically, this is the case for molecules that have no net charge. Consider, for example, a molecule of HCN, where C is the central atom. From Table 2-1, we know that H prefers one bond and no lone pairs, so the hydrogen–carbon bond must be a single bond. We also know that C prefers four bonds and no lone pairs, so it must have three bonds in addition to the one to H. Because N is the only other atom to which C is bonded, the carbon–nitrogen bond must be a triple bond. Finally, we know from Table 2-1 that N prefers three bonds and one lone pair.

TABLE 2-1 Preferred Numbers of Bonds and Lone Pairs for Common Atoms

Atom	Preferred Number of Bonds	Preferred Number of Lone Pairs
H	1	0
C	4	0
N	3	1
O	2	2
F	1	3
Ne	0	4

Because N is triply bonded to C, all that remains is to give N one lone pair of electrons. Thus, H—C≡N: is the final Lewis structure.

TIME TO TRY

Using the complete five-step procedure from Section 2.2, construct the Lewis structure of HCN, and verify it is the same as what we arrived at above.

The benefits of completing Lewis structures in this way are fully realized when the molecule is significantly larger. For example, consider a molecule with the following connectivity. (The C atoms are numbered in order to distinguish them from one another.)

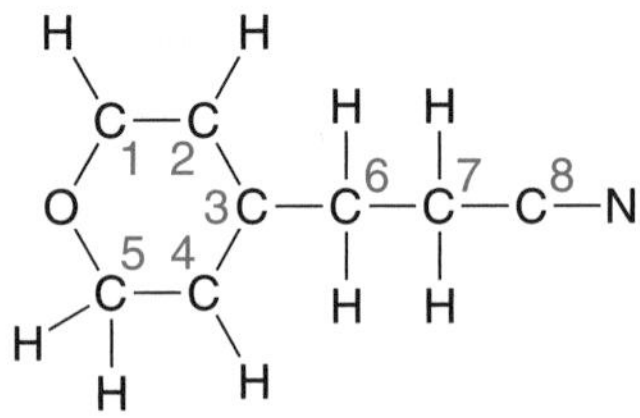

Notice that all the H atoms are already singly bonded, so they should receive no additional bonds or lone pairs. The O atom already has two bonds, but no lone pairs. Therefore, according to Table 2-1, it receives an additional two lone pairs of electrons. The C-1 atom has three bonds currently, so it prefers one more bond. The same is true of C-2. We can satisfy the preferences of both C-1 and C-2 by connecting the two atoms by a double bond. Similarly, both C-3 and C-4 currently have three bonds, so they should be connected by a double bond. By contrast, C-5, C-6, and C-7 already have four bonds each, so they are satisfied. This is not true of C-8, which has only two bonds. We must add two more bonds to C-8, but the bonds to C-7 must be left alone—otherwise, C-7 would exceed its octet. The only choice is to convert the bond to N into a triple bond. That would give N the three bonds that it prefers, leaving only to add a lone pair to N. The resulting Lewis structure appears as below.

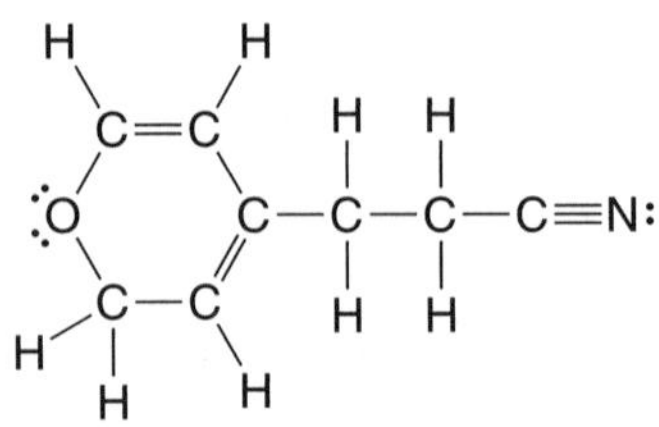

✔ QUICK CHECK

Complete the Lewis structure (including all lone pairs) of the compound whose connectivity is given below, assuming that each atom's formal charge is 0.

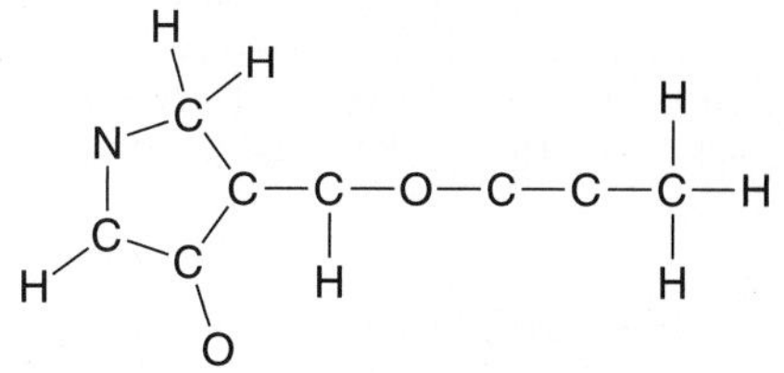

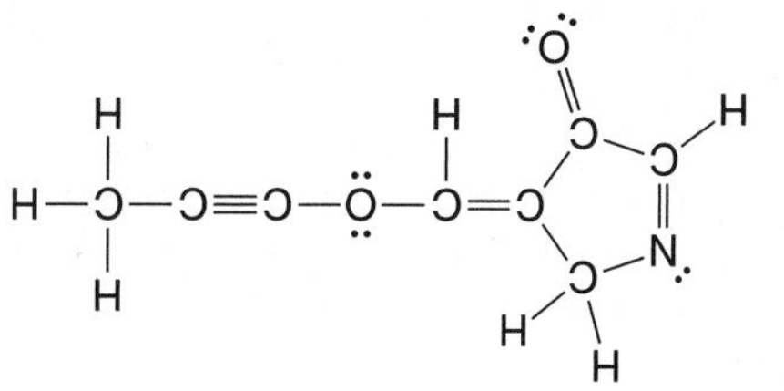

Answer:

2.6 Application: Draw All Resonance Contributors of . . .

Being able to draw all valid resonance structures of a species is important for two reasons. First, a species that has resonance is described by its resonance hybrid, which has contributions from *all* of its resonance contributors. Second, and perhaps more importantly, the number of resonance structures that can be drawn correlates with that species' stability—something that is critical to know in order to understand its reactivity.

All else being equal, the more resonance structures that can be drawn, the more stable the species.

(We touch more on this idea in Chapter 6, and it will be developed more fully in your traditional textbook.)

Drawing all resonance contributors can be straightforward if we use a systematic approach. We begin by applying what we already know about resonance contributors:

1. A resonance contributor must be a valid Lewis structure.
   a. If possible, all atoms should have an octet.
   b. Some atoms in the second row, such as B, C, and N, may have less than the octet, but cannot have more.
2. Resonance contributors are related by moving around electrons within the species, while the atoms remain frozen in place.

Examining once again Figure 2-8 (which shows the conversion of one resonance contributor of NO_2^- into the second) provides insight into the systematic way of drawing all resonance contributors of a given species. Notice in Figure 2-8 which types of electrons are involved—lone pairs and pairs of electrons from double bonds. Specifically, a lone pair of electrons from one O atom is converted into a bonding pair, thereby converting a single bond into a double bond. Simultaneously, a pair of electrons from a different double bond is converted into a lone pair on another O atom, thereby converting that second double bond into a single bond. The lessons we gain from such an interconversion can be generalized as follows:

The types of electrons that are involved in drawing different resonance contributors are lone pairs of electrons and pairs of electrons from multiple bonds—that is, double bonds and triple bonds.

The reason is that, as we saw in the case of NO_2^-, if a pair of electrons from a double bond is shifted elsewhere, a single covalent bond remains. Similarly, if a pair of electrons from a triple bond is shifted elsewhere, a double bond remains. However, if the pair of electrons from a single bond is shifted elsewhere, there are no electrons left to keep the atoms bonded together!

Although we now know that resonance involves only electrons from multiple bonds and/or lone pairs, simply having these types of electrons present in a particular species does not automatically guarantee that more than one resonance structure exists. *The relative positioning of such electrons is also crucial.* Looking back at NO_2^- in Figure 2-8, notice that what enables a second resonance structure to exist is the fact that the O=N double bond is attached to another atom that has a lone pair of electrons—the singly bonded O atom. This allows the lone pair on the singly bonded O atom to become an additional bond to N, simultaneously allowing a pair of electrons from the O=N double bond to be converted to a lone pair.

These lessons apply to resonance structures of other species as well.

In general, whenever an atom with a lone pair of electrons is attached to a multiple bond, an additional resonance structure can be drawn.

To draw the other resonance structure, a lone pair of electrons from the atom attached to the double bond is converted to an additional bond, and a pair of electrons from the multiple bond is converted to a lone pair.

Two examples are shown in Figure 2-10. In Figure 2-10a, the N atom has a lone pair of electrons and is attached to a triple bond. Thus, to draw the other resonance structure, a lone pair of electrons from N is converted to an additional bond to C, and a pair of electrons from the C≡C triple bond is converted to a lone pair on the leftmost C. In Figure 2-10b, the uncharged O has a lone pair of electrons and is attached to a C=O double bond. To draw the other resonance structure, the lone pair from the uncharged O becomes an additional bond to C, and a pair of electrons from the C=O double bond is converted to an additional lone pair on O.

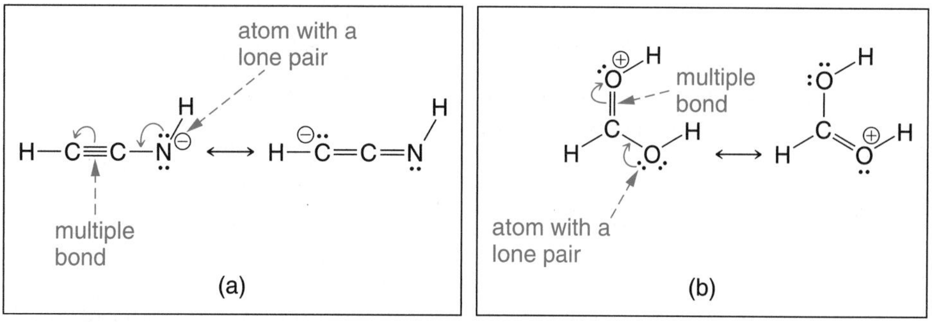

FIGURE 2-10 Resonance Structures in Which an Atom with a Lone Pair Is Attached to a Multiple Bond. (a) The N atom has a lone pair and is attached to the C≡C triple bond. To draw the additional resonance structure, the lone pair is converted to an additional bond to C, and a pair of electrons from the triple bond is converted to a lone pair on the leftmost C. (b) The uncharged O atom has a lone pair and is attached to the C=O double bond. To draw the additional resonance structure, the lone pair is converted to an additional bond to C, and a pair of electrons from the double bond is converted to an additional lone pair on O.

LOOK OUT

When drawing resonance structures that involve the interconversion of a bond and a lone pair, care must be taken to make sure that an atom with a lone pair of electrons is in fact *attached* to a multiple bond. For example, consider the species below. Even though it has a double bond and several lone pairs, it does not have any resonance structures. The O atom on the left has three lone pairs, but is attached to a C atom that is involved in only single bonds. The N and O atoms on the right also have lone pairs, but instead of being *attached* to a multiple bond, they are *part of* a multiple bond.

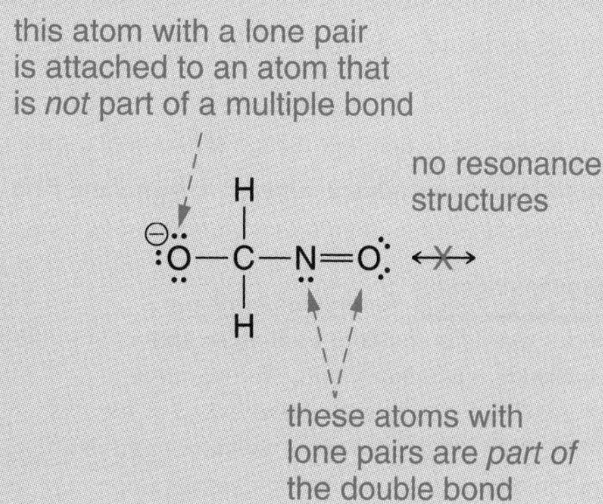

✔ QUICK CHECK

Consider the structure on the left of Figure 2-10b, shown again below. Explain why a resonance structure cannot be obtained by the electron movement indicated by the curved arrows below.

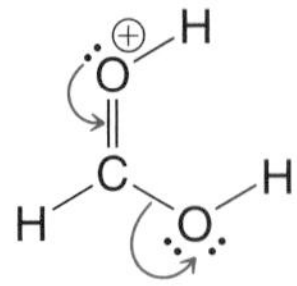

Answer: The curved arrow at the top depicts the involvement of an atom with a lone pair that is *part of* a multiple bond, not *attached to* a multiple bond. The curved arrow at the bottom indicates the involvement of a pair of electrons that constitute a single bond.

In addition to the previous examples, resonance structures can be drawn for species that don't have any lone pairs at all. An example is shown in Figure 2-11. What enables an additional resonance structure to be drawn in this case is the fact that an atom lacking an octet (the one with the positive formal charge) is attached to a double bond.

In general, whenever an atom lacking an octet is attached to a multiple bond, an additional resonance structure can be drawn. ■

To draw the other resonance structure, a pair of electrons from the multiple bond becomes an additional bond to the atom initially lacking an octet. ■

Finally, some cyclic molecules can undergo resonance even without involving a lone pair of electrons or an atom lacking an octet. An example is shown in Figure 2-12a. What allows an additional resonance structure to be drawn in this case is the fact that the ring is composed *entirely* of alternating single and double bonds. Thus, a pair of electrons from each double bond can be shifted to an adjacent position around the ring. (It doesn't matter if this shifting of electrons takes place clockwise or counterclockwise.)

This molecule is a specific case of a more general scenario.

In general, whenever a species has a ring consisting entirely of alternating single and multiple bonds, an additional resonance structure can be drawn. ■

To draw the other resonance structure, a pair of electrons from each multiple bond is shifted to an adjacent bond around the ring. ■

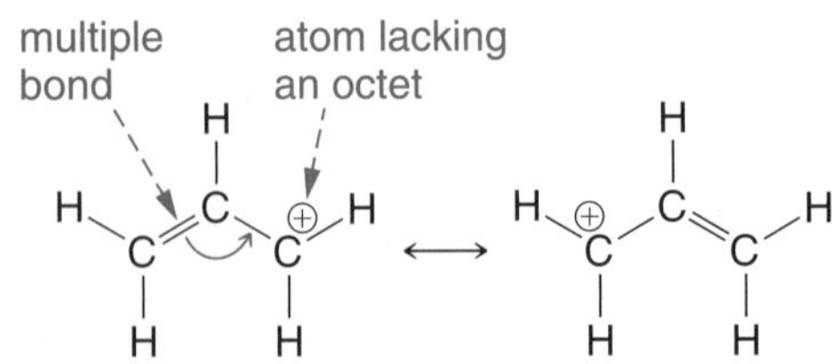

FIGURE 2-11 Resonance Involving a Species that Has an Atom Lacking an Octet Attached to a Multiple Bond. The positively charged C lacks an octet and is attached to the C=C double bond. Another resonance structure can be drawn by converting a pair of electrons from the double bond into an additional bond to the atom lacking an octet. The result is a resonance structure in which an atom from the initial double bond lacks an octet.

✔ QUICK CHECK

For each species below, determine whether another resonance structure exists. If another one does, use curved arrows to indicate the electron movement necessary to get to the other resonance structure, and draw that resonance structure.

1.

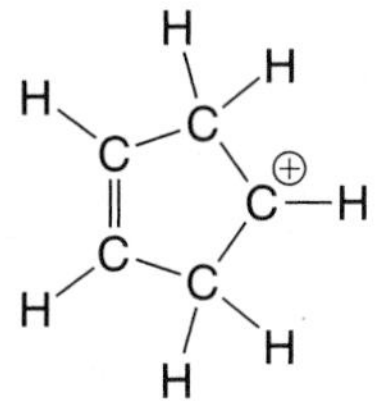

2.

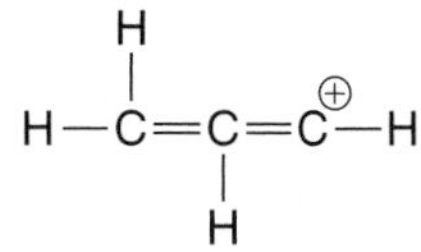

Answers: 1. No other resonance structure exists. Although the species has a multiple bond and an atom lacking an octet, that atom is not attached to the multiple bond. 2. The C with the positive charge lacks an octet and is attached to the C$=$C bond involving the two C atoms on the left. A curved arrow can be drawn from the double bond on the left toward the bonding region involving the atom lacking the octet. As shown below, the resulting resonance structure has a single bond between the C atoms at the left and a triple bond between the C atoms at the right. The leftmost C atom is left without an octet and has a formal charge of +1.

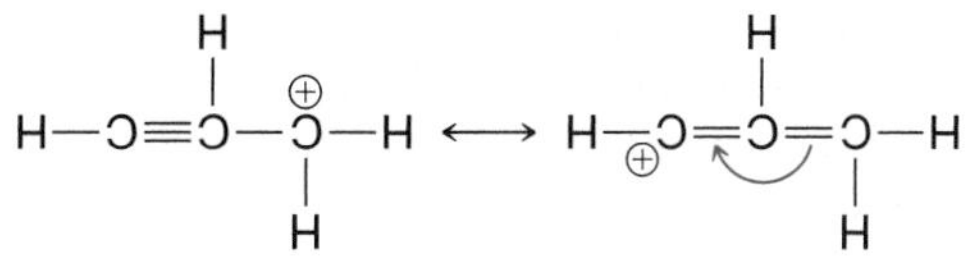

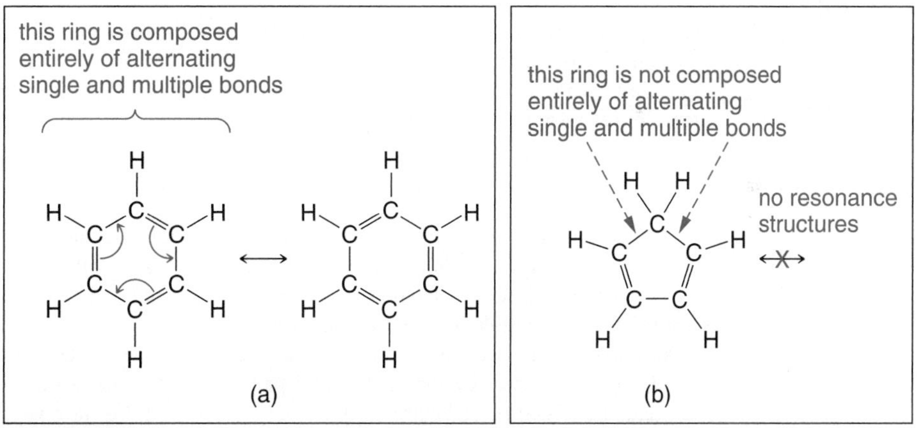

FIGURE 2-12 Resonance Involving a Species with a Ring of Alternating Single and Multiple Bonds. (a) The ring is composed entirely of alternating single and multiple bonds, so an additional resonance structure can be drawn in which a pair of electrons from each multiple bond is shifted to an adjacent position around the ring. (b) The ring has some alternating single and multiple bonds, but the ring is not composed *entirely* of alternating single and multiple bonds. As indicated, there are two adjacent single bonds in the ring. Therefore, this molecule has no additional resonance structures.

In order to draw resonance structures for molecules like this, the alternation of the single and multiple bonds is critical. Figure 2-12b, for example, shows a cyclic molecule that has some alternating single and double bonds, but the alternating single and double bonds do not go *entirely* around the ring. Instead, as indicated, the ring has two adjacent single bonds. Therefore, this molecule does not have any additional resonance structures.

So far, the examples we have seen have had only two total resonance structures. Many species, however, will have more than two resonance structures, and you will be expected to draw *all* of them. To do so, it is best to begin by determining if the species has one (or more) of the three structural features we have discussed in this section—that is, an atom with a lone pair attached to a multiple bond, an atom lacking an octet attached to a multiple bond, or a ring of alternating single and multiple bonds. Once you identify which structural feature is present, draw the new resonance structure by shifting the electrons according to the directions that correspond to the specific structural feature. Then examine the new resonance structure to determine which of the structural features are present, and draw the next resonance structure accordingly. Continue this procedure until you have reached the point where any feasible shifting of electrons results in a resonance structure you have already drawn.

Consider the species in Figure 2-13a. Notice that it has a ring of alternating single and double bonds. Therefore, to draw the next resonance structure, we can shift a pair of electrons from each double bond around the ring, yielding the resonance structure in Figure 2-13b. The new resonance structure also has a ring of alternating single and double bonds, but if we shift the electrons around the ring one more time, we arrive back at the initial resonance structure. (See for yourself!)

TIME TO TRY

Consider the structure from Figure 2-13b. Use curved arrows to indicate moving the electrons from the multiple bonds completely around the ring (in the same direction indicated in Figure 2-13a), and draw the structure that would result. How does that resulting structure compare to the one in Figure 2-13a?

Notice that the new resonance structure in Figure 2-13b also has an atom with a lone pair (oxygen) attached to a double bond (C=C). So to draw an additional resonance structure, we can convert a lone pair from O into an additional bond and convert a pair of electrons from the C=C double bond into a lone pair on C, yielding the structure in Figure 2-13c. That new structure also has an atom with a lone pair (carbon) attached to a C=C double bond, so we can shift the lone pair and a pair of electrons from the C=C double bond to yield the structure in Figure 2-13d. Doing this one more time yields the structure in Figure 2-13e. Finally, notice that the structure in Figure 2-13e also has an atom with a lone pair (carbon) attached to a double bond (C=O), but shifting the lone pair and the pair of electrons from the double bond one more time yields the structure in Figure 2-13f, which is the same as the structure in Figure 2-13a. Therefore, this does not count as an additional resonance structure, and we are done.

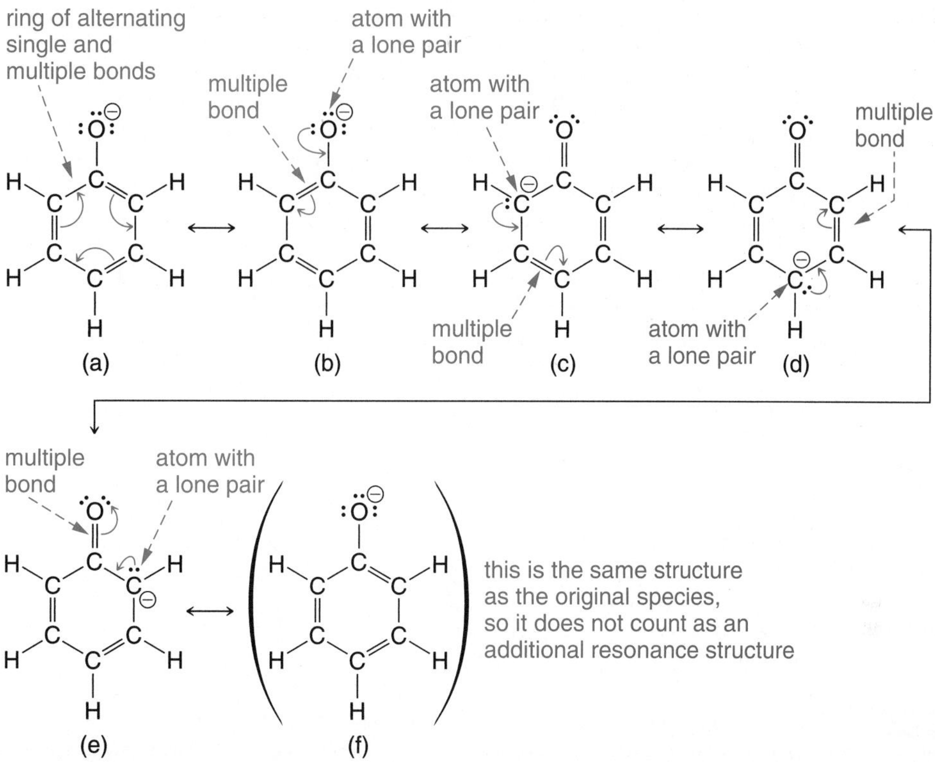

FIGURE 2-13 **Drawing Multiple Resonance Structures of $C_6H_5O^-$.** (a) In the initial species given, there is a ring of alternating single and double bonds, so to get to the next resonance structure, we shift a pair of electrons from each double bond around the ring. (b) This structure has an O atom with a lone pair attached to a C=C double bond, so a lone pair is converted to an additional bond, and a pair of electrons from the double bond is converted to a lone pair. (c) This structure has a C atom with a lone pair attached to a C=C double bond, so a lone pair is converted to an additional bond, and a pair of electrons from the double bond is converted to a lone pair. (d) This structure has a C atom with a lone pair attached to a C=C double bond, so a lone pair is converted to an additional bond, and a pair of electrons from the double bond is converted to a lone pair. (e) This structure has a C atom with a lone pair attached to a C=O double bond. However, if a lone pair is converted to an additional bond and a pair of electrons from the double bond is converted to a lone pair, we arrive at a structure that is the same as in (a). Therefore, there are no additional resonance structures.

Let's do one more example, shown in Figure 2-14. In Figure 2-14a, notice that there is an atom lacking an octet (the C with the positive charge) attached to a double bond (C=C), so we can shift a pair of electrons from the double bond toward the atom lacking an octet, yielding the resonance structure in Figure 2-14b. We can do this one more time to yield the resonance structure in Figure 2-14c. The structure in Figure 2-14c has none of the three features discussed in this section, but it does have an atom with a lone pair attached to an atom lacking an octet. According to step 5 of the rules for drawing Lewis structures (Section 2.2b), a lone pair of electrons is converted to an additional bond to yield the structure in Figure 2-14d. At this point, there is no other feasible way to shift electrons to produce another resonance structure.

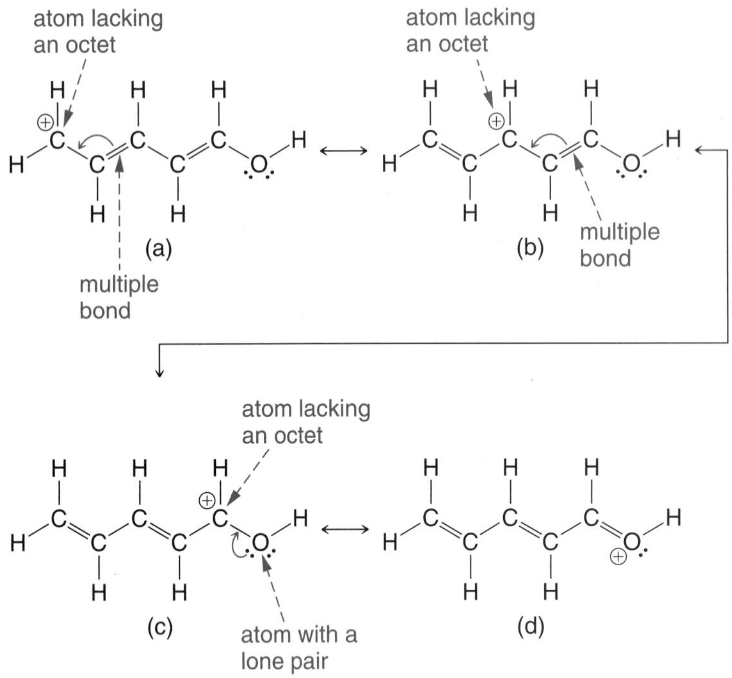

FIGURE 2-14 **Drawing All Resonance Structures of H_2C^+—CH=CH—CH=CH—OH.** (a) The initial structure has an atom lacking an octet (the positively charged C) attached to a C=C double bond, so an additional resonance structure can be drawn by shifting a pair of electrons from the double bond toward the atom lacking an octet. (b) The new resonance structure also has an atom lacking an octet attached to a double bond, so a pair of electrons from the double bond can again be shifted toward the atom lacking an octet. (c) An atom with a lone pair (oxygen) is adjacent to an atom lacking an octet, so a lone pair can be converted to an additional bond. (d) No additional resonance structures can be drawn.

LOOK OUT

When attempting to draw all resonance structures, you should try to use as few curved arrows as possible to arrive at a new resonance structure.

- For an atom with a lone pair attached to a multiple bond, two curved arrows are necessary.
- For an atom lacking an octet attached to a multiple bond, one curved arrow is necessary.

If you use more curved arrows, you will likely skip over some resonance structures. For example, consider the structure from Figure 2-13b, repeated below. The four curved arrows that are shown can be used to arrive at another resonance structure—specifically, the one shown previously in Figure 2-13e. But in doing so, we would skip over two resonance structures—those in Figures 2-13c and 2-13d.

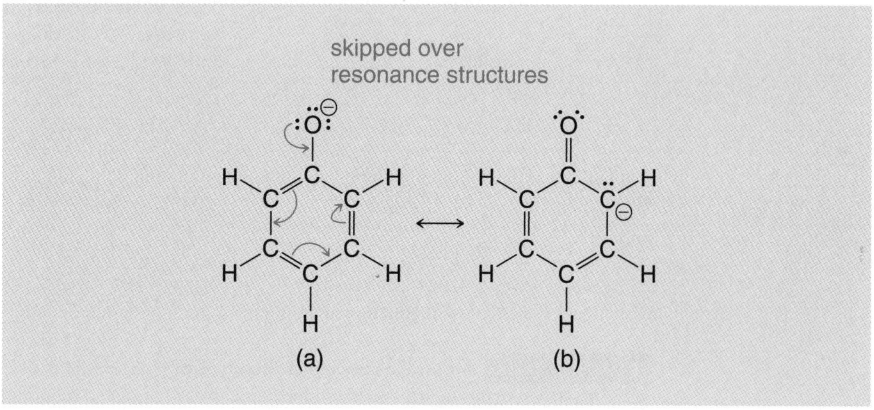

QUICK CHECK

The structure from Figure 2-14a is shown below. We could arrive at another resonance structure using the three curved arrows indicated. Draw the resulting resonance structure. To arrive at this resonance structure, how many resonance structures were skipped over?

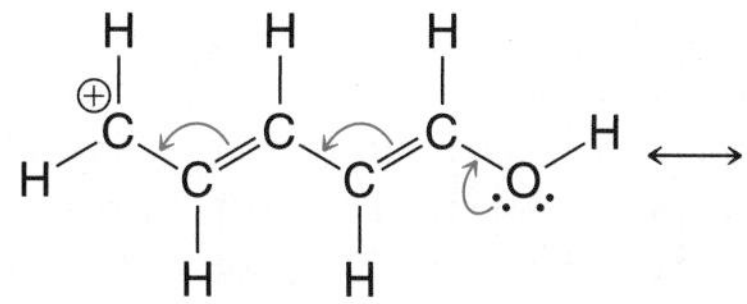

Answer: The new resonance structure is the same as the one in Figure 2-14d. Thus, two resonance structures are skipped over—the ones in Figures 2-14b and 2-14c.

There is one more thing you should be aware of when drawing all resonance structures of a particular species. Notice in Figure 2-14a that, in addition to what we have already called attention to, there is an atom with a lone pair (oxygen) attached to a multiple bond (C=C). So you might imagine drawing an additional resonance structure by converting a lone pair on O into an additional bond and converting a pair of electrons from the C=C double bond to a lone pair, as shown in Figure 2-15a. However, the resulting structure in Figure 2-15b is not an acceptable resonance structure. The reason is that the new structure has introduced two additional formal charges. As we will learn in greater detail in Chapter 6, this is quite bad!

This lesson can be generalized as follows.

In general, a resonance structure is not valid if it is the result of having introduced additional formal charges.

Some exceptions to this rule do exist. You will learn about them, as necessary, in your traditional organic chemistry course.

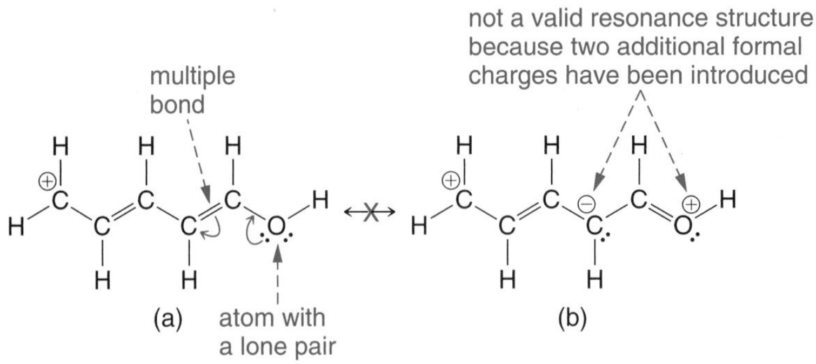
not a valid resonance structure
because two additional formal
charges have been introduced
multiple
bond
(a)
atom with
a lone pair
(b)

FIGURE 2-15 **An Invalid Resonance Structure of H_2C^+—CH=CH—CH=CH—OH.** (a) The structure has an atom with a lone pair (oxygen) attached to a double bond (C=C), so it appears that another resonance structure can be drawn by converting a lone pair into an additional bond and converting a pair of electrons from the double bond into a lone pair, as indicated. However, the resulting structure (b) is not a valid resonance structure because two additional formal charges have been introduced.

2.7 Application: Shorthand Notations

Learning organic chemistry requires drawing numerous molecules, and it would be extremely cumbersome if we had to draw the complete, detailed Lewis structure of a species each and every time. For this reason, organic chemists have devised various shorthand notations that save time, but do not entail any loss of structural information.

We will use these shorthand notations throughout the rest of the book. You should therefore take the time now to become comfortable with them.

2.7A LONE PAIRS AND CHARGES

As we mentioned earlier, lone pairs of electrons are frequently omitted. However, as we continue to study organic chemistry, we will see that lone pairs of electrons have vital roles in molecular structure as well as in many organic reactions; you must therefore be able to put them back in as necessary. Doing so requires knowledge of how formal charge relates to the numbers of bonds and lone pairs on various atoms, as illustrated in Table 2-2. Pay particular attention to the fact that in all of the scenarios shown, the atoms have octets except in the case of C^+.

✔ QUICK CHECK

For each species below, complete the Lewis structure by adding all lone pairs.

1. 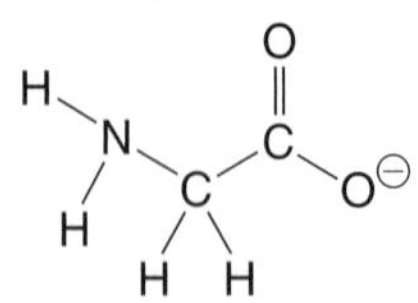

3. 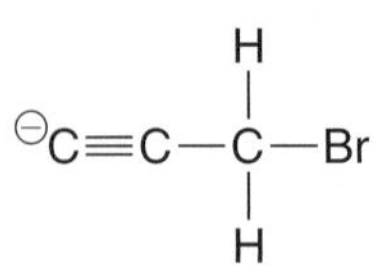

2. 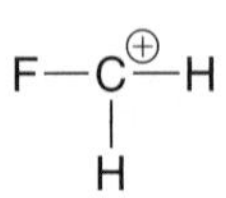

Answers: 1. 2. 3.

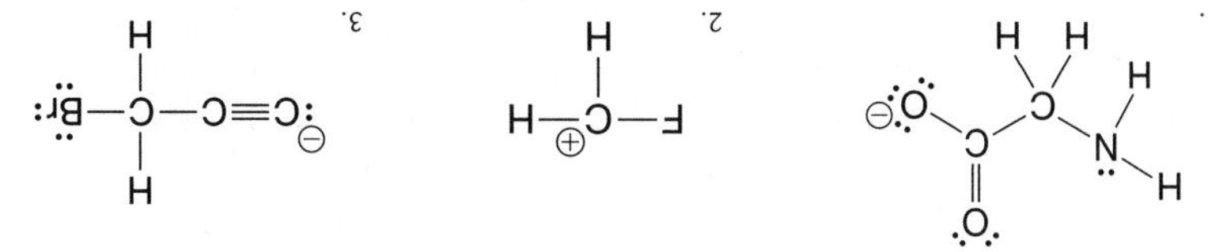

TABLE 2-2 **Formal Charges on Atoms with Various Bonding Scenarios**

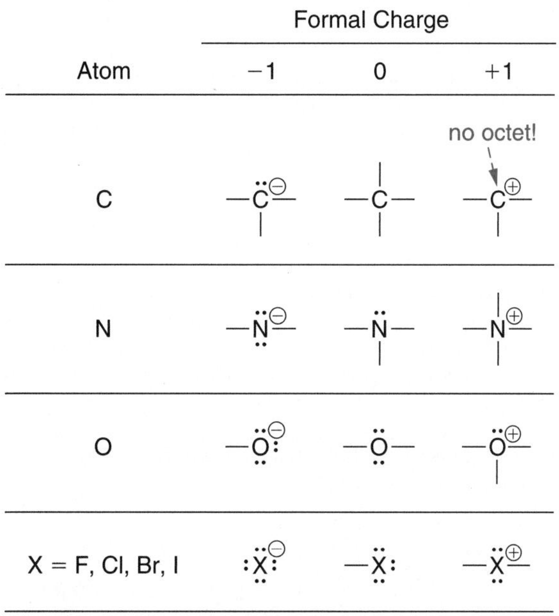

Formal Charge
Atom
−1
0
+1
no octet!
C
N
O
X = F, Cl, Br, I

2.7B CONDENSED FORMULAS

Condensed formulas allow us to include molecules and molecular ions as part of text. To do so, we adhere to these general rules.

RULES FOR CONDENSED FORMULAS

- Each non-hydrogen atom is written explicitly, followed immediately by the number of hydrogen atoms that are bonded to it.

Adjacent non-hydrogen atoms are understood to be bonded to each other.

For example, the condensed formula CH_3CHN^- indicates that there is a central carbon atom bonded to another carbon atom and to a nitrogen atom, giving rise to the skeleton that appears on the left below. The structure is completed by adding the electrons necessary to give each C a formal charge of 0 (four bonds, no lone pairs) and N a formal charge of −1 (two bonds, two lone pairs).

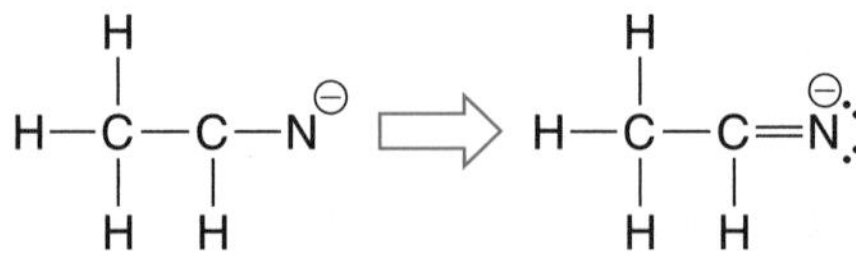

QUICK CHECK

Draw the complete Lewis structure of $CH_3CHCHCHO$.

Answer: According to the condensed formula, four C atoms are bonded sequentially, followed by an O atom. Three H atoms are bonded to the first C atom, and one H atom is bonded to each of the other C atoms, as shown below at the left. The formal charge on each atom is 0, so each C atom must have four bonds and the O atom must have two bonds and two lone pairs. The resulting structure is shown below at the right.

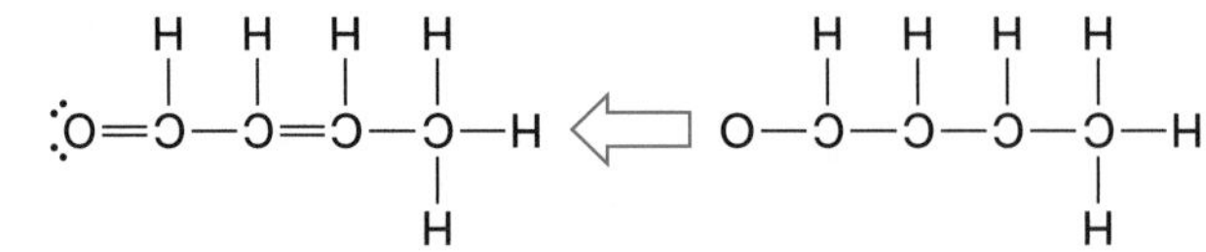

Condensed formulas are a little trickier when three or four groups are attached to a given atom. Two examples are shown below.

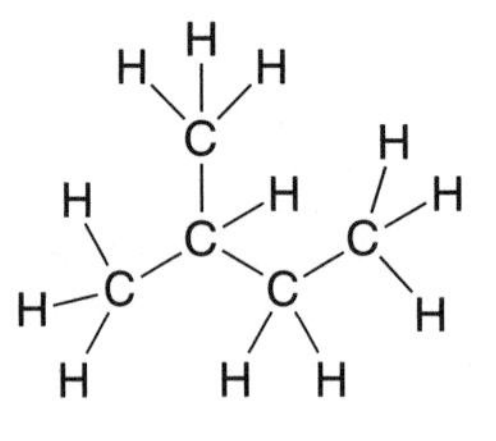

$CH_3CH(CH_3)CH_2CH_3$

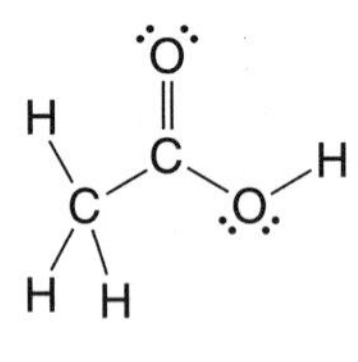

CH_3CO_2H

In the first example, notice that the second carbon atom from the left is bonded to two CH_3 groups and to a CH_2CH_3 group. To take into account additional groups, parentheses can be used. In this case, the condensed formula would be written as $CH_3CH(CH_3)CH_2CH_3$.

The second example is a common scenario in which a carbon atom is bonded to one O atom by a double bond and to a second O atom by a single bond. Using parentheses, the condensed formula could be written $CH_3C(O)OH$. However, to simplify it futher, it is almost always written CH_3CO_2H.

Condensed formulas for cyclic structures are problematic. Consider the molecule below at the left. If the formula is written on a single line of text, the leftmost carbon atom must be bonded to the rightmost one in order to complete the ring. This would appear as follows: $\lceil CH_2CH_2CH_2CH_2CH_2CH_2 \rceil$. Because this is fairly awkward, we generally do not represent rings in their fully condensed forms. Instead, rings are often shown in their *partially* condensed form, as illustrated below at the right.

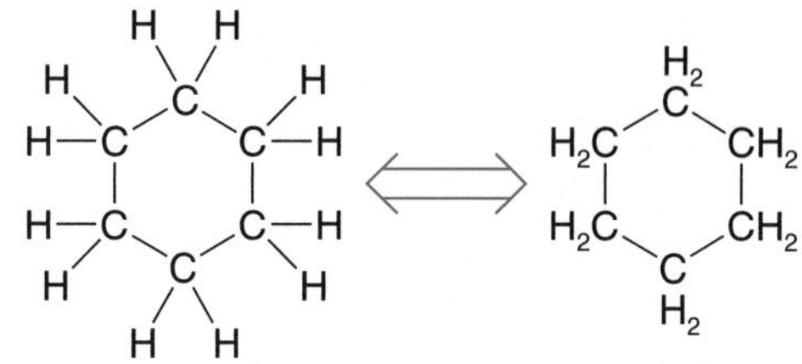

2.7C LINE STRUCTURES

Line structures, like condensed formulas, can be drawn quickly and easily using very little space. Unlike condensed formulas, however, they are not intended to be written as part of text. The rules for drawing line structures are as follows.

RULES FOR LINE STRUCTURES

- Carbon atoms are not drawn.
- Hydrogen atoms bonded to C are not drawn. H atoms bonded to other atoms are drawn.
- All non-C and non-H atoms are drawn.
- Bonds to H are not drawn. All other bonds are drawn explicitly.
- Carbon atoms are implied at the end of every bond that is drawn unless another atom is written there.
- Several C atoms bonded in a single chain are represented by a zig-zag structure.
- Enough H atoms are assumed to be bonded to each C atom to fulfill the C atom's octet, paying attention to the formal charge on C (see Table 2-2).

As an example, the line structure of $CH_3CH_2CH_2CH_2CH_2CH_2CH_2NH_2$ is shown below at the right. Notice that the intersection of each bond represents a carbon atom, as does the end of the bond on the left side of the molecule. The NH_2 group is written in explicitly.

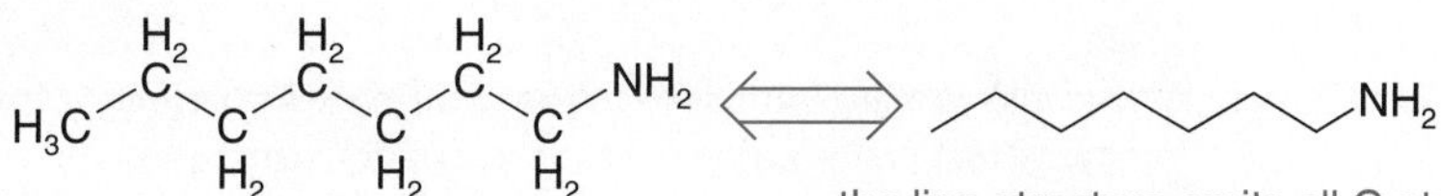

the line structure omits all C atoms and all H atoms bonded to C atoms

QUICK CHECK

Draw the line structure of the first species below. Write the condensed formula of the second species.

1. 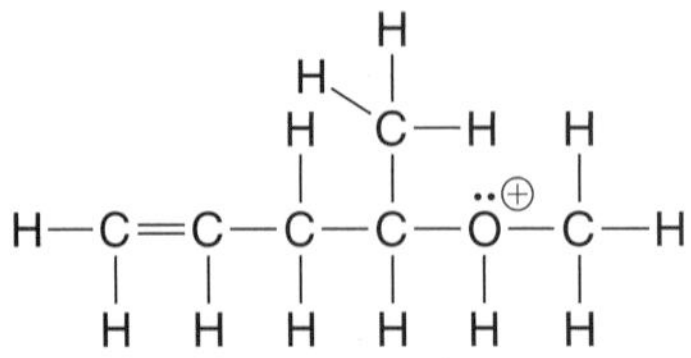 ______________________

2. 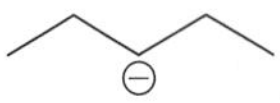 ______________________

Answers:
1. 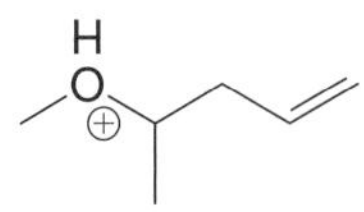
2. $CH_3CH_2CH^-CH_2CH_3$.

LOOK OUT

Line structures can be used when working with resonance. However, if you are not used to "seeing" the H atoms and lone pairs that are not shown, you can end up making some serious mistakes. For example, many students early on will see the following two structures as resonance structures, but they are not.

The reason becomes obvious when we draw the *complete* Lewis structures, as shown below. As we can see, in order to convert one structure into the other, atoms would have to move, and this violates the rules of resonance.

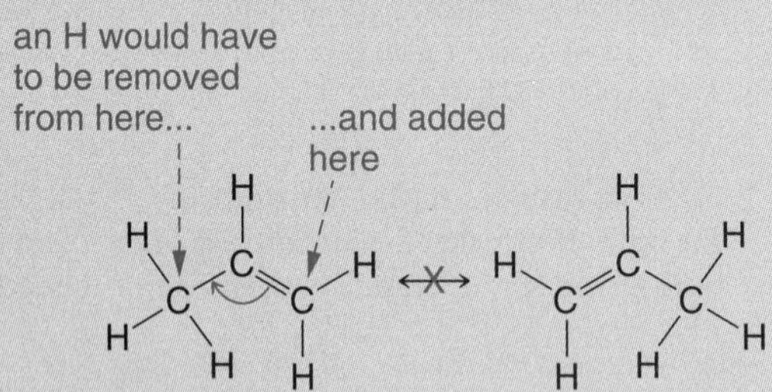

Because of these pitfalls associated with line structures, it is highly recommended that, when first starting to draw resonance structures, you work with the complete Lewis structures, with all lone pairs and all bonds to H added back.

WHAT DID YOU LEARN?

2.1 How many valence electrons are there in C^-, O^+, and F, respectively? How many core electrons are there? Draw each species' Lewis structure.

2.2 Assuming that all atoms have a formal charge of 0, complete the Lewis structure (including all lone pairs) of each of the following molecules with the given connectivity.

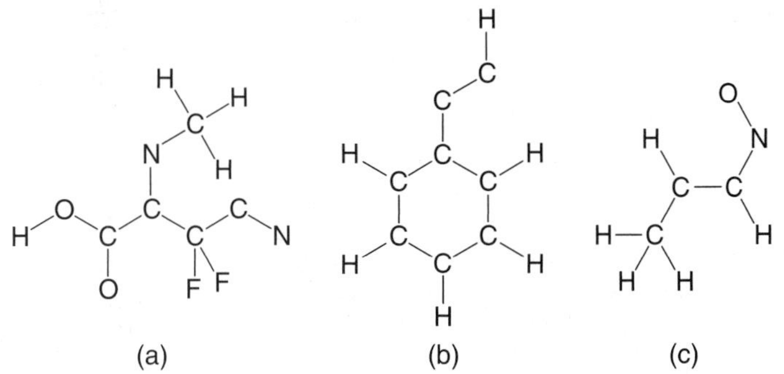

2.3 Draw the Lewis structure of each of the following molecules and calculate the formal charge of every atom. Include all lone pairs in the Lewis structures.

(a) $[CH_3O]^-$ (the C is the central atom)
(b) HNO_3 (the N is bonded to three O atoms, and the H is bonded to one O)
(c) C_2H_6 (the two Cs are bonded together, and each C is bonded to three Hs)
(d) C_2H_4 (the two Cs are bonded together, and each C is bonded to two Hs)
(e) O_3 (a central O atom is bonded to two other Os)

2.4 Using Table 2-1, quickly identify the atoms in the following species that will *not* have a formal charge of 0, and compute their formal charges.

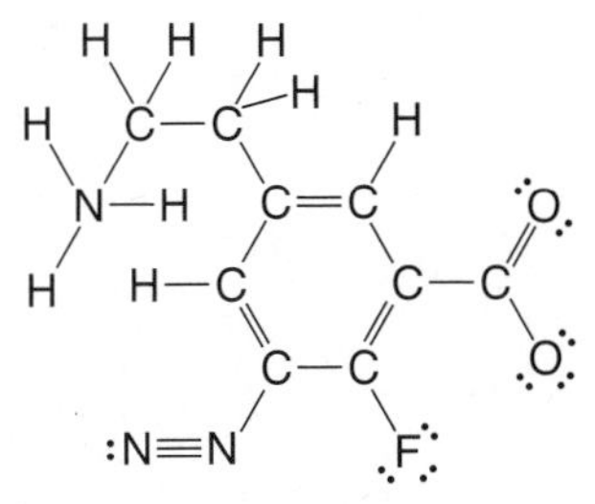

2.5 Draw the molecule in the previous problem as a condensed formula and as a line structure.

2.6 Draw each line structure as a complete Lewis structure and as a condensed formula.

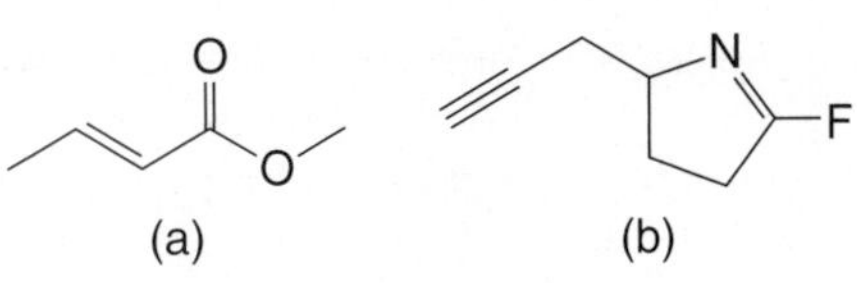

2.7 Draw all resonance contributors of the following species.

(a) O_3 (a central O atom is bonded to each of the others)
(b) HCO_2^- (the C is the central atom)
(c) CO_3^{2-} (the C is the central atom)

(d) 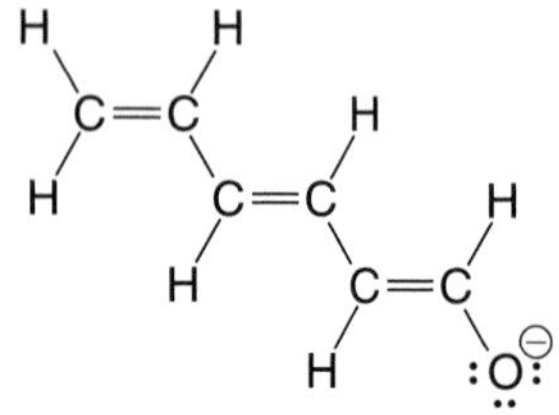

2.8 For each species in the previous problem, draw the resonance hybrid.

2.9 Which of the following pairs are *not* related by resonance (i.e., resonance contributors of the same overall species)? For each pair that is not, explain. (*Note:* All lone pairs of electrons might not be shown.)

(a) 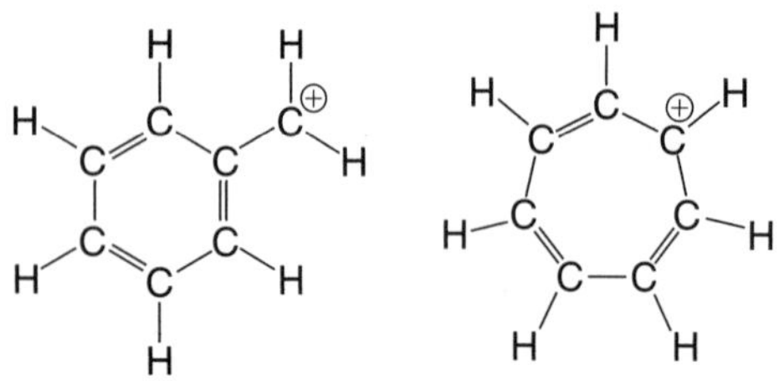

(b) 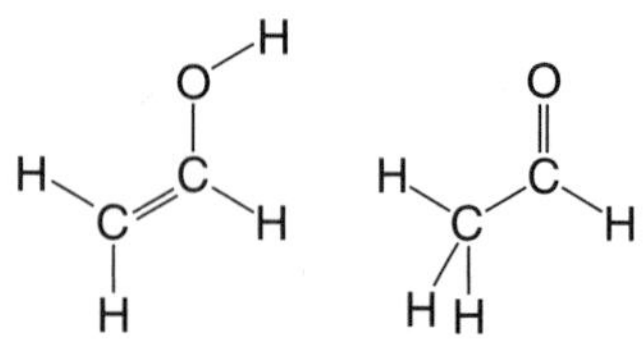

(c) 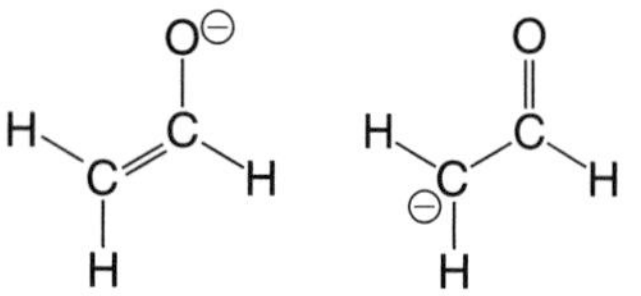

(d) 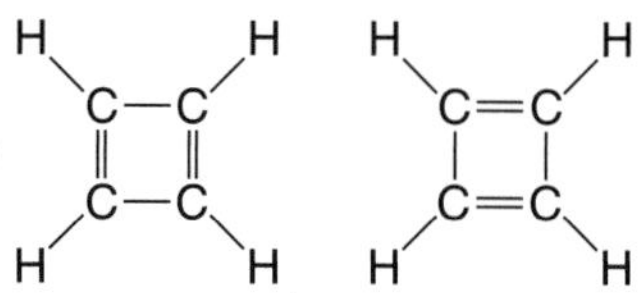

(e) 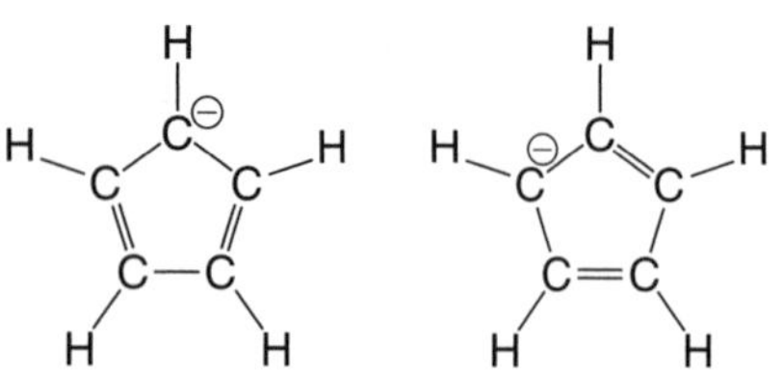

(f) 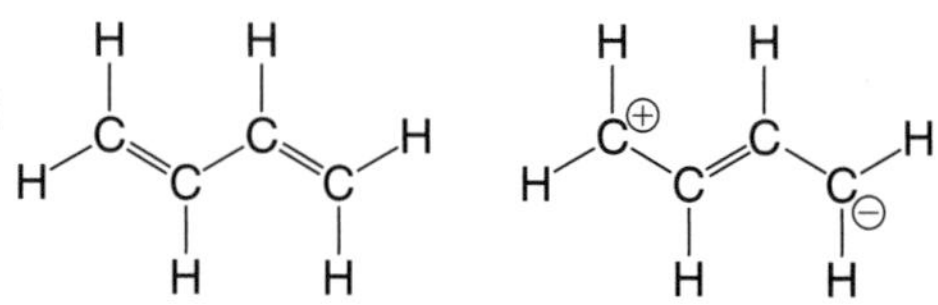

g) 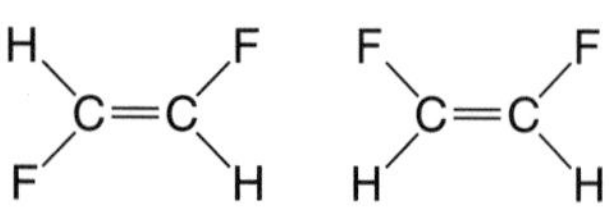

2.10 Draw all resonance structures for each of the following species. Determine which of the species is more stable. Explain why.

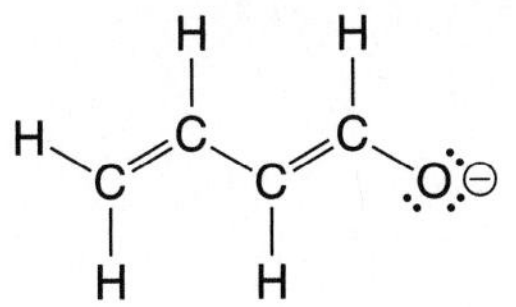

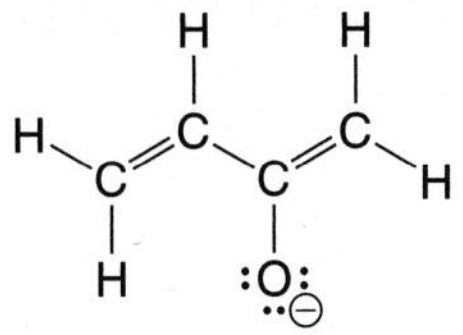

3 Molecular Geometry, Dipole Moments, and Intermolecular Interactions

When you complete this chapter, you should be able to:

- Explain the basis of VSEPR theory.
- Predict the electronic geometry, molecular geometry, and bond angle about an atom in a molecule, given only a Lewis structure.
- Represent a three-dimensional molecule accurately using dash-wedge notation.
- Recognize the various types of symmetry that a tetrahedral atom has.
- Specify the direction and relative magnitudes of bond dipoles within a molecule.
- Determine the direction and magnitude of the net molecular dipole moment of a molecule.
- Describe the various types of intermolecular interactions that can exist between molecules, and rank them according to their strength.
- Identify the specific intermolecular interactions that exist in pair of molecules, given only their Lewis structures.
- Predict the stronger of two dipole–dipole interactions, two hydrogen-bonding interactions, and two induced dipole–induced dipole interactions, given only the Lewis structures.
- Rank the melting points and boiling points of compounds, given only their Lewis structures.
- Predict whether a particular solute should be soluble in a specified solvent, given only their Lewis structures.

Your Starting Point

Answer the following questions to assess your knowledge about VSEPR theory, polarity, and intermolecular interactions.

1. According to VSEPR theory, what are the electron groups about an atom trying to achieve? ______________________

2. What are the electronic geometries called when at atom has two, three, and four electron groups, respectively? What bond angles are associated with these geometries?
____

3. When a covalent bond connects atoms that differ in their electronegativity, toward which atom does the bond dipole point? ______________________

4. When a molecule has polar covalent bonds, how do you know whether the molecule is polar or nonpolar? ______________________

5. Name as many types of intermolecular interactions as you can, and rank them in order of their relative strength. ______________________
____

6. Given that a hydrogen bond consists of atoms in the arrangement X—H·····Y, what types of atoms can X and Y be? ______________________

7. As the strength of intermolecular interactions increases, does a compound's boiling point increase or decrease? ______________________
____

8. If the strength of intermolecular interactions is substantially stronger among two pure substances than it is among the mixture of those substances, will those substances readily mix? ______________________

Answers: 1. They are trying to get as far away from each other as possible. 2. Linear (180°), trigonal planar (120°), and tetrahedral (109.5°). 3. Toward the more electronegative of the two atoms. 4. The molecule is nonpolar if all of the bond dipoles perfectly cancel. Otherwise, the molecule will be polar. 5. Ion–ion interactions > ion–dipole interactions > hydrogen-bonding > dipole–dipole interactions > dipole–induced dipole interactions > induced dipole–induced dipole interactions. 6. The X and Y atoms can be F, O, or N. 7. Increase. 8. No, the substances will not mix readily.

3.1 Introduction

The specific shape of a molecule dictates a number of important properties. As indicated by the chapter title, one of those properties is the dipole moment, or polarity. Polarity, as we will see, plays a major role in *intermolecular interactions*—and hence the physical properties of a compound. Also, as you will see in Chapter 4, the three-dimensional nature of organic compounds gives rise to an interesting phenomenon, called *stereoisomerism*. Both stereoisomerism and intermolecular interactions contribute to a given species' specific type of chemical reactivity.

Water is perhaps the best example with which to highlight the importance of molecular geometry. Water is essential for life, largely because of its unique physical and chemical behavior. Many of those properties, in turn, are a direct result of the fact that the water molecule, H_2O, has a strong dipole moment. From its Lewis structure, we know that the connectivity of the atoms is H—O—H. But *the Lewis structure does not tell us the specific shape*. As it turns out, the water molecule is bent, with an H—O—H angle of around 105°. If, instead, the molecule was linear (i.e., with an angle of 180°), then it would have no dipole moment whatsoever. Without a dipole moment, the boiling point of water would be much lower than 100°C, and water would be incapable of dissolving important species in the body, including the Na^+ and K^+ ions. Life would be very different.

3.2 VSEPR Theory and Three-Dimensional Molecular Geometry

Lewis structures are quite convenient in that they show which atoms are connected by covalent bonds and by what type of covalent bond (single, double, or triple). However, we must be aware of an important limitation.

Lewis structures do *not* provide explicit information about the three-dimensional geometry of a molecule.

This was previously mentioned for H_2O. Another example is NH_3. The Lewis structure of NH_3 in Figure 3-1 implies that all of the atoms are in the same plane (the plane of the paper) and that the atoms are arranged in a T-shape. In actuality, neither of these is the case.

Because of this shortcoming of Lewis structures, chemists often turn to **valence shell electron pair repulsion** (VSEPR) theory to predict the three-dimensional geometry about an atom. Although you probably covered VSEPR theory to some depth in general chemistry, I have found that many students fail to appreciate the power of such a simple model. Much of this is because general chemistry textbooks tend to provide a large table that summarizes the results of VSEPR theory in which various Lewis structure scenarios are correlated with

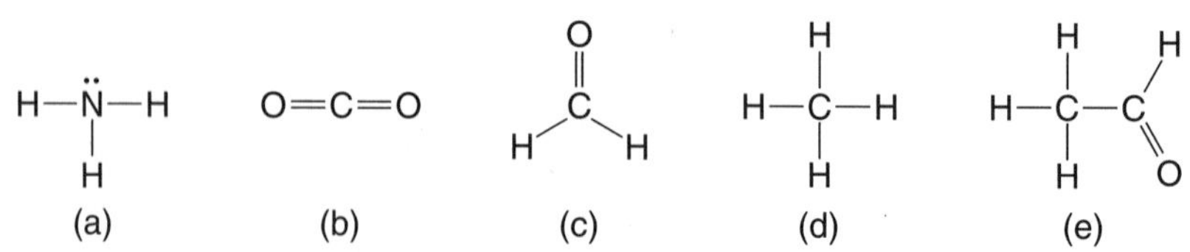

FIGURE 3-1 **Lewis structures of various molecules.** (a) Ammonia, NH_3; (b) carbon dioxide, CO_2; (c) formaldehyde, H_2CO; (d) methane, CH_4; and (e) acetaldehyde, $H_3CCH{=}O$.

TABLE 3-1 Electron groups in VSEPR theory

Electron Class	Number of e^- Total	Number of e^- "Groups"
1 lone pair	2	1
1 single bond	2	1
1 double bond	4	1
1 triple bond	6	1

molecular geometries (e.g., AX_4 = "tetrahedral," AX_2E_2 = "bent"). The problem is that those tables promote memorization; students tend to want to memorize the dozen or so results rather than understand where they came from. This is particularly problematic because those tables are typically not provided on exams.

VSEPR theory is quite straightforward and requires no memorization. The idea is that all electrons in a molecule (as bonds or lone pairs) exist as "groups" of electrons (e.g., a triple bond is one "group" of six electrons). Table 3-1 explicitly shows the four different groups of electrons found in organic molecules.

With this in mind, VSEPR theory stems from the fact that electrons are negatively charged and therefore repel each other. Thus,

Electron groups prefer to be as far away from each other as possible.

As an example, we can consider CO_2 (whose Lewis structure is provided in Figure 3-1b) and ask what the geometry is about the central C atom. The electrons about C appear as two double bonds. According to Table 3-1, that is two groups of electrons that repel each other. To get as far away from each other as possible, one group will be opposite to (180° apart from) the other group. It turns out, then, that the Lewis structure in Figure 3-1b is an accurate representation of the molecule's shape.

LOOK OUT

It is a common mistake for students to interpret a double bond as two electron groups, and a triple bond as three electron groups. Whereas these represent two and three electron *pairs*, respectively, they each count as one electron *group* in VSEPR theory. This is why the C atom in CO_2 is said to have two electron groups. As another example, the C atom in H—C≡N also has two electron groups—one single bond and one triple bond.

TIME TO TRY

In Figure 3-1b, circle each electron group surrounding the C atom.

Once we know the accurate three-dimensional structure of a species, we can describe it using one of two different types of geometries. **Electronic geometry** describes the orientation of the *electron groups* about a particular atom. **Molecular geometry** describes the orientation of *atoms* about a particular atom.

In the case of CO_2, because the electron groups are 180° apart, we say that the electronic geometry about the C atom is **linear.** To determine the molecular geometry about that atom, our goal is to describe the arrangement of the O atoms about the C atom. Realize that the O atoms must be at the ends of the bonds from C, and we have already determined that those bonds are arranged in a linear fashion. Thus, in this case, the molecular geometry is the same as the electronic geometry—linear.

The electronic and molecular geometries about a particular atom must be the same if every electron group belonging to that atom is in the form of a bond.

The reason is simply that, in such a case, an atom must be found at the end of every bond.

Next, let's look at a molecule of formaldehyde, $H_2C{=}O$ (Figure 3-1c). Surrounding the central C atom, we find two single bonds (attached to H atoms) and one double bond (attached to the O atom). According to Table 3-1, that is three total groups of electrons.

TIME TO TRY

In Figure 3-1c, circle each electron group surrounding the C atom.

For those groups to be as far away from each other as possible, they will end up pointing roughly to the corners of an equilateral triangle, being about 120° apart (give or take, depending on the effective sizes of the different groups). We therefore say that the electronic geometry is **trigonal planar.** Again, because there is an atom at the end of each group of electrons, we also say that the molecular geometry is trigonal planar, and the Lewis structure provided in Figure 3-1c accurately represents the molecule's geometry. (Recognize that *trigonal planar* refers to the bond angles about the central C atom and also emphasizes the fact that the O atom and two H atoms are in the same plane as that C atom.)

Methane, CH_4, is slightly more complex. In Figure 3-1d, the Lewis structure makes it appear as if the entire molecule is in one plane, but that is not true. There are four groups of electrons surrounding the carbon atom—the four single bonds. If all of them were to be constrained to the same plane, then they would be 90° apart, pointing to the corners of a square. However, the electron groups are *not* constrained in this way; they can relax into a formation that is three-dimensional, allowing them to be up to 109.5° apart. In such a case, each group points toward the corner of a tetrahedron, shown in Figure 3-2. The C atom is at the center of that tetrahedron, and an H atom is at each corner. Both the electronic and the molecular geometries are **tetrahedral.** Notice that, because of a tetrahedron's symmetry, all four hydrogen atoms are equivalent. That is, they are all the exact same distance from the C atom, and every H—C—H angle is exactly the same at 109.5°.

The aforementioned three molecules encompass the most common electronic geometries that appear in organic chemistry—linear, trigonal planar, and tetrahedral (the two other common

FIGURE 3-2 Representation of methane in its tetrahedral geometry. In the tetrahedral geometry of methane, the C atom is at the center of a tetrahedron, and the H atoms are at the tetrahedron's four corners. This geometry allows the four C—H bonds to be as far apart as possible, 109.5°.

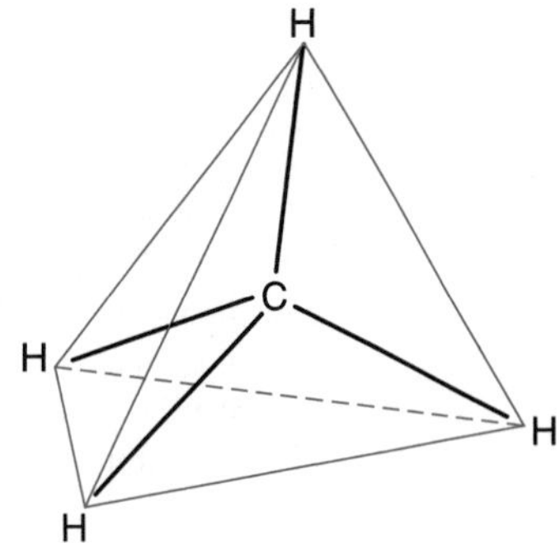

ones you learn in general chemistry, trigonal bipyramidal and octahedral, are not common in organic molecules). The lesson you should learn from the exercises we just went through is the *rigorous correlation among the number of electron groups, electronic geometry, and bond angle.* Specifically, if you know any one of the three, you know the other two. For example, if you know that the electronic geometry about an atom is trigonal planar, there must be a total of three electron groups, and any bond angle must be roughly 120°. Likewise, if you know that a bond angle is roughly 109.5°, then there must be a total of four electron groups about the central atom, and the electronic geometry must be tetrahedral. This idea is summarized in Table 3-2.

✔ **QUICK CHECK**

For each molecule, determine the number of electron groups surrounding the central atom, as well as the electronic and molecular geometries about that atom. (In each case, the central atom is underlined.)

1. $\underline{C}Br_4$ ____________________
2. $\underline{Be}H_2$ ____________________
3. $F_2\underline{C}{=}O$ ____________________

Answers: 1. Four electron groups (four single bonds); tetrahedral; tetrahedral. 2. Two electron groups (two single bonds); linear; linear. 3. Three electron groups (two single bonds and one double bond); trigonal planar; trigonal planar.

TABLE 3-2 Correlations in VSEPR theory

Number of Electron Groups	Electronic Geometry	Approximate Bond Angle
2	Linear	180°
3	Trigonal planar	120°
4	Tetrahedral	109.5°

In each of the molecules we just examined with VSEPR theory, there are no lone pairs of electrons on the central atom. In other words, in those examples, there is an atom at the end of each group of electrons. That's why the geometry that describes the electron groups about the central atom (the electronic geometry) is the same as that which describes the positioning of the atoms about the central atom (the molecular geometry). However, *molecules often contain lone pairs of electrons.* In those circumstances, the electronic geometry is determined no differently than before (Table 3-1). The molecular geometry, however, must be considered further.

Ammonia, NH_3, is an example of a molecule that contains a lone pair of electrons on the central atom. The Lewis structure (Figure 3-1a) shows that there are four groups of electrons on the N atom—three single bonds and one lone pair. Therefore, the electronic geometry must be tetrahedral (Table 3-2). There are H atoms at the ends of three of those four groups, as shown in Figure 3-3a. Consequently, the molecular geometry cannot be tetrahedral—that would require an atom at the end of the fourth group.

In general, if an atom has at least one lone pair of electrons, the electronic and molecular geometries about that atom must be different.

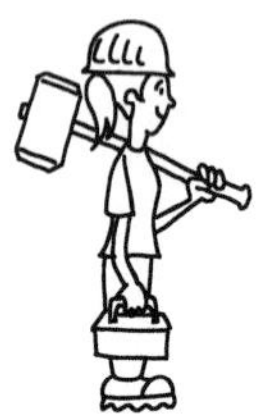

TIME TO TRY

Circle each electron group surrounding the N atom in Figure 3-1a.

The molecular geometry that is chosen for NH_3 is one that accurately describes the positioning of the atoms about the central N atom. Focusing only on the atoms (Figure 3-3b), it appears that the N atom is sitting on top of the three H atoms and those H atoms are located at the corners of a triangle. Therefore, the N atom and the three H atoms form a pyramid, with a triangular base. The molecular geometry for NH_3 is said to be **trigonal pyramidal.**

The main difference between NH_3 and CH_4, as far as the application of VSEPR theory is concerned, is the molecular geometry. The rest is essentially the same. Specifically, both molecules have the same number of electron groups about the central atom. From Table 3-2, we see

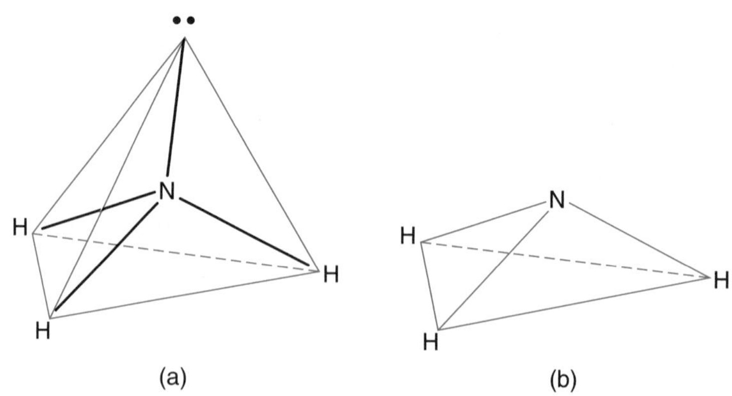

FIGURE 3-3 **VSEPR geometries of ammonia.** (a) The N atom of NH_3 is surrounded by four electron groups—three single bonds and a lone pair. Thus, its electronic geometry is tetrahedral. (b) Focusing only on the atoms, the N atom appears to sit atop a trigonal pyramid and thus has a trigonal pyramidal molecular geometry.

that they both have the same tetrahedral electronic geometry and the angle between groups of electrons is roughly 109.5°. Furthermore, since the bond angle is the same as the angle between bonding electron groups, the H—C—H angle in CH_4 is about the same as the H—N—H angle in NH_3 (they are actually a few degrees different).

QUICK CHECK

For each molecular species, determine the number of electron groups surrounding the central atom, as well as the electronic and molecular geometries about that atom. (In each case, the central atom is underlined.)

1. $H_3\underline{C}^-$ ______________________
2. $H_3\underline{O}^+$ ______________________
3. $H\underline{N}{=}O$ ______________________

Answers: 1. Four groups (three single bonds and a lone pair); tetrahedral; trigonal pyramidal. 2. Four groups (three single bonds and a lone pair); tetrahedral; trigonal pyramidal. 3. Three groups (one single bond, one double bond, and one lone pair); trigonal planar; bent (or angular).

So far, the molecular species we have examined have just one central atom. However, it may be that in a given molecule there are two or more atoms about which we can define a geometry. A good example is acetaldehyde (Figure 3-1e). One carbon atom has four single bonds, or four groups of electrons around it. Its electronic and molecular geometries are therefore tetrahedral. The other carbon atom has three groups of electrons surrounding it—two single bonds and one double bond. Its electronic and molecular geometries are therefore both trigonal planar.

TIME TO TRY

In Figure 3-1e, label the tetrahedral carbon atom and the trigonal planar carbon atom.

Realize that an atom can have an electronic geometry without a molecular geometry. Such is the case with either of the oxygen atoms on CO_2 (Figure 3-1b). Each oxygen atom has three groups of electrons—one double bond and two lone pairs (the lone pairs are not shown in the Lewis structure). According to VSEPR theory, then, the electronic geometry about that atom should be trigonal planar, and the angle between any two groups should be roughly 120°. However, there is no bond angle that exists with the oxygen atom at the center. Therefore, there is no molecular geometry that can be defined about that oxygen atom.

TIME TO TRY

Without reviewing the above discussion, specify the electronic geometry, molecular geometry, and bond angle associated with each non-hydrogen atom in Figure 3-1.

3.3 Tetrahedral Geometry and the Dash-Wedge Notation

Tetrahedral geometry introduces three-dimensionality into organic molecules. You must therefore be able to draw and interpret their three-dimensional representations. This can often be difficult, given that the representations of three-dimensional structures are drawn in two dimensions. **Dash-wedge notation** was developed to make such a task straightforward. It consists of three aspects.

CONVENTIONS OF DASH-WEDGE NOTATION

1. A straight line (—) represents a bond that is parallel to the paper, such that the atoms at either end of the bond are both in a plane parallel to the paper.
2. A wedge (◄) represents a bond that points toward you and comes out of the plane of the paper. If the atom at the thinner end of the wedge is in the plane of the paper, then the atom bonded at the thicker end is in front of the page. The idea is that, if a tube (representing the bond) is pointed toward you, the end that is closer appears to be bigger.
3. A dashed line (IIIIII or IIII···) represents a bond that points away from you. If the central atom is in the plane of the paper, then the atom at the other end of the dashed line is behind the plane of the paper.

Using dash-wedge notation, there are two common ways of representing a tetrahedral C atom. This is illustrated in Figure 3-4, which shows two representations of CH_4—a molecule whose geometry is a tetrahedron. It is important to realize that *both representations are of exactly the same molecule*, with the same 109.5° bond angles. The only difference is the vantage point from which you are viewing the molecule.

TIME TO TRY

To better understand the representations in Figure 3-4, build a molecule of CH_4 using a molecular modeling kit. Hold that molecule up to each of the two dash-wedge representations, and rotate the model until its orientation looks exactly like one representation. Then rotate the model again until it looks exactly like the other.

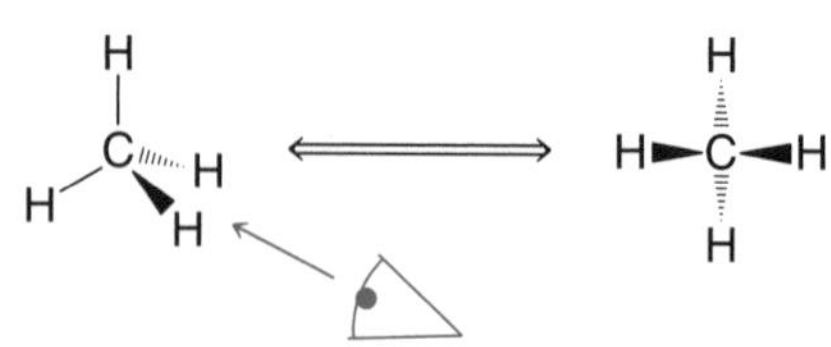

FIGURE 3-4 **Dash-wedge representations of CH_4.** Two different representations of CH_4 using the dash-wedge notation. These representations imply viewing the molecule from different vantage points. The representation on the right is exactly how CH_4 appears if the representation on the left is viewed from where indicated with the eyeball.

Being able to draw and to fully interpret the two different dash-wedge representations of a tetrahedral atom in Figure 3-4 is so important that we will go through some additional exercises that provide insight into the nature of a tetrahedral structure. The exercises specifically demonstrate two different ways to obtain a molecule of CH_4 and should help you understand the characteristics of a tetrahedron much better.

The first way to construct a CH_4 molecule is to begin with a *fictitious* molecule of CH_4 that is square planar—that is, all five atoms are in the same plane (Figure 3-5). In Figure 3-5a, we view the planar molecule from the side, such that two H atoms and the C atom are in the plane perpendicular to the paper—one of those H atoms is pointing toward you, while the other is pointing away from you (pay close attention to the dash-wedge notation here). The H atoms on the left and right are then bent upward until the H—C—H bond angle is 109.5°, whereas the H atoms in front and in back are bent downward until that H—C—H bond angle is 109.5°. The result is a dash-wedge representation that is essentially the same as the one on the left in Figure 3-4, where the C and two Hs are in the plane of the paper, one H is in front, and one H is behind.

In Figure 3-5b, we view the original planar molecule from the top (so that it begins entirely in the plane of the paper) and repeat the process of bending the C—H bonds. In this case, one pair of opposite H atoms is bent toward you, resulting in wedges, while the other pair of opposite H atoms is bent away from you, resulting in dashed lines. The result is the dash-wedge representation on the right in Figure 3-4.

TIME TO TRY

Carry out the "experiment" shown in Figure 3-5a using four pencils to represent the four bonds. To do so, you might need a friend to hold two pencils, while you hold the other two. Begin by having the four pencils touch in the middle, and make sure they are all parallel to the ground. Then bend two pencils upward, and two downward, until the structure appears to be a tetrahedron. Repeat this exercise to reproduce what is shown in Figure 3-5b.

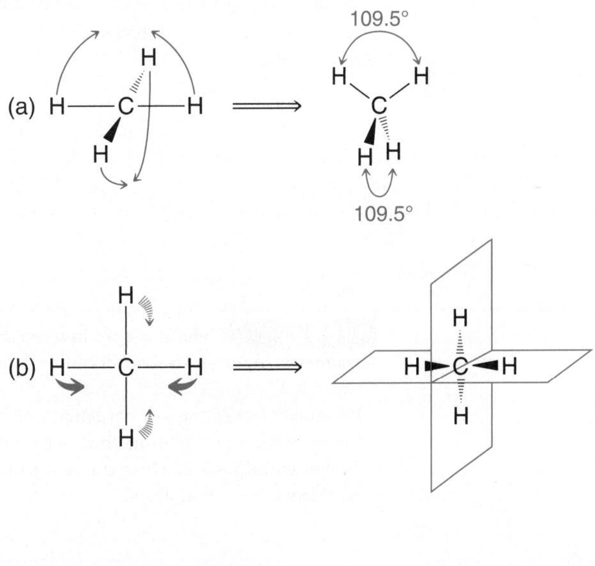

FIGURE 3-5 Tetrahedral geometry from square planar. The tetrahedral geometry of CH_4 can be obtained by beginning with a fictitious molecule of CH_4 that is square planar. (a) The planar molecule is viewed from the side. (b) The planar molecule is viewed from the top. In this representation, we can most readily see the two perpendicular planes in a tetrahedral structure.

One of the main reasons for going through that exercise is to realize an important aspect of symmetry.

In a tetrahedral molecule like methane, there are two perpendicular planes that are defined. One plane is defined by the C atom and a pair of H atoms (three points are needed to define a plane). The second plane is defined by the C atom and the other pair of H atoms.

The fact that those planes are perpendicular to each other is most easily seen at the right in Figure 3-5b. When viewing the tetrahedral structure from that vantage point, it appears that the line connecting one pair of Hs is perpendicular to the line connecting the other pair. Therefore, the plane that contains the C atom and the first pair of H atoms must be perpendicular to the plane that contains the C atom and the second pair of H atoms.

In light of these perpendicular planes, you can envision constructing a molecule of CH_4 using two V-shaped pieces, each with an angle of 109.5° (Figure 3-6). Specifically, orient these pieces so that the plane of one is perpendicular to the plane of the other and the openings of the Vs are pointed toward each other. Bring the two pieces together, and fuse them at the vertices of the Vs. The fusion at the vertices becomes the tetrahedral C. (It is worth noting that one brand of molecular modeling kit takes advantage of this. Two V-shaped pieces, snapped together in exactly the way described here, yield a single tetrahedral atom.)

Having gone through this exercise, you should always think about these V shapes when using dash-wedge notation to depict a tetrahedral atom:

If an atom with tetrahedral geometry is drawn with two bonds in the plane of the paper, forming a V that opens in one direction, then the other two bonds—one dash and one wedge—must form a V that opens in the *opposite* direction.

QUICK CHECK

Use dash-wedge notation to draw two representations of a molecule of NH_3, one similar to the representation depicted in Figure 3-4a and one similar to the representation depicted in Figure 3-4b.

Answer:

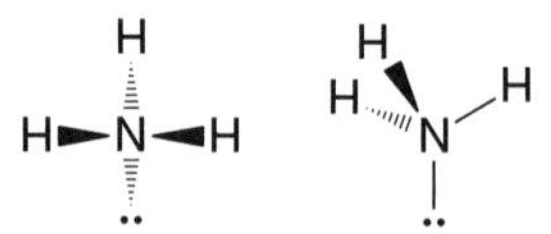

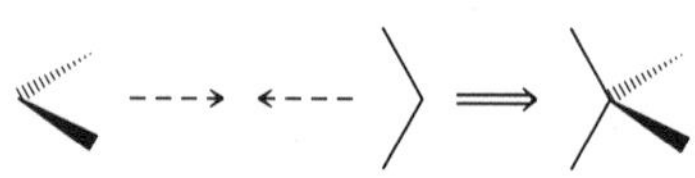

FIGURE 3-6 The V shapes in a tetrahedral geometry. A tetrahedral atom can be constructed from two V-shaped pieces, each with an angle of 109.5°. One V-shaped piece is oriented perpendicular to the second before they are brought together. In the resulting structure, the point at which the Vs are fused together represents the central atom.

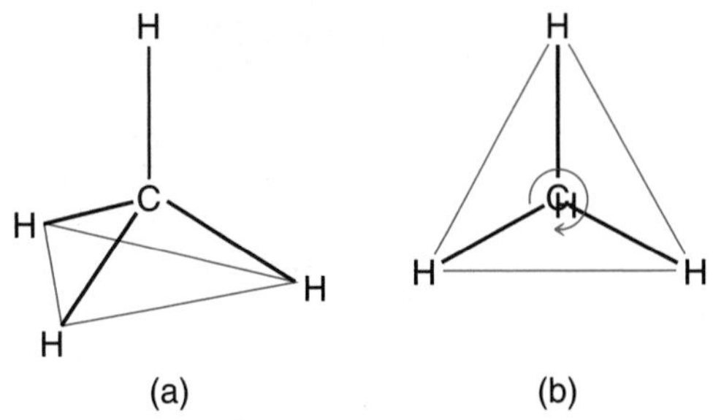

FIGURE 3-7 Threefold symmetry of a tetrahedron. (a) Construction of CH_4 by placing an H atom on top of a trigonal pyramid containing a C atom and three H atoms. (b) View of CH_4 from the top, explicitly showing the H atoms at the corners of an equilateral triangle. Therefore, if the CH_4 molecule is rotated 120° about the topmost C—H bond (indicated by the curved arrow), the molecule will end up looking no different.

The second way to construct CH_4 demonstrates a different aspect of the molecule's symmetry. Realize from Figure 3-2 that a tetrahedron has four faces and each face is an equilateral triangle (all three sides and all three angles are identical). The C atom is at the center of the tetrahedron. Therefore, if we ignore the topmost H atom and focus only on the C atom and the bottom three H atoms, we appear to have a pyramidal structure (Figure 3-7). The C atom sits on top of the three H atoms that make up the bottom face of the tetrahedron. The molecule of CH_4 can therefore be viewed as a pyramid constructed of one C atom and three H atoms, with a fourth H atom placed directly on top of that pyramid.

There are two lessons to learn from the last exercise. One is that *the CH_4 molecule is perfectly symmetric*. That is, we could have chosen any set of three H atoms to use as the pyramid's base because all four faces of a tetrahedron are identical. The second, and perhaps more important, lesson is to understand that a molecule of CH_4 has what is called *threefold rotational symmetry*. What this means is that, as we rotate the molecule 360° about a certain axis, it appears to be the same at three different angles of rotation—120°, 240°, and 360°. This rotational symmetry can be seen in Figure 3-7b, where we view the molecule from the top. The axis about which we rotate the molecule is the top C—H bond, as indicated in the figure.

TIME TO TRY

Use a molecular modeling kit to construct a molecule of CH_4. Hold one of the C—H bonds between your thumb and finger, and rotate that bond in your hand. Observe how the other three C—H bonds rotate like a propeller.

WHY SHOULD I CARE?

The threefold rotational symmetry is important because, as we will see later in this chapter, a portion of a molecule can rotate freely about a single bond. If that portion is a CH_3 group, then rotation will allow all three H atoms to exchange locations in space rapidly, meaning that those H atoms are indistinguishable.

3.3A PITFALLS WITH THE DASH-WEDGE NOTATION

One of the most common mistakes pertaining to the dash-wedge notation is to take it for granted. A student might intend to draw a tetrahedral atom using only straight lines, omitting the dashes and wedges, as is shown in Figure 3-8a. The intention might be to imply, simply by

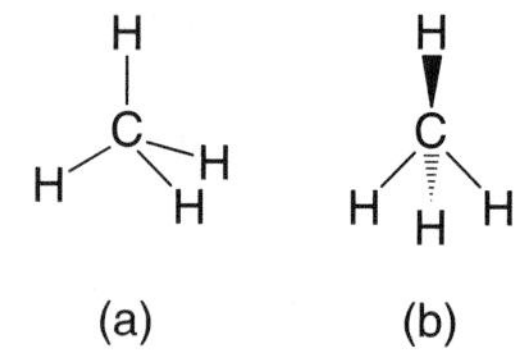

FIGURE 3-8 Common mistakes with dash-wedge notation. (a) The bonds seem to imply tetrahedral geometry, but dash-wedge notation is not used. (b) Dash-wedge notation is used incorrectly. To correct the mistake, both of the bonds not in the plane of the paper—the dash and the wedge—should be pointing upward.

the positioning around the central atom, that certain atoms or groups of atoms are in front of the plane of the paper, while others are behind. However, if the dash-wedge notation is not explicitly shown, then the information provided to anyone interpreting the structure is no different than that provided by the Lewis structure—that is, only the connectivity (which atoms are connected together and by what types of bonds) is actually provided.

By the same token, when interpreting a structure that does not contain dashes and wedges explicitly, a student might infer that, because an atom or a group of atoms appears at a certain location around the central atom, it must be in front of the plane of the paper. The student might also infer that another atom is behind the plane of the paper. As before, unless a tetrahedral atom is drawn explicitly with the dashes and wedges, those inferences are not valid.

Another common mistake with the dash-wedge notation arises despite good intentions by the student. That is, the student includes the dashes and wedges, but does so incorrectly, as shown in Figure 3-8b. If we recall (see Figure 3-6) that the tetrahedral C atom is the result of fusing together two perpendicular Vs that open in opposite directions, then we should see right away that the representation in Figure 3-8b is invalid. In Figure 3-8b, the Vs are not opening in opposite directions. The V that is in the plane of the paper opens downward. The other two bonds, comprising the second V (perpendicular to the first), *should* open in the upward direction. In other words, both of the bonds that are not in the plane of the paper should be pointing upward. In Figure 3-8b, one points upward, and the other points downward, resulting in a representation that does not accurately depict the true tetrahedral structure. Because of this, such a structure is, once again, no more useful than a simple Lewis structure and thus does not actually provide any three-dimensional information.

WHY SHOULD I CARE?

Proper dash-wedge notation is critical to *stereochemistry*, a topic we encounter in Chapter 4. When stereochemistry is relevant to a particular molecule, changing the dash-wedge representation can lead to an entirely different molecule! In order to depict the correct molecule, dash-wedge representation must be accurate. Therefore, when you make a mistake such as those in Figure 3-8, your structure will not have the proper information, and you will lose points on exams!

3.4 Rotations about Single and Double Bonds

The aspects of geometry we have examined so far have dealt only with the relative locations of groups about a given central atom. But what about aspects of geometry that extend further throughout the molecule—namely, aspects of bond rotation? When two atoms are bonded together—each with their separate geometries—can they (along with their attached groups) rotate relative to each other about that bond?

The answer is that it depends on the type of bond.

Groups that are connected by a single bond can freely rotate relative to each other.

Groups that are connected by a double bond cannot freely rotate—they are locked in place.

A complete understanding of why this is so requires an in-depth discussion of the various types of orbitals that exist in molecules, which will be left to your full-year course. For now, it will suffice just to know that these rotational characteristics of bonds are built into molecular modeling kits.

Trust what your molecular modeling kit tells you.

Figure 3-9, for example, shows molecules of $H_3C{-}CH_3$ and $H_2C{=}CH_2$ constructed from a modeling kit. As can be seen in these models, the single bond allows the CH_3 groups in the first molecule to rotate freely relative to each other, but the double bond prevents rotation of the CH_2 groups in the second molecule.

TIME TO TRY

Using a molecular modeling kit, construct the two molecules in Figure 3-9, and verify that the modeling kit allows for free rotation about single bonds, but not double bonds.

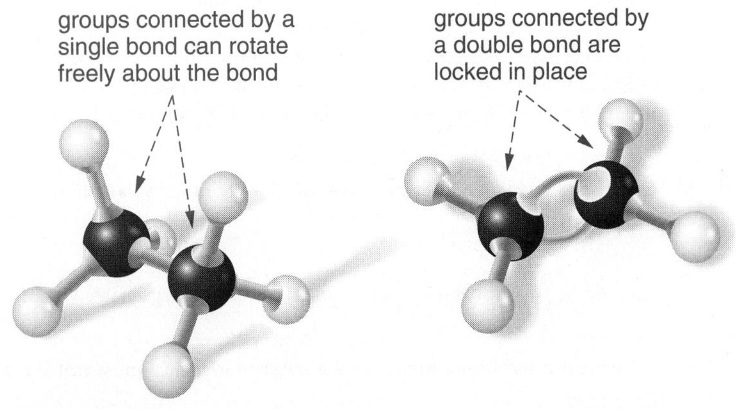
groups connected by a single bond can rotate freely about the bond
groups connected by a double bond are locked in place

FIGURE 3-9 **Modeling kits and rotations about single and double bonds.** Molecules of (a) $H_3C{-}CH_3$ and (b) $H_2C{=}CH_2$ built from a molecular modeling kit. The single bond in (a) allows the CH_3 groups to rotate freely relative to each other, but the double bond in (b) prevents rotation of the CH_2 groups.

FIGURE 3-10 Planarity of atoms involved in a double bond. A molecule of 2-methyl-2-heptene. The five C atoms and the H atom encompassed by the dotted rectangle must all be in the same plane.

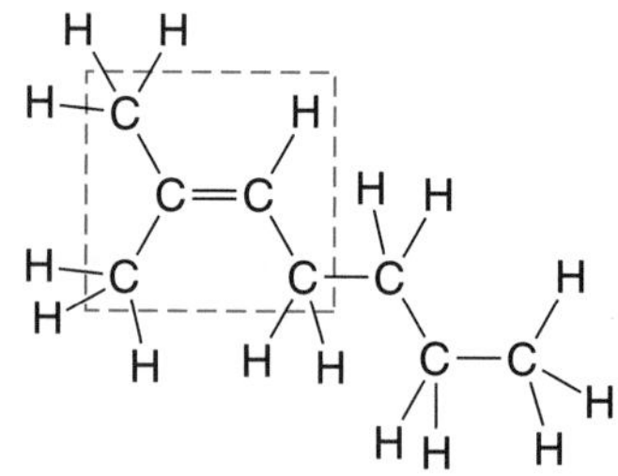

Notice that the CH_2 groups in Figure 3-9b are not locked in just any position relative to each other, but rather all six atoms are in the same plane. In general,

Two atoms that are connected by a double bond, along with any atoms to which they are *directly* bonded, must lie in the same plane.

This idea is emphasized with a more complex molecule, 2-methyl-2-heptene, in Figure 3-10. Specifically, the two C atoms that are doubly bonded are required to be in the same plane as the other three C atoms and the H atom to which they are directly bonded.

3.5 Bond Dipoles and Polarity

We learned earlier that atoms can be bonded through the sharing of a pair of electrons—that is, a covalent bond. If a bond is perfectly covalent, the pair of electrons is shared equally. Such is the case in a molecule of H_2. In the H—H bond, the pair of electrons is shared between identical atoms. Therefore, there is no reason for the electrons to prefer one H atom over the other.

The pair of electrons in the H—H bond is shared equally because the H atoms have exactly the same electronegativity. **Electronegativity** can be thought of as an atom's *pull of electrons toward itself when that atom is part of a covalent bond.* The electronegativity of an atom is quantified by a scale that goes up to a value of about 4.0, and those values are listed for some key atoms in the periodic table on the inside front cover of this book. Recall, however, that electronegativity is one of the periodic table trends that you learned in general chemistry.

Electronegativity generally increases up the periodic table and also from left to right across the periodic table.

Because no two atoms have exactly the same electronegativity, we arrive at an important result.

A covalent bond between atoms that are not identical must not share the pair of electrons equally.

A molecule of H—F, for example, must not share the electrons equally between the H and F atoms. According to the periodic table trend, the F atom has a much higher electronegativity than the H atom and, as a result, has a much stronger pull, drawing the shared electrons toward itself.

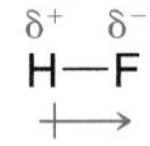

FIGURE 3-11 **The bond dipole of the H—F bond.** (*top*) The F atom is more electronegative than the H atom, so F receives more of a share of the bonding electrons than does H. Thus, a partial negative charge develops on F. Because the molecule is uncharged overall, a partial positive charge is left on H, equal in magnitude to the partial negative charge on F. (*bottom*) The resulting bond dipole is depicted with an arrow that points from the atom that has the partial positive charge to the atom that has the partial negative charge.

Consequently, the F atom has more than its fair share of the pair of electrons and bears a partial negative charge (δ^-, read as "delta minus"), as shown in Figure 3-11. Since the molecule's total charge is zero, there must be a balance of positive charge on the H atom (δ^+, read as "delta plus"). Such *a separation of charge, caused by the unequal sharing of electrons in a covalent bond*, is called a **bond dipole moment,** or **bond dipole** for short.

To get a better feel for a given bond dipole, an arrow is drawn that explicitly shows the direction in which the bonding electrons are being pulled (Figure 3-11). The head of the arrow is therefore analogous to the δ^-, and the tail of the arrow is analogous to the δ^+. To remind you of this, a short line is drawn through the tail of the arrow to resemble a "+."

✔ **QUICK CHECK**

Determine which of the indicated covalent bonds below are polar. For those that are, draw in the dipole arrow, and add "δ^+" and "δ^-."

1. Cl—Cl ______
2. H_3C—F ______
3. Br—I ______
4. HC≡N ______
5. N≡N ______
6. O=CH_2 ______

Answers: 1. Nonpolar. 2. Arrow points from C to F; δ^+ on left and δ^- on right. 3. Arrow points from I to Br; δ^+ on right and δ^- on left. 4. Arrow points from C to N; δ^+ on left and δ^- on right. 5. Nonpolar. 6. Arrow points from C to O; δ^+ on right and δ^- on left.

If there are bond dipoles present in a molecule, each contributing a δ^+ and a δ^-, then it is possible for one end of the overall molecule to bear a δ^+, while the other end bears an equal and opposite δ^-. In other words, there can be a **net molecular dipole moment,** also called a **permanent dipole moment,** making the entire molecule **polar;** a trivial example is the case of HF mentioned previously, given that its only bond is a polar covalent bond. On the other hand, it is also possible for the bond dipoles of a molecule to add up in a certain way so as to completely cancel one another—that is, the molecule can be **nonpolar.**

The net molecular dipole is determined by adding up all of the bond dipoles within the molecule.

Such bond dipoles can be added up "vectorially" to determine the overall molecular dipole moment. That is, each bond dipole can be split into its component vectors along the *x*, *y*, and *z* axes, allowing you to add up all of the component vectors along one axis separate from the ones along the other axes. Two vectors that point in the same direction along an axis combine to make a larger vector in that direction, whereas two vectors that point in opposite directions will partially or fully (if they are the same magnitude) cancel each other. Once the component vectors are summed along each of the three axes, the resulting vectors can be added together.

PICTURE THIS

If you are not comfortable adding vectors, think about a tug-of-war. Each bond represents a rope tying two atoms together, and each bond dipole represents a tug-of-war between those two atoms; the bond dipole arrow is drawn in the direction of the atom that is pulling harder. In a molecule with several bonds, each with a bond dipole of its own, think about several ropes and several tugs-of-war going on at the same time. If all of those imaginary tugs-of-war will result in the overall molecule being pulled in one direction, then the molecule is polar. *The direction of the net molecular dipole is the same as the direction in which the molecule will move if each bond dipole is treated as a tug-of-war.* Conversely, if all of the imaginary tugs-of-war will result in no movement of the overall molecule, then the molecule is nonpolar.

To help you better understand this idea, let's determine the net molecular dipole for each molecule in Figure 3-12. The first is CO_2 (Figure 3-12a). From VSEPR theory, we know that the geometry is linear about the C atom and that the O atoms are on opposite sides. There are two bond dipoles in this molecule—one for each C═O bond—each represented by a thin blue dipole arrow. One is pointed from the C atom toward one O atom, and the other is pointed from the C atom toward the second O atom. Treating each bond dipole as a tug-of-war between an O and the C, it appears that the C atom is being pulled equally in opposite directions. The result of our imaginary tugs-of-war is that the molecule does not move, and we associate this with the molecule being nonpolar.

✔ QUICK CHECK

Draw an accurate three-dimensional structure of BeF_2, include the bond dipoles, and add them to determine the direction of the net dipole.

Answer: The molecule is linear, F—Be—F, so the Be—F bond dipoles point in opposite directions (from Be to each F). The bond dipoles therefore cancel, making the molecule nonpolar.

Next, we have F—C≡N (Figure 3-12b), which is also a linear molecule. Both the F and the N atoms are more electronegative than C, such that the C—F and C—N bond dipoles originate from the C atom and point to the F and N atoms, respectively. In terms of our imaginary

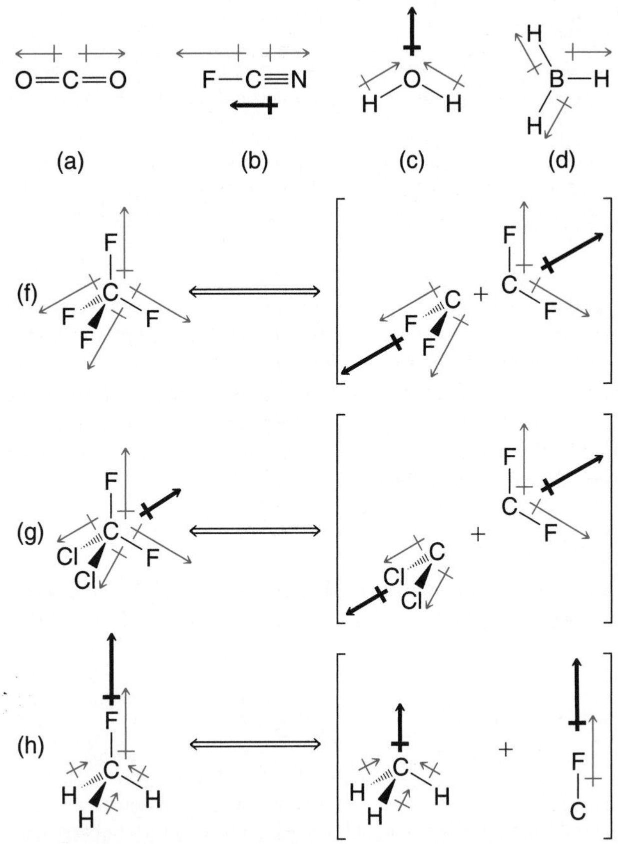

FIGURE 3-12 **Net molecular dipole moments.** (a) The two C=O bond dipoles (thin blue arrows) are equal in magnitude and point in opposite directions, so they cancel, leaving the molecule nonpolar. (b) The C—F and C—N bond dipoles sum to the net dipole (thick black arrow), which points toward F. (c) The two O—H bond dipoles sum to the net molecular dipole that bisects the H—O—H bond angle and points from the H atoms to the O atom. (d) The three B—H bond dipoles cancel out to leave no net dipole moment. (e) The three N—H bond dipoles sum to the net dipole moment that points from the center of triangle defined by the H atoms toward the N atom. (f) The four C—F bond dipoles at the left are split into two perpendicular Vs inside the brackets. The net dipoles of the Vs (thick arrows) cancel out, leaving the molecule nonpolar. (g) The net dipoles of the Vs sum to the overall net dipole moment shown at the right, which bisects the F—C—F angle and points toward the F atoms. (h) The four bond dipoles that are shown at the right are split into a CH_3 trigonal pyramid and a C—F bond inside the brackets. The two net dipoles point upward, so the overall net dipole also points upward.

tugs-of-war, the F atom is pulling the C atom to the left, and the N atom is pulling the C atom to the right. Since the F atom is more electronegative than the N atom, it is pulling the C atom harder, and the entire molecule is pulled to the left. Therefore, we say that the molecule is polar, and the net molecular dipole points to the left.

✔ **QUICK CHECK**

Draw an accurate three-dimensional structure of OCS (C is the central atom), include the bond dipoles, and add them to determine the direction of the net dipole.

Answer: The molecule is linear, S=C=O, with the C=O and C=S bond dipoles pointing in opposite directions (from C to O and C to S, respectively). The bond dipoles partially cancel, but because O has a greater electronegativity than S, the C=O bond dipole is larger and gives rise to a net dipole that points toward O.

From the discussion at this chapter's beginning, we already know that H_2O (Figure 3-12c) is a polar molecule. This is because it is a bent molecule, with two bond dipoles. Both originate from an H atom and point toward the O atom. Rather than thinking about it in terms of tug-of-war, consider it a "push-of-war," where both H atoms are pushing on the O atom. One H atom is pushing the O atom up and to the right, and the other is pushing it up and to the left. The result is the movement of the entire molecule directly upward, which is the same as the direction of the net molecular dipole.

✔ **QUICK CHECK**

Draw an accurate three-dimensional structure of FNO (N is the central atom), include the bond dipoles, and add them to determine the direction of the net dipole.

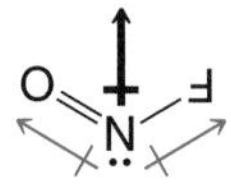

Answer: The N atom has three electron groups, giving it a trigonal planar electronic geometry and a bent molecular geometry. The N—F and N=O bond dipoles do not completely cancel. The net dipole lies between those two bonds. (Because F is more electronegative than O, the net dipole should be tilted slightly toward F. However, in this book, we won't be concerned with that level of detail.)

BH_3 (Figure 3-12d) is a trigonal planar molecule, with the H atoms at the corners of an equilateral triangle and the B atom at the center. There are three B—H bond dipoles to consider. It turns out that hydrogen has a greater electronegativity than boron (this is difficult to tell by the periodic table trend due to the placement of H in the periodic table, but you can verify this by finding the values on the inside front cover), so those bond dipoles originate from the B atom and point toward the different H atoms. Now consider those bond dipoles as tugs-of-war where each H atom is tied to the B atom and pulls it toward the corners of the equilateral triangle. Because of the symmetry of an equilateral triangle, the end result is no movement of the molecule whatsoever, and we say that there is no net molecular dipole.

Next, we examine NH_3 (Figure 3-12e), which is a pyramidal molecule. The H atoms form an equilateral triangle, and the N atom is sitting on top of them, at the triangle's center. Each of the N—H bond dipoles points from an H atom to the N atom. As with H_2O, think of these as pushes-of-war where each H atom is pushing on the N atom. That is, each H atom simultaneously pushes inward and upward. The horizontal (inward) forces cancel out because each

H atom pushes from the corner of an equilateral triangle. As a result, there is no motion in the horizontal direction. What is left is the upward force from each H atom. Therefore, the overall resulting motion of the entire molecule is straight upward, indicating the direction of the net molecular dipole.

✔ QUICK CHECK

Draw an accurate three-dimensional structure of NF_3, include the bond dipoles, and add them to determine the direction of the net dipole.

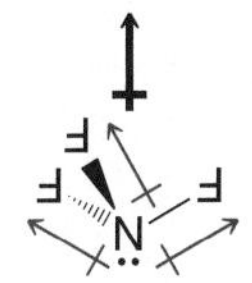

Answer: The N atom has four electron groups, giving it a tetrahedral electronic geometry, but a trigonal pyramidal molecular geometry. The F atoms pull electrons toward the corners of the base, giving a net pull directly along the central axis of the pyramid.

In the final examples, we consider three molecules with a tetrahedral central C atom. The first is CF_4 (Figure 3-12f), where each of the F atoms is at the corner of a tetrahedron and the C atom is at the center. Because F is more electronegative than C, each F atom can be thought of as pulling on the C atom. Is there any resulting movement of the molecule? The answer is no because of the symmetry of a tetrahedron. We can see this better if we split CF_4 into two identical V-shaped portions (similar to Figure 3-6) where C is at the vertex and an F atom is at each end—this is what is shown inside the brackets. For each V-shaped piece, then, the net dipole bisects the F—C—F angle and points from C toward the F atoms. Because CF_4 is composed of two identical Vs, pointing in opposite directions, the net dipole from each V-shaped piece cancels, leaving no net molecular dipole.

Next, let's examine CF_2Cl_2 (Figure 3-12g). Again, we can split this molecule into two V-shaped pieces—a CF_2 and a CCl_2—as shown inside the brackets. As before, the net dipoles from the Vs point in opposite directions. However, because the bond dipole of a C—F bond is stronger than that of a C—Cl bond, the V formed by the CF_2 portion pulls more strongly than that formed by the CCl_2 portion. As a result, the net molecular dipole points in the direction of the opening of the V formed by the CF_2 portion.

Finally, let's consider CH_3F (Figure 3-12h). As shown inside the brackets, we can separate this molecule into an F atom and a pyramidal CH_3 portion (with C on top of the three Hs). Each of the C—H bond dipoles points toward the C atom, so let's think of each H atom as pushing on the C atom. The resulting movement of the CH_3 group is directly upward, similar to what we saw with NH_3. If we then tack on the F atom, we simply add on the bond dipole of the C—F bond, which points toward the F atom. The pull of the F atom is in the same direction as the movement of the CH_3 group, so the overall movement of the CH_3F molecule is directly upward. Therefore, the net molecular dipole is along the C—F bond, pointing in the direction of the F atom.

✔ **QUICK CHECK**

Draw an accurate three-dimensional structure of CH_2Cl_2, include the bond dipoles, and add them to determine the direction of the net dipole.

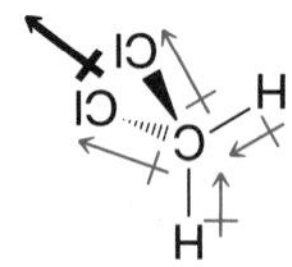

Answer: The molecule is tetrahedral. The two C—H bond dipoles point toward the C atom, and the two C—Cl bond dipoles point toward the Cl atoms. The result is a net dipole that points in the direction of the opening of the V formed by the CCl_2 group.

3.6 Intermolecular Interactions

Thus far in the chapter, we have dealt with molecular shapes and the consequences for the distribution of charge within the molecule. As we will see here, such distribution of charge dictates the type and the strength of the interaction that exists *between* two separate molecules—either two of the same or two different molecules. In other words, these **intermolecular interactions** (or **intermolecular forces**) dictate how well two molecules "stick together"—or, alternatively, how much energy it would take to pull the two molecules apart.

WHY SHOULD I CARE?

Later in this chapter, we see how such intermolecular interactions govern physical properties such as boiling point, melting point, and solubility. In Chapter 6, we also see how intermolecular interactions lead to specific roles that solvents can play in the reactivity of certain types of reactants–so-called *solvent effects*.

The important intermolecular interactions we examine here are (1) ion–ion interactions, (2) ion–dipole interactions, (3) dipole–dipole interactions, (4) hydrogen-bonding, (5) dipole–induced dipole interactions, and (6) induced dipole–induced dipole interactions (also called dispersion forces). The names of these interactions are descriptive of the species that are involved. For example, ion–ion interactions are those that keep two ions stuck together; dipole–induced dipole interactions are those that keep a neutral molecule that has a net molecular dipole stuck to one that has an *induced dipole moment* (which we will introduce later). It is important to keep in mind, however, that even though these interactions involve different types of species, all intermolecular interactions originate from the same physical phenomenon:

Opposite charges attract, and the stronger the concentration of one or both of those charges is, the stronger the two species stick together.

REALITY CHECK

The attraction between opposite charges is analogous to the attraction between the opposite poles of magnets—north and south. Magnets are stabilized when they are attracted by their opposite poles, which is why it takes energy to separate them. In much the same way, opposite charges in close proximity to one another undergo stabilization and therefore require energy to be separated.

Despite the relatively simplistic origin of these intermolecular interactions, many students want to just memorize the names of the interactions and their relative strengths. It is very important that you do *not* do this! If you try to memorize various aspects of intermolecular interactions, along with the numerous other bits and pieces throughout the course, then intermolecular interactions will be among the earliest to fall by the wayside. But intermolecular interactions are important throughout all of organic chemistry, and you will likely revisit them on your final exam or on the MCAT—either directly or (worse) indirectly.

3.6A ION–ION INTERACTIONS

Ion–ion interactions are conceptually the simplest of the intermolecular interactions and therefore do not require a substantial discussion. As the name suggests,

Ion–ion interactions exist between two ions.

For example, a positive ion like Na^+ is attracted to a negative ion like Cl^-. In fact, as you are probably already aware, strong ionic bonds are formed between those two ions, making a very stable ionic compound—NaCl, or table salt.

Beyond simply recognizing the presence of the ion–ion interactions, it is also important to know the relative strength of such interactions.

Ion–ion interactions are the strongest of all the intermolecular interactions.

This is evidenced by the extremely high melting points and boiling points we often see with ionic compounds (see Section 3.7). Furthermore, this is consistent with the earlier statement that the stronger the concentration of charge, the stronger the interaction. Ions bear full charges—positive or negative—and therefore have the strongest concentration of charge of any organic chemistry species.

3.6B ION–DIPOLE INTERACTIONS

As with other intermolecular interactions, the name *ion–dipole interaction* is descriptive of the types of species involved.

Ion–dipole interactions are the attractive interactions that exist between an ion—either positive or negative—and a neutral molecule that has a net molecular dipole.

One example is the interaction between an Na^+ ion and a molecule of water. Another is between a molecule of water and a Cl^- ion.

WHY SHOULD I CARE?

As you may have guessed, and as you will see at the end of this chapter, ion–dipole interactions are what are primarily responsible for water's ability to dissolve ionic compounds such as NaCl.

Before we continue, let's recall that a net molecular dipole, also called a permanent dipole, exists in an uncharged molecule when one end bears a partial positive charge (δ^+) and the other bears an equal and opposite partial negative charge (δ^-). In Section 3.4, we saw that the direction and magnitude of a dipole moment are represented by an arrow ($+\!\!\longrightarrow$), where the arrow's head represents the δ^- and the tail represents the δ^+. Therefore, a positive ion like Na^+ will stick to the head of the arrow, and a negative ion like Cl^- will stick to the tail (Figure 3-13).

TIME TO TRY

In Figure 3-13, label the molecular partial charges δ^+ and δ^- for each water molecule, and draw a circle around the pair of opposite charges that undergo attraction.

The concentrations of charge at the head or the tail of a dipole moment are much smaller than the concentration of charge of an ion with a full negative or positive charge; the charges within a dipole are only *partial* charges. Therefore,

Ion–dipole interactions are weaker interactions than ion–ion interactions.

Realize, however, that the magnitude of a dipole moment can vary, depending on the specific makeup of the molecule. Consequently, the strength of an ion–dipole interaction can vary as well.

In general, the strength of an ion–dipole interaction increases as the magnitude of the dipole moment increases.

For example, consider two different interactions—one between an Na^+ ion and the dipole moment of a molecule of $H_3C{-}F$ and the second between that Na^+ ion and the dipole moment of a molecule of $H_3C{-}Cl$. Because F has a greater electronegativity than Cl, we expect that

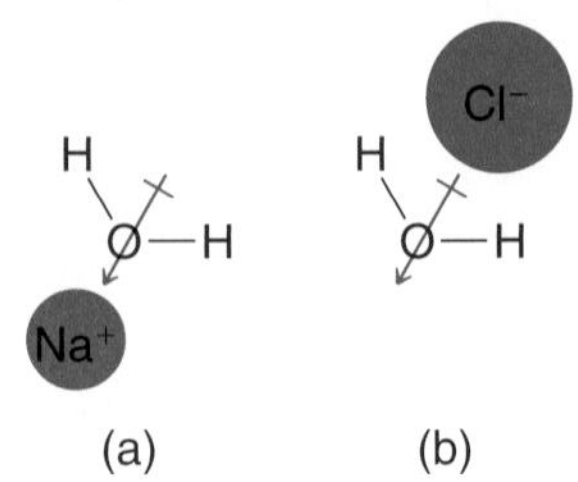

FIGURE 3-13 Ion–dipole interactions. (a) Shows the attraction between the positive charge of Na^+ and the partial negative charge of the dipole moment in a molecule of water. (b) Shows the attraction between the negative charge of Cl^- and the partial positive charge of the dipole moment in a water molecule.

the dipole moment of $H_3C—F$ is of greater magnitude than that of $H_3C—Cl$. Therefore, we expect the Na^+ ion to stick better to a molecule of $H_3C—F$ than to a molecule of $H_3C—Cl$. That is, we expect the ion–dipole attractive interactions to be greater.

QUICK CHECK

Show how F^- sticks to a molecule of $H_3C—Cl$. Label each partial and full charge, and circle the attractive interaction.

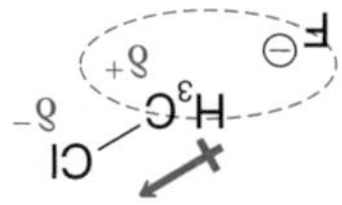

Answer: F^- is an ion and has a full negative charge. H_3CCl is a neutral molecule with a permanent dipole that points from C to Cl. Therefore, δ^+ is at the end with the C atom, and δ^- is at the end with the Cl atom. F^- sticks to the end with the C atom.

3.6C DIPOLE–DIPOLE INTERACTIONS

Just as with the other intermolecular interactions, the name *dipole–dipole interaction* is descriptive of the types of species involved.

Dipole–dipole interactions are the attractive forces between two neutral molecules, each with a permanent dipole moment.

They can be two of the same molecule, or they can be different molecules. Furthermore, these molecules can line up in two ways in order to give an overall attractive interaction (Figure 3-14). In both orientations, note that the δ^+ of one dipole moment is near the δ^- of the other—opposite charges attract. Not surprisingly, both of these orientations are called **head-to-tail orientations.**

TIME TO TRY

In Figure 3-14, label each δ^+ and δ^-. Circle each instance where opposite charges attract.

As mentioned in Section 3.6B, the concentrations of charge at the head and the tail of each dipole moment are smaller than that of an ion. Therefore,

FIGURE 3-14 **Head-to-tail orientations.** Two ways in which two molecules can be oriented so as to provide an overall attractive interaction. In each case, the partial positive charge of one dipole is adjacent to a partial negative charge of the other.

Dipole-dipole interactions are weaker than ion-dipole interactions.

For instance, a molecule of acetone, $(CH_3)_2C{=}O$, will stick better to a Cl^- ion than it will to another molecule of acetone (Figure 3-15).

In addition, dipole–dipole interactions become stronger if one or both of the dipole moments are increased in magnitude. Therefore, because oxygen has a greater electronegativity than nitrogen, the dipole–dipole interactions are stronger between two molecules of H_2O than between two molecules of NH_3 (Figure 3-16).

✔ QUICK CHECK

Which do you think will stick together more strongly via dipole-dipole interactions—two molecules of $H_2C{=}CH_2$ or two molecules of $H_2C{=}O$? Why?

Answer: Two molecules of $H_2C{=}O$. That's because $H_2C{=}O$ has a strong permanent dipole that points from C to O, giving rise to dipole-dipole interactions between two such molecules. By contrast, $H_2C{=}CH_2$ is non-polar, so $H_2C{=}CH_2$ molecules cannot stick together via dipole-dipole interactions.

3.6D HYDROGEN-BONDING

Despite the fact that *hydrogen-bonding* (often abbreviated "H-bonding") is classified as a different intermolecular interaction, it is very closely related to the dipole–dipole interactions we examined in the previous section. In fact, many organic chemists treat hydrogen bonding essentially as a subcategory of dipole–dipole interactions. However, there is one major difference between the two: Whereas dipole–dipole interactions involve the *net molecular dipole*

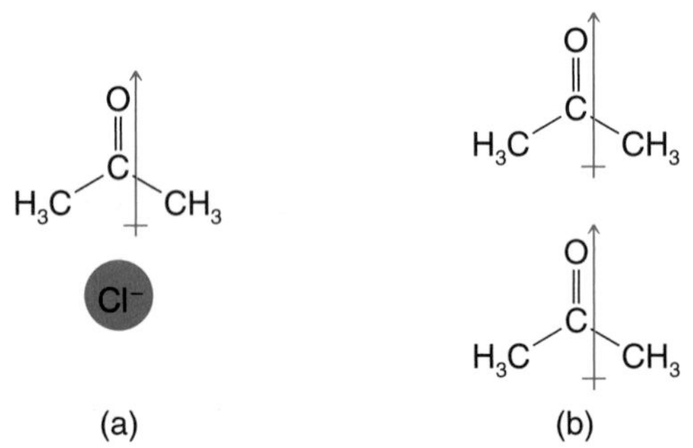

FIGURE 3-15 **Relative strengths of ion-dipole and dipole-dipole interactions.** (a) Ion-dipole interactions lead to the attraction between a Cl^- ion and the dipole moment of a molecule of $(CH_3)_2C{=}O$. (b) Dipole-dipole interactions lead to the attraction between two molecules of $(CH_3)_2C{=}O$. The attractive forces in (a) are stronger than those in (b) because the δ^+ on $(CH_3)_2C{=}O$ is attracted more strongly to a full negative charge on Cl^- than it is to a δ^- on another molecule of $(CH_3)_2C{=}O$.

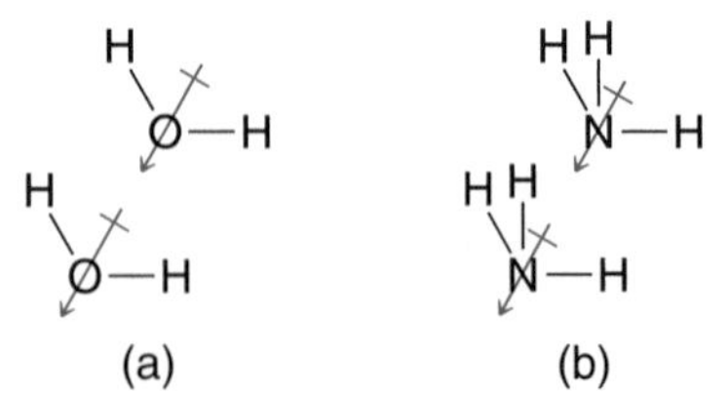

FIGURE 3-16 **Relative strengths of dipole-dipole interactions.** (a) Dipole-dipole interaction between two molecules of H_2O. (b) Dipole-dipole interaction between two molecules of NH_3. With the greater electronegativity of O than N, the dipole moment of H_2O is larger than that of NH_3, so we expect that the dipole-dipole interaction is stronger in (a) than in (b).

moment of one molecule and the *net molecular dipole moment* of another, hydrogen-bonding involves a *bond dipole* in one molecule and a *bond dipole* in another. Moreover, hydrogen bonding requires a specific arrangement and identity of atoms.

A **hydrogen bond** is formed when atoms are arranged in the order X—H·····Y, where X and H are covalently bonded together and Y is either an atom on a different part of the same molecule or a part of an entirely separate molecule.

The X and Y atoms must be F, O, or N.

Notice, in particular, that the H atom is between X and Y; the presence of the hydrogen-bonding interaction (i.e., the hydrogen bond) is represented by the dotted line and implies an attraction between H and Y.

LOOK OUT

A common mistake is to say that, in the arrangement X—H·····Y, the X—H bond is the hydrogen bond. It is not. Instead, *the hydrogen bond is the entire set of those three atoms in that arrangement*. The X—H portion is a *covalent bond* between X and H.

F, O, and N are the required atoms because of their high electronegativities. As a result, the X—H bond dipole is quite large in magnitude—that is, there is a large concentration of negative charge on the X atom, leaving a large concentration of positive charge on the H atom. In addition, there should be a large concentration of negative charge on the Y atom, given that Y also has a high electronegativity. Therefore, when the two molecules come together in a fashion yielding the arrangement X—H·····Y, the large partial positive charge of the H atom is adjacent to the large partial negative charge of the Y atom, leading to a very stable situation. In other words, the molecules are stuck together rather strongly. Figure 3-17a demonstrates this point with a hydrogen bond between two molecules of H_2O, and Figure 3-17b demonstrates this with a hydrogen bond between a molecule of F—H and a molecule of $(CH_3)_2C{=}O$.

In every hydrogen bond that is formed, there is a hydrogen-bond donor and a hydrogen-bond acceptor.

A **hydrogen-bond donor** is the part of a molecule that contains the formal covalent bond between the highly electronegative X atom and the H atom.

The **hydrogen-bond acceptor** is the electronegative atom Y.

In Figure 3-17a, the donor is the covalent O—H bond in the H_2O molecule in the upper left, and the hydrogen-bond acceptor is the oxygen atom in the molecule in the lower right. In Figure 3-17b, the hydrogen-bond donor is the covalent F—H bond, and the acceptor is the O atom of the other molecule.

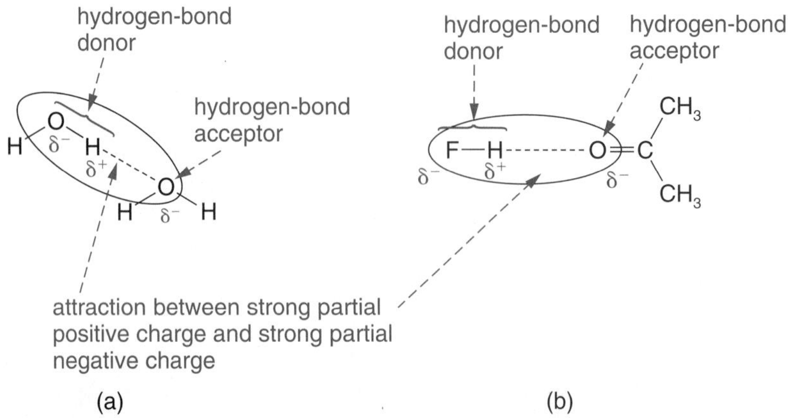

FIGURE 3-17 **Hydrogen bonds.** A hydrogen bond between (a) two molecules of water, and (b) a molecule of HF and a molecule of $(CH_3)_2C{=}O$. In (a), O—H on one molecule is the hydrogen-bond donor, and O on the other molecule is the hydrogen-bond acceptor. In (b), F—H is the hydrogen-bond donor, and O is the hydrogen-bond acceptor. In each case, the oval encompasses the complete hydrogen bond, and the attraction is between the partial positive charge on the hydrogen-bond donor and the partial negative charge on the hydrogen-bond acceptor.

✔ QUICK CHECK

Draw the hydrogen bond that forms between two molecules of HF. Label the hydrogen-bond donor and the hydrogen-bond acceptor.

Answer: H—F·····H—F; the hydrogen-bond acceptor is the F atom on the left, and the hydrogen-bond donor is the H—F bond on the right.

One reason we must discuss hydrogen-bond donors and acceptors is that not all molecules that contain F, O, or N can form hydrogen bonds.

If there are only hydrogen-bond acceptors, then no hydrogen bonds can be formed.

For example, no hydrogen bonds can form between two molecules of $H_3C{-}F$, between two molecules of $(CH_3)_2C{=}O$, or between one molecule of $H_3C{-}F$ and one molecule of $(CH_3)_2C{=}O$. In all three of these scenarios, each molecule has a hydrogen-bond acceptor only. By contrast, hydrogen bonds can be formed between a molecule of $H_2C{=}O$ and a molecule of water (that is, $H_2C{=}O\cdots\cdots H{-}OH$), given that there is a donor-acceptor pair.

Another reason to discuss hydrogen-bond donors and acceptors is to gain a sense of how strong hydrogen-bonding can be. We do so by counting the total number of hydrogen-bond donors and hydrogen-bond acceptors of the species involved.

For two molecules that can stick together via hydrogen-bonding, the greater the number of hydrogen-bond donors and/or acceptors, the stronger the hydrogen-bonding interaction.

Consider, for example, two molecules of H_2O versus two molecules of ethylene glycol ($HO—CH_2CH_2—OH$), which is a product sold as antifreeze for your car. A pair of water molecules has a total of four hydrogen-bond donors (each O—H bond) and two hydrogen-bond acceptors (each O atom). A pair of ethylene glycol molecules has four hydrogen-bond donors (each O—H bond) and four hydrogen-bond acceptors (each O atom). Because two molecules of ethylene glycol have a greater number of donors and acceptors than two molecules of water, we expect the former to stick together much more strongly.

QUICK CHECK

Which will undergo stronger hydrogen-bonding with water?

Answer: The second molecule. A molecule of water and the first molecule will have a total of three H-bond donors and four H-bond acceptors. A molecule of water and the second molecule will have a total of five H-bond donors and four hydrogen-bond acceptors.

Before moving on, the last thing we must consider about hydrogen-bonding is the relative strength of this type of interaction compared to the other ones we have discussed so far.

In general, hydrogen bonding is weaker than ion–dipole interactions, but stronger than dipole–dipole interactions.

Hydrogen-bonding is weaker than an ion–dipole interaction because a hydrogen bond is really between a *partial* positive charge and a *partial* negative charge, whereas an ion–dipole interaction involves a *partial* charge and a *full* charge. Hydrogen-bonding is stronger than a dipole–dipole interaction in part because the X and Y atoms of the hydrogen bond are very highly electronegative. Therefore, in the H·····Y interaction, the Y atom has a large partial negative charge, and the H atom has a large partial positive charge. Moreover, the H atom is very small, so the partial positive charge on H and the partial negative charge on Y can get very close together.

3.6E DIPOLE–INDUCED DIPOLE INTERACTIONS

The next type of intermolecular interaction we will examine is the dipole—induced dipole interaction.

Dipole–induced dipole interactions exist between two uncharged species–one that has a net molecular (permanent) dipole and another that does not.

For example, dipole–induced dipole interactions are responsible for the attraction between a molecule of HF, which has a net molecular dipole, and a molecule of H_2, which is nonpolar. But how could two molecules like these stick together? You might expect that there is no attraction that could exist because a neutral molecule that has no permanent dipole has no region that bears an excess charge—either positive or negative. Therefore, the partial positive charge on the molecule that has the permanent dipole doesn't see any excess negative charge on the other molecule. Likewise, the partial negative charge on the molecule with the permanent dipole doesn't see any excess positive charge on the other molecule. And this is in fact the case—at least initially. However, it turns out that the molecule with a permanent dipole *induces* a dipole (although a very weak and temporary one) on the molecule with no permanent dipole, and the subsequent interaction *becomes* a type of dipole–dipole interaction. There is, however, a major difference between this interaction and the dipole–dipole interactions we saw before.

In dipole–dipole interactions, the interaction is between two permanent dipoles, whereas in dipole–induced dipole interactions, the interaction is between a permanent dipole of one molecule and an induced (or temporary) dipole of the other.

How does this induced dipole happen? The molecule with a permanent dipole approaches the other molecule in one of two ways—either with the positive end of the dipole pointed toward the nonpolar molecule or with the negative end pointed toward the nonpolar molecule. If the positive end is pointed toward the nonpolar molecule, the electrons (which are negatively charged and are relatively free to move around) surrounding the nuclei of the neutral molecule will subsequently be attracted toward the positive end of the dipole. Therefore, the distribution of electrons in the nonpolar molecule has changed! There is a buildup of electrons (δ^-) on one side of the nonpolar molecule (the side closer to the partial positive charge of the molecule with the permanent dipole) and a deficiency of electrons (δ^+) on the other side (Figure 3-18a). In other words, the net dipole from the polar molecule has *induced* a small, temporary dipole on the nonpolar molecule, and the nonpolar molecule has become **polarized.** The two molecules can then stick together, although weakly. A similar situation exists when the polar molecule approaches with its negative end (Figure 3-18b) except that, instead of attracting electrons to its side, it repels electrons to the other side.

TIME TO TRY

In Figure 3-18, write "δ^+" and "δ^-" on each of the four HF molecules. On the right-hand sides of Figures 3-18a and 3-18b, circle each attractive interaction.

As with the other intermolecular interactions we have examined, it is important to know the relative strength of dipole–induced dipole interactions.

Dipole–induced dipole interactions are typically weaker than dipole–dipole interactions.

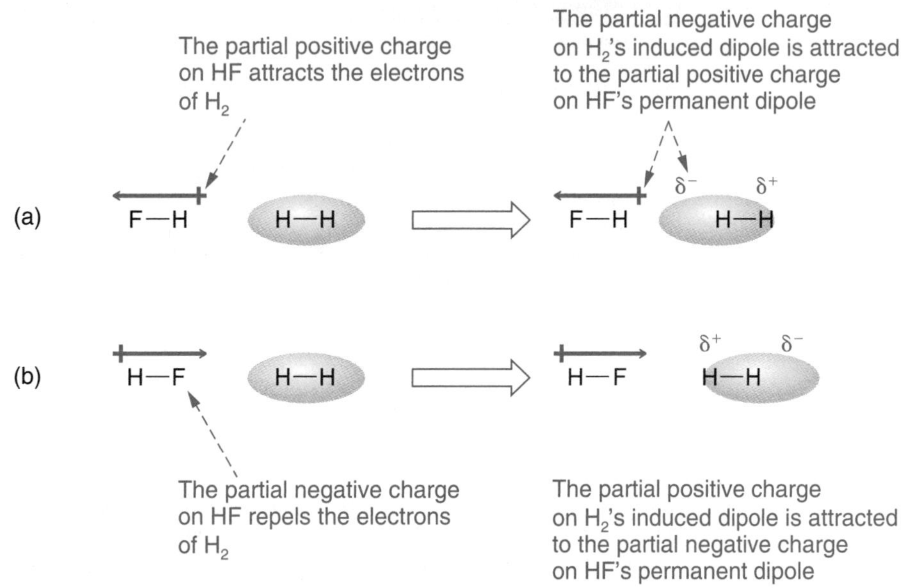

FIGURE 3-18 **Dipole–induced dipole interactions.** (a) At the left, the H_2 molecule initially has an equal distribution of charge about the two nuclei, and the positive end of the HF molecule's dipole points toward H_2. The electrons of H_2 are therefore attracted toward HF. At the right, an induced dipole on H_2 is shown to have appeared as a result of electrons building up to the left of the molecule, leaving an excess of positive charge to the right. As a result, the two molecules stick together due to the attraction of the opposite charges. (b) At the left, the H_2 molecule initially has an equal distribution of charge about the two nuclei, and the negative end of the HF molecule's dipole points toward H_2. The electrons of H_2 are therefore repelled by HF. At the right, an induced dipole on H_2 is shown to have appeared as a result of electrons building up to the right of the molecule, leaving an excess of positive charge to the left. As a result, the two molecules stick together due to the attraction of the opposite charges.

This should make sense because dipole–induced dipole interactions are very similar to dipole–dipole interactions, but the dipole that is induced on the nonpolar molecule is generally very weak in magnitude. And as we saw in Section 3.5C, the magnitude of the dipoles in such an interaction dictates the strength of the interaction.

3.6F INDUCED DIPOLE–INDUCED DIPOLE INTERACTIONS (ALSO KNOWN AS LONDON DISPERSION FORCES)

To understand induced dipole–induced dipole interactions and their consequences, we must first better understand the behavior of electrons in a molecule that has no net dipole, such as H_2. In a molecule of H_2, there are two total electrons, which are shared between the two H nuclei. They are shared equally because the H atoms are identical and therefore have identical electronegativities. On *average*, then, those electrons spend an equal amount of time around each of the two nuclei, which is what gives rise to the H_2 molecule having no permanent dipole.

However, it is *not* true that at every instant in time the two electrons are exactly opposite each other in the molecule. If we were to take a snapshot of the H_2 molecule at some instant in

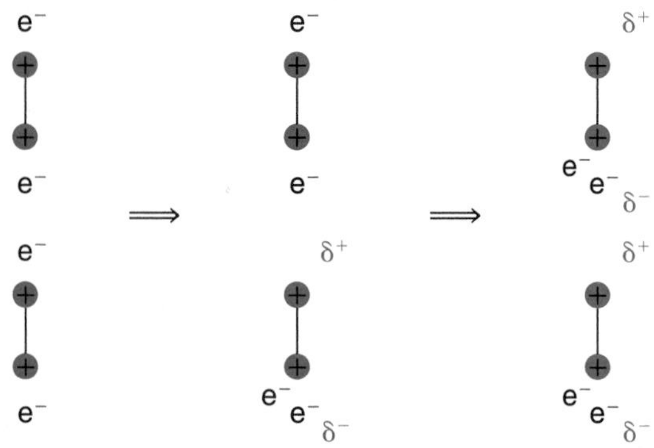

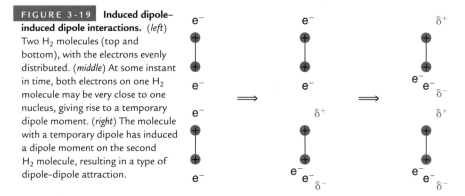

FIGURE 3-19 Induced dipole–induced dipole interactions. (*left*) Two H_2 molecules (top and bottom), with the electrons evenly distributed. (*middle*) At some instant in time, both electrons on one H_2 molecule may be very close to one nucleus, giving rise to a temporary dipole moment. (*right*) The molecule with a temporary dipole has induced a dipole moment on the second H_2 molecule, resulting in a type of dipole-dipole attraction.

time, the electrons might indeed be opposite each other (Figure 3-19, left). However, at some later instant in time, both electrons might be very close to one nucleus and far away from the other nucleus (Figure 3-19, middle). At that moment, there exists an **instantaneous dipole,** reminiscent of the induced dipole we saw in the previous section.

Such an induced dipole is the basis of induced dipole–induced dipole interactions.

Induced dipole–induced dipole interactions arise when an instantaneous dipole *induces* a dipole on another molecule, resulting in an attraction that resembles dipole–dipole interactions.

This is illustrated on the right in Figure 3-19.

Keep in mind two things about induced dipole–induced dipole interactions. The first is how prevalent such interactions are.

Induced dipole–induced dipole interactions exist between any two atomic or molecular species that contain electrons.

This is because of what gives rise to induced dipole–induced dipole interactions—the distortion of those electrons relative to positively charged nuclei. And every species that we will encounter has both nuclei and electrons.

The second thing to keep in mind is the relative strength of induced dipole–induced dipole interactions.

Induced dipole–induced dipole interactions are usually very weak—in general, weaker than dipole–induced dipole interactions.

This is because, in dipole–induced dipole interactions, the dipole that does the inducing is a permanent dipole, which can be rather large in magnitude. By contrast, in induced dipole–induced dipole interactions, the dipole that does the inducing of the second dipole is a temporary dipole, which is often much smaller in magnitude than a permanent dipole.

WHY SHOULD I CARE?

Although induced dipole–induced dipole interactions are typically quite weak, they are significant enough to worry about. As we will learn later, if it were not for these interactions, nonpolar molecules would exist only as gases and could never turn to liquids or solids, even at the coldest of temperatures—nothing would allow those molecules to stick together. However, we know that nonpolar molecules like H_2, O_2, and N_2 *do* liquefy if the temperature is cold enough. For example, N_2 liquefies at around −200°C. (The cold temperatures that are required are a testament to how weak this type of interaction really is.)

We must be careful when comparing the strength of induced dipole–induced dipole interactions to that of other intermolecular interactions. This is because induced dipole–induced dipole interactions are highly variable in strength.

The strength of induced dipole–induced dipole interactions in general increases with the number of electrons in each species.

This is because, with a greater number of electrons, it is easier for the electrons to be pushed around. Therefore, whereas the induced dipole–induced dipole interactions are quite weak for the interaction between two molecules of CH_4 (with only 10 total electrons), they are much stronger for the interaction between two molecules of $CH_3(CH_2)_{16}CH_3$ (with 146 total electrons)—in fact, even stronger than the hydrogen bonding that exists between two molecules of water!

QUICK CHECK

Which will stick together more strongly via induced dipole–induced dipole interactions—two molecules of Br_2 or two molecules of $CH_3(CH_2)_4CH_3$?

Answer: A molecule of Br_2 has 70 total electrons, whereas one of $CH_3(CH_2)_4CH_3$ has 50 total electrons. Because the strength of induced dipole–induced dipole interactions generally increases with the number of total electrons, such interactions will be stronger with two molecules of Br_2.

3.6G SUMMARY OF INTERMOLECULAR INTERACTIONS

Each of the intermolecular interactions we have learned about in this chapter was introduced independent of the other intermolecular interactions. Rarely, however, can we say that one and only one intermolecular interaction is in effect. Under most circumstances where there are attractive forces between two molecules, there is more than one intermolecular interaction present. In fact, typical organic chemistry problems often expect you to consider multiple intermolecular interactions simultaneously. Some ask you to determine exactly which of the six intermolecular interactions exist between a pair of given molecules. Others have you determine the relative strengths of the intermolecular interactions that are active.

As an example, suppose we are asked to determine what intermolecular interactions exist between a molecule of NH_3 and a molecule of H_2O. First, we see that hydrogen bonding can exist between them. Either the O—H of the H_2O molecule can be the H-bond donor with the N atom of the NH_3 molecule as the H-bond acceptor, or the N—H of the NH_3 molecule can be the donor with the O atom of the H_2O molecule as the acceptor. Second, both molecules have permanent dipoles; therefore, dipole–dipole interactions should also be present. Third, realize that the permanent dipole of the H_2O molecule can push around the electrons of the NH_3 molecule, giving rise to dipole–induced dipole interactions. (*Note:* Because NH_3 already has a permanent dipole, we should think about this induced dipole as being a *modification* to the permanent dipole.) Similarly, the permanent dipole of the NH_3 molecule can induce a temporary dipole moment on the H_2O molecule. Fourth, both neutral molecules have electrons—therefore, induced dipole–induced dipole interactions should also be active. In all, then, there are four types of intermolecular interactions that are simultaneously active to help a molecule of H_2O and a molecule of NH_3 stick together.

Another type of question may have you go one step further, asking you to determine how strongly two molecules stick together, in a relative sense. For example, you might be asked which pair of molecules sticks together better, either H_2O with NH_3 or H_2O with CH_4. To answer this question, just do two things. First, for each pair of molecules, determine exactly which intermolecular interactions are active. Second, figure out which of those interactions is going to be the most important—that is, the strongest. We can assume that the strongest active intermolecular interaction will overshadow the weaker ones that are present. In other words,

How well a pair of molecules sticks together is determined largely by the strongest intermolecular interaction that is present.

We have already identified the four intermolecular interactions active in the first pair of molecules; of those, the hydrogen-bonding interaction is the most important to consider. In the second pair of molecules, there can be no hydrogen-bonding. Furthermore, because CH_4 is a nonpolar molecule, there can be no dipole–dipole interactions—those interactions that do exist are only dipole–induced dipole interactions and induced dipole–induced dipole interactions. The more important of the interactions between CH_4 and H_2O is therefore the dipole–induced dipole interaction, which is much weaker than hydrogen bonding. Therefore, we can say that a molecule of H_2O will stick to a molecule of NH_3 better than it will to a molecule of CH_4.

3.7 Application: *cis* and *trans* Isomers

As we saw in Section 3.4, rotation about a double bond is not allowed. Therefore, it is possible to isolate two molecules that differ only by an *imaginary* 180° rotation about a double bond—the two molecules do not interconvert. Because of this unique relationship, such molecules receive a special classification.

Two molecules that differ only by an *imaginary* 180° rotation about a double bond constitute a pair of *cis/trans* isomers.

They are called *isomers* because they are different molecules that have the same formula (the topic of isomers is discussed more fully in Chapter 4). One of the molecules has what is called a *cis* configuration about that double bond, and the other has what is called a *trans* configuration.

In a ***cis* configuration,** two non-H groups reside on the same side of the double bond.

In a ***trans* configuration,** two non-H groups reside on opposite sides of the double bond.

One of the simplest pairs of *cis*/*trans* isomers is that of 2-butene, $CH_3CH{=}CHCH_3$ (Figure 3-20a). We can see that the molecule on the left of the figure has a *cis* configuration, given that the two CH_3 groups reside on the same side of the double bond, and the molecule on the right has a *trans* configuration, given that the CH_3 groups are on opposite sides of the double bond. Because of the different locations of the CH_3 groups relative to the double bond, the molecules are indeed different. Furthermore, as shown in the figure, an imaginary rotation of 180° about the double bond does indeed convert one configuration into the other.

TIME TO TRY

Obtain a molecular modeling kit, and follow its directions to construct both molecules in Figure 3-20a. Hold the two molecules up to each other, and verify that they are indeed different molecules.

Certainly, the surefire way of determining whether a molecule has a pair of *cis*/*trans* isomers is to construct one molecule with a molecular modeling kit, construct a separate molecule that differs from the first only by a 180° rotation about the double bond, and then hold the two molecules together in various orientations in space to see whether or not they are the same. Doing this with several different molecules, however, you will notice a trend.

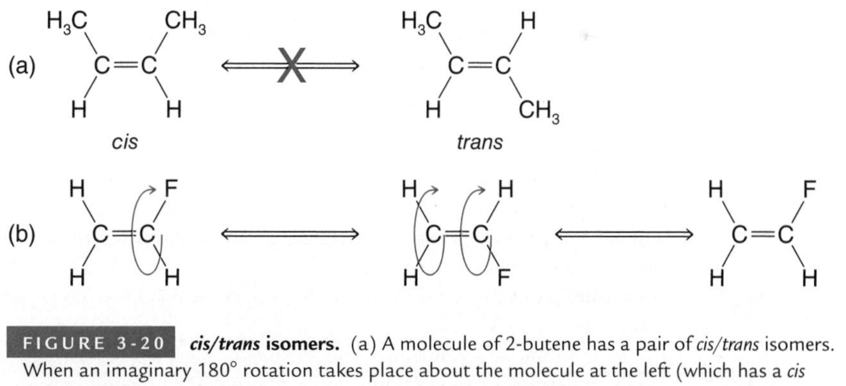

FIGURE 3-20 ***cis/trans* isomers.** (a) A molecule of 2-butene has a pair of *cis*/*trans* isomers. When an imaginary 180° rotation takes place about the molecule at the left (which has a *cis* configuration), a different molecule is produced (which has a *trans* configuration). (b) This molecule does not have a pair of *cis*/*trans* isomers because an imaginary 180° rotation about the double bond produces exactly the same molecule. Specifically, the imaginary 180° rotation of the CHF group (leaving the CH_2 group frozen in place) produces the molecule in the middle, but when that molecule is flipped over, the result is a molecule that is exactly the same as the one on the far left.

A molecule will have a pair of *cis/trans* isomers if each atom that is part of the double bond is itself singly bonded to two different groups.

Otherwise, the molecule will not have a pair of *cis/trans* isomers.

Looking back at Figure 3-20a, you can see that this is consistent with the fact that 2-butene has a pair of *cis/trans* isomers. The groups attached to one of the doubly bonded Cs are H and CH_3, which are different from each other, and the same is true for the other doubly bonded C.

By contrast, according to the above rules, a molecule of $H_2C{=}CHF$ should not have a pair of *cis/trans* isomers because one of the doubly bonded Cs is singly bonded to two of the same group—H atoms. To convince ourselves that the molecule indeed does not have a pair of *cis/trans* isomers, we can examine the molecules that differ only by a 180° rotation about the double bond, as is indicated in the first two structures of Figure 3-20b. As they appear in the figure, the two structures do not look identical. However, if the entire species shown in the middle of the figure is flipped over, yielding the structure at the right, we can see that it is exactly the same species as the initial molecule at the far left.

Before moving on, we should say a few more words about labeling a species as having a *cis* or *trans* configuration. Notice that, according to the definitions above, applying such labels requires that an H atom is attached to each doubly bonded atom. If at least one of the doubly bonded atoms has two non-H groups, then the *cis* and *trans* labels do not apply. (Instead, as you will see in your year-long organic chemistry course, we use the labels *Z* and *E*.) However, in such cases, it is still possible for the molecules to differ after a 180° rotation about the double bond. Consider $CH_3CF{=}CHCH_3$. It has a pair of *cis/trans* isomers (convince yourself of this), but the doubly bonded C atom on the left is not directly attached to any H atoms.

3.8 Application: Melting Point and Boiling Point Determination

You will invariably face homework and exam problems that give you the Lewis structures of several molecules and ask you to rank those molecules in order of boiling point or melting point (the respective temperatures at which boiling or melting occurs). One way to attack such a problem is to apply the various tricks that you may have memorized from somewhere else. Wrong! The better way is to apply what you have learned in the previous sections of this chapter as well as what you know about the processes of melting and boiling.

Let's first write out what we know about melting and boiling. Both processes are nothing more than *changes of phase*—melting is from solid to liquid, and boiling is from liquid to gas (or vapor). They are *not* chemical reactions. Covalent bonds are *not* broken and/or formed. So, for example, when ice melts, we say that solid H_2O turns into liquid H_2O, but *all* of the H_2O molecules remain intact during that process. We also know from experience that both of these processes require heat.

In addition, we know something about what it means to be a solid, a liquid, and a gas. In a solid, molecules are in very close contact, essentially touching one another. They are also in a very strictly ordered pattern—what we call either a lattice or a crystal—and are *frozen in place*. The molecules do not **translate** (i.e., move from one place to another), nor do they rotate in place.

Liquids, like solids, have molecules that are in very close contact—essentially touching one another. Unlike molecules in a solid, however, molecules in a liquid can translate and rotate.

Gases are essentially isolated molecules—they basically don't see or feel any other molecules. Like liquids, gaseous molecules can translate and rotate.

Understanding this and what we learned about intermolecular interactions, we can explain why it is that the process of melting requires heat energy. Solids are composed of molecules that are frozen in place in a highly ordered arrangement, which is the best scenario in which to maximize the attractive intermolecular interactions between molecules. Figure 3-21 demonstrates this with dipole–dipole interactions specifically. The left side of the figure represents the solid, where the molecules are very nicely ordered to give the most head-to-tail interactions that could possibly exist. When that solid substance is melted, the resulting liquid on the right side of the figure has less order, due to the translation and the rotation of the molecules. With less order, the intermolecular interactions are no longer maximized. In other words,

Melting causes the destruction of some of the good intermolecular interactions that exist in the solid.

That costs energy! And in order to supply that energy, heat must be provided.

If we go one step further and boil the substance, converting it from liquid to gas, then all of the molecules become isolated. This is shown in Figure 3-22. Therefore,

Boiling causes the destruction of *all* of the intermolecular interactions that exist in the liquid.

That requires even more energy in the form of heat, and that is why boiling points are higher than melting points!

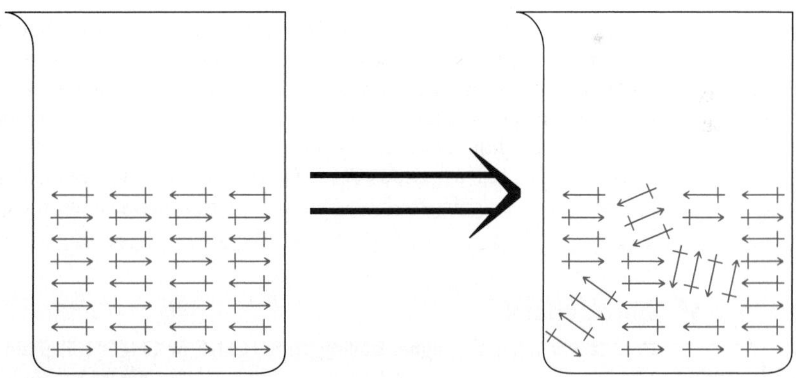

FIGURE 3-21 **Melting and intermolecular interactions.** (*left*) Representation of a solid crystal substance in which the intermolecular interactions (in this case, dipole–dipole interactions) are maximized. (*right*) Representation of the same substance in liquid form; it has less order, and therefore, the intermolecular interactions are no longer maximized. Thus, melting causes some of the intermolecular interactions to be destroyed.

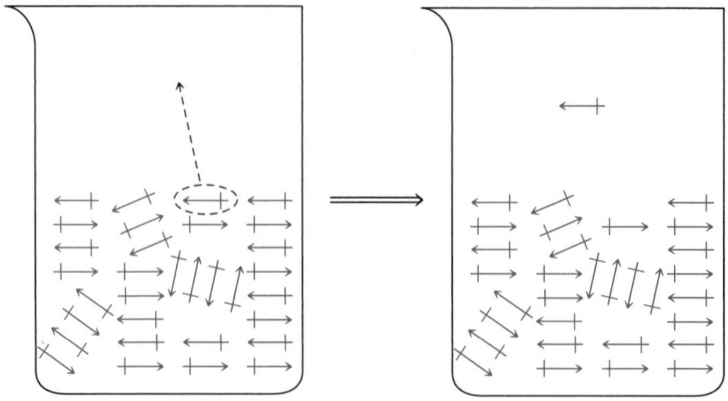

FIGURE 3-22 **Boiling and intermolecular interactions.** (*left*) In a liquid, some of the intermolecular interactions are intact. To become part of the gas, the indicated molecule leaves the others behind. (*right*) In a gas, molecules are effectively isolated from one another. Thus, boiling causes the destruction of all of the intermolecular interactions that exist in the liquid.

Now that we know melting involves the destruction of some percentage of the good interactions between molecules and boiling involves the destruction of all of the rest, we can rank melting points and boiling points, given only the Lewis structures of a set of molecules. The key idea that we need to apply is this: *The stronger the intermolecular interactions that are responsible for the molecules sticking together, the larger the amount of heat energy that is required to destroy those interactions.* Therefore,

In general, the stronger the intermolecular interactions that are present between molecules of a particular substance, the higher the melting point and the boiling point of that substance.

The only other piece of the puzzle is to apply what you know about the relative strength of attraction from the various intermolecular interactions, which we examined in Section 3.6G.

Let's look at some examples. Suppose we are asked to determine which molecule has the higher boiling point, CH_4 or CH_3F. We must determine the relative amount of heat energy it takes to convert liquid CH_4 to gaseous CH_4 and compare that to the heat energy required to convert liquid CH_3F to gaseous CH_3F. For the former, we realize that the interactions that must be overcome upon going from liquid to gas are those between two molecules of CH_4, which are

QUICK CHECK

Predict which has the higher boiling point, CH_3F or CH_3OH. Explain.

Answer: For CH_3F and CH_3OH to boil, the strongest intermolecular interactions that must be overcome are dipole–dipole interactions and hydrogen-bonding, respectively. Because hydrogen-bonding is generally more important than dipole–dipole interactions, it takes more energy to remove molecules of CH_3OH from one another, giving CH_3OH the higher boiling point.

just induced dipole–induced dipole interactions (CH_4 is nonpolar). For the latter, the interactions that must be overcome are those between two molecules of CH_3F, the most important of which are dipole–dipole interactions. Because dipole–dipole interactions are much stronger, we would expect CH_3F to have a significantly higher boiling point than CH_4.

Suppose, now, that we are asked to determine the relative melting points of CH_4 and C_2H_6. Because CH_4 is nonpolar, the only intermolecular interactions that exist between two molecules of CH_4 are induced dipole–induced dipole interactions. The same is true for two molecules of C_2H_6. The difference is that C_2H_6 has more electrons. As a result, induced dipole–induced dipole interactions are stronger between two molecules of C_2H_6, which therefore gives rise to a higher melting point for C_2H_6 than CH_4. This idea can be generalized as follows.

All else being equal, more massive molecules (because they have a greater number of electrons) have higher melting and boiling points.

QUICK CHECK

Predict which has the higher boiling point, CF_4 or CCl_4. Explain.

Answer: Both CF_4 and CCl_4 are nonpolar, so the only intermolecular interactions that must be overcome in order to boil are induced dipole–induced dipole interactions. CCl_4 has many more electrons and therefore has stronger induced dipole–induced dipole interactions than CF_4. Consequently, CCl_4 has the higher boiling point.

3.9 Application: Solubility

As with boiling points and melting points, let's first make sure that we understand the process of dissolving before we try to say anything about solubility. When one pure substance (the **solute**) dissolves into another pure substance (the **solvent**), the first simply mixes with the second to yield a solution. A **solution** is a homogeneous mixture of the solute and the solvent. (Although the solvent can technically be in any of the three phases—solid, liquid, or gas—your undergraduate organic chemistry course will deal exclusively with the solvent being a liquid.)

Given that the process of dissolving involves the mixing together of two substances, we must discuss the driving force behind mixing in general. Consider an example of salt and pepper. Suppose that some salt is added to a container and then some pepper is added on top of the salt. Initially, there are two separate layers, but if the container is then shaken, you would observe that the salt and pepper become mixed together. Why should this happen? The answer has to do with **entropy,** which can be thought of as a measure of *disorder*. In the mixture, there is more entropy than there is in the unmixed samples of salt and pepper. Therefore, it appears that systems tend to prefer greater entropy. In terms of solubility, we can say the following.

In general, a solute tends to dissolve in a solvent in order to achieve an increase in entropy.

Realize, however, that the process of dissolving is not always favorable. A classic example involves oil and water. If you shake a container that contains oil and water, it might initially appear that a solution has formed. But given a little time, the oil and water will *unmix*, producing two separate layers—the more dense water layer will be on the bottom, and the less dense oil layer will be on the top. What causes the two substances to unmix—and therefore overcome the driving force toward increased entropy? This is where intermolecular interactions enter the picture. Consider the fact that there are intermolecular interactions at play in the solute, the solvent, and the solution. And in general, these do not all have the same strength. If the strength of the intermolecular interactions is greater in the pure solute and solvent than it is in the solution, then the pure solute and solvent will be favored instead of the solution, and the solution will unmix. Looking at this in another way,

A solute tends not to dissolve in a solvent if the intermolecular interactions that are at play in the separated solute and solvent are substantially stronger than those in the solution.

With this in mind, suppose that we are asked if water dissolves in hexane, $CH_3(CH_2)_4CH_3$. Without considering intermolecular interactions, we know that entropy provides a driving force for the two substances to mix. Now let's consider the various intermolecular interactions. In H_2O, the most important intermolecular interactions are hydrogen-bonding. In $CH_3(CH_2)_4CH_3$, the predominant intermolecular interactions are induced dipole–induced dipole interactions. In a mixture of the two substances (Figure 3-23), we must consider the intermolecular interactions that exist between the two different types of molecules. Because H_2O is polar and $CH_3(CH_2)_4CH_3$ is nonpolar, the most important intermolecular interactions between the molecules are dipole–induced dipole interactions, which we know are significantly weaker than hydrogen bonding. In summary, then, there are significantly stronger intermolecular interactions in the separated solute and solvent (hydrogen bonding) than there are in the mixture (dipole–induced dipole interactions). Therefore, according to the general rule above, water should not dissolve in hexane.

Let's now look at what happens if we try to mix CH_2Cl_2 with $(CH_3)_2C{=}O$. Both molecules are polar. In pure CH_2Cl_2, the most important intermolecular interactions are dipole–dipole interactions, and the same is true for pure $(CH_3)_2C{=}O$. If the two substances are mixed

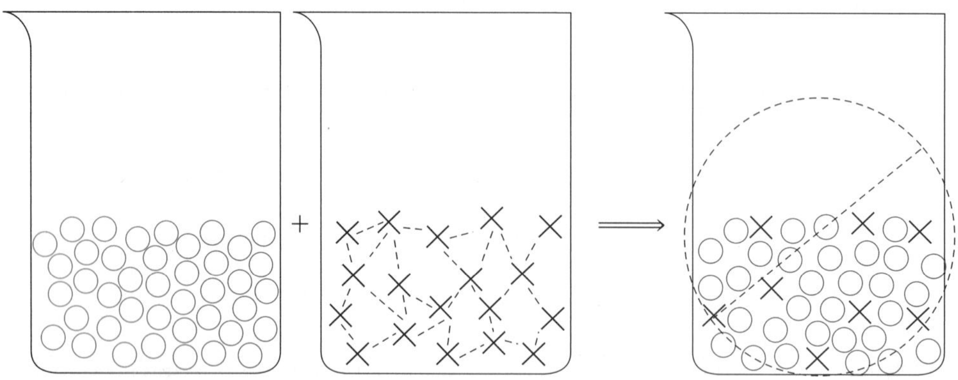

FIGURE 3-23 **Mixing a nonpolar substance with a hydrogen-bonding substance.** (*left*) A nonpolar substance such as hexane, $CH_3(CH_2)_4CH_3$, and a hydrogen-bonding substance such as water are in separate containers. The hydrogen bonds are indicated by the dashed lines. (*right*) The two substances are mixed. The strongest intermolecular interactions in the separated substance are hydrogen bonding, whereas the strongest interactions in the mixture are dipole–induced dipole interactions. Because there are substantially stronger intermolecular interactions in the separated substances than in the mixture, the two substances will not dissolve in each other.

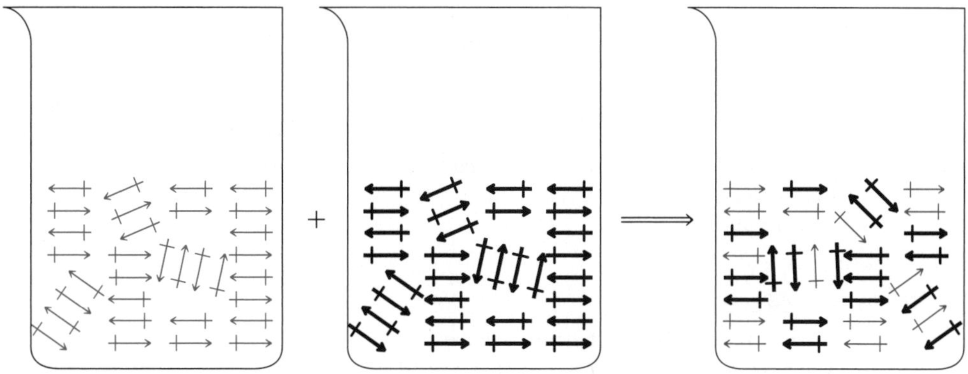

FIGURE 3-24 **Mixing of two polar substances.** (*left*) Two polar substances—one represented by thin dipole arrows and the other by thick arrows—are in separate containers. (*right*) The substances are mixed. In both the mixture and the separated substances, the strongest intermolecular interactions are dipole-dipole interactions, represented by the head-to-tail orientations of the dipole arrows. Because the intermolecular interactions are about the same strength in both the mixture and the separated substances, the intermolecular interactions do not prevent the substances from mixing.

(Figure 3-24), molecules of CH_2Cl_2 become interspersed with molecules of $(CH_3)_2C{=}O$, and the most important intermolecular interactions between those two molecules are also dipole-dipole interactions. Therefore, the situation in which the two substances are mixed appears to be no better and no worse than that when the two substances are pure and separated. Without intermolecular interactions that prevent mixing, the increase in entropy will cause the two substances to mix. Thus, we should expect CH_2Cl_2 to dissolve in $(CH_3)_2C{=}O$.

In another scenario, we can mix together two nonpolar substances (substances whose molecules possess no net molecular dipole) such as hexane, $CH_3(CH_2)_4CH_3$, and carbon disulfide, $S{=}C{=}S$. In this case, there are relatively weak induced dipole–induced dipole interactions in each of the pure substances and, likewise, in the mixture (Figure 3-25). Thus, just as in the

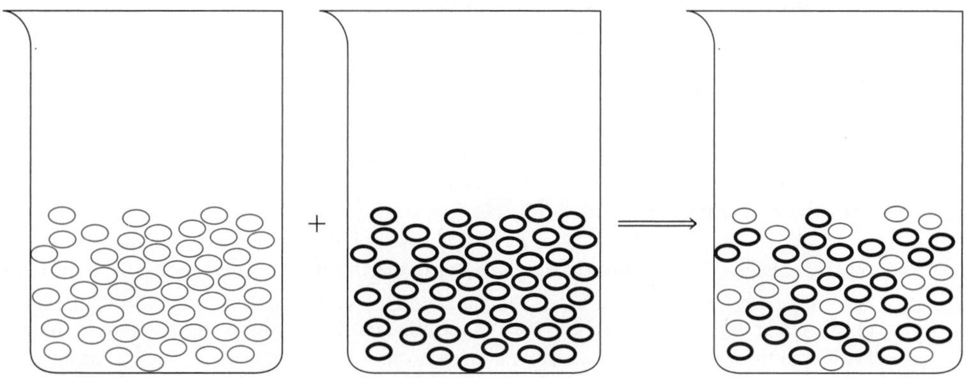

FIGURE 3-25 **Mixing of two nonpolar substances.** (*left*) Two nonpolar substances—one substance represented by thin ovals and the other by thick ovals—are in separate containers. (*right*) The substances are mixed. In both the mixture and the separated substances, the strongest intermolecular interactions are induced dipole–induced dipole interactions. Because the intermolecular interactions are about the same strength in both the mixture and the separated substances, the intermolecular interactions do not prevent the substances from mixing.

previous example, the intermolecular interactions in the mixture are about as strong as those in the separated substances, so the intermolecular interactions do not prevent mixing. In other words, we expect the two substances to dissolve readily in each other.

In yet another example, we can predict that H_2O and $HOCH_2CH_2OH$ are readily soluble in each other. The reason here is that significant hydrogen-bonding can exist between the two different molecules, just as in the pure separated substances. This is another situation in which the interactions in the mixture are about as good as when separated, so increased entropy becomes the important factor.

QUICK CHECK

1. Which of the following is the most soluble in water, H_2O?
2. Which is most soluble in hexane, $CH_3(CH_2)_4CH_3$?

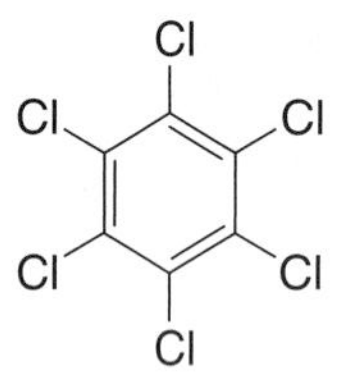

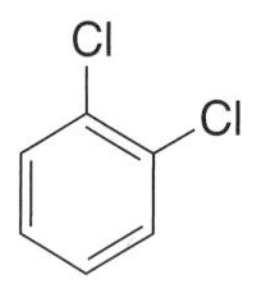

$H_2NCH_2CH_2NH_2$

Answers: 1. The third compound is the most soluble in water. When any compound dissolves in water, hydrogen bonding among the water molecules must be destroyed. Of all three compounds, the third one is the only one that can undergo substantial hydrogen bonding with water. 2. The first compound is the most soluble in hexane. The intermolecular interactions that can exist between hexane and any of the solutes involve an induced dipole—and thus will be relatively weak. The most-soluble compound will therefore be the one that, when pure, has the weakest interactions to be overcome. That compound is the first one because, when pure, the only intermolecular interactions that exist are induced dipole-induced dipole interactions. The second compound is polar, so it has dipole–dipole interactions, and the third compound has hydrogen-bonding.

Unfortunately, the rule above is not perfect. Consider, for example, that CH_2Cl_2 and CH_3CH_2OH are quite soluble in each other. In this example, hydrogen-bonding is the strongest intermolecular interaction that exists in the separated substances, whereas in the mixed state, dipole–dipole interactions are the strongest. Therefore, the intermolecular interactions are stronger when the substances are separated, suggesting that the two substances should not mix. They do mix, however, because hydrogen-bonding is not tremendously stronger than dipole–dipole interactions—not enough to overcome the driving force for mixing that is provided by the increase in entropy.

Understanding all of this, we can then ask, "In what kinds of solvents do ionic compounds typically dissolve?" For ionic compounds to be soluble, the ion–ion interactions (from the pure substance) must be overcome, which, as we previously argued, requires a *large* amount of energy. Therefore, whatever interactions exist in the mixed solution must be sufficient to overcome the loss of those ionic bonds. Ion–dipole interactions are sufficient; the dipole can be either a net molecular dipole or a strong bond dipole from a hydrogen-bond donor, such as O—H.

The reason ion–dipole interactions can overcome ion–ion interactions is not because the strength of an individual ion–dipole interaction is greater than that of an ion–ion interaction (ionic bond)—in fact, we know that the reverse is true. Instead, it is because of the *number* of ion–dipole interactions that can be formed.

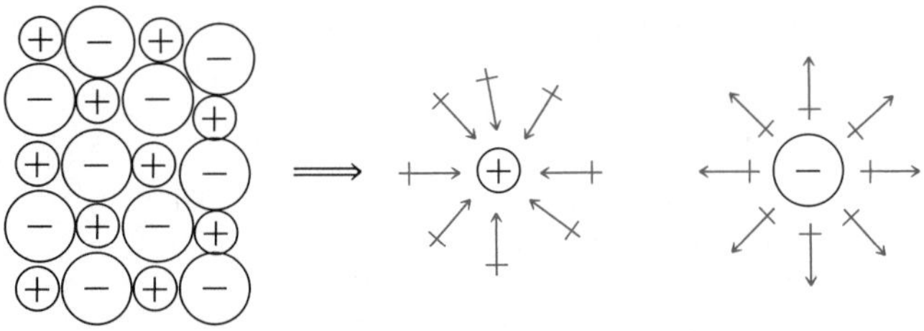

FIGURE 3-26 **Dissolving an ionic compound in a polar solvent.** (*left*) The ions of an ionic compound are held together by very strong ion–ion interactions. (*right*) The dissolved ions are surrounded by the dipoles of the solvent, leading to ion–dipole interactions. The multitude of ion–dipole interactions that result, called *solvation*, constitute an overall stronger interaction than the ion–ion interactions in the pure ionic compound. Thus, ionic compounds tend to dissolve in polar solvents.

For each ion–ion interaction that must be overcome to dissolve an ionic compound, *several* ion–dipole interactions involving a polar solvent can be formed.

This is illustrated in Figure 3-26.

With enough of these relatively strong ion–dipole interactions, the total strength of the intermolecular interactions in the mixed state can be substantially greater than that of the intermolecular interactions in the separated state. Therefore,

In general, polar solvents (like H_2O and CH_3CN) can dissolve ionic compounds.

Such a phenomenon that exists in the mixed solution (the surrounding of ions by the dipoles of solvent molecules) is called **solvation** and is important in *solvent effects*—the topic of Section 6.9.

When ionic compounds do dissolve, they dissolve as their constituent ions. Most of us are familiar with NaCl dissolving as $Na^+ + Cl^-$ and with NaOH dissolving as Na^+ and OH^-. However, students commonly have difficulty dealing with compounds such as $KOCH_3$ and $NaNH_2$ dissolved in solution. But the last two are no different from the first two—they are ionic compounds that dissolve as $K^+ + CH_3O^-$ and as $Na^+ + H_2N^-$, respectively.

WHAT DID YOU LEARN?

3.1 For each non-hydrogen atom in the following species, specify the electronic geometry. If appropriate, specify each such atom's molecular geometry and bond angle.

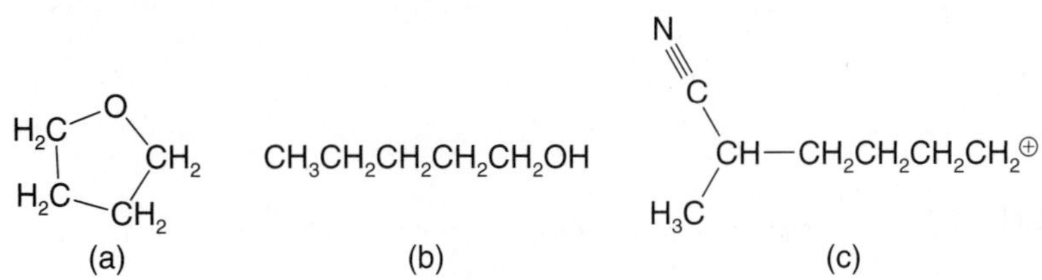

3.2 For each non-hydrogen atom in the following species, specify the electronic geometry. If appropriate, specify each such atom's molecular geometry and bond angle.

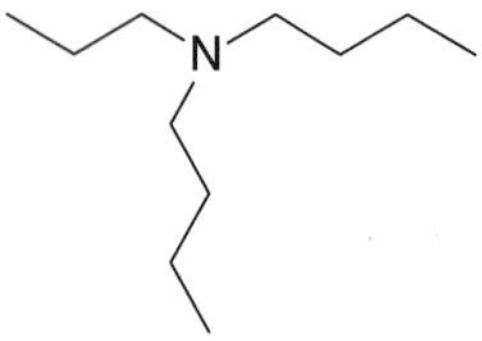
(a)

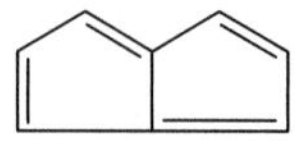
(b)

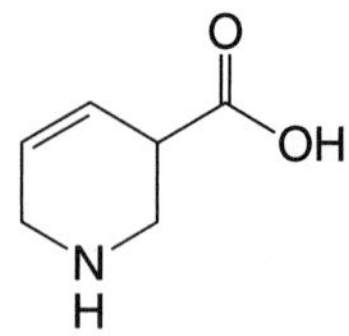
(c)

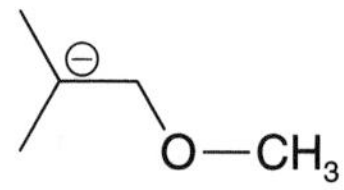
(d)

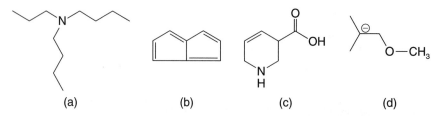

3.3 For each molecule below, add all bond dipoles, and determine whether the molecule is polar or nonpolar. For those that are polar, determine the direction in which the net molecular dipole points.

(a)
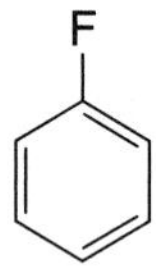

(b)

(c)
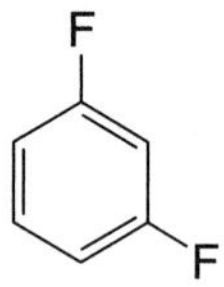

(d)

(e) CF_3Cl

(f) $AlCl_3$

(g) SF_2

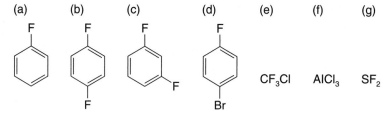

3.4 Determine all intermolecular interactions between pairs of each of the following molecules, and identify which is the most important.

(a)
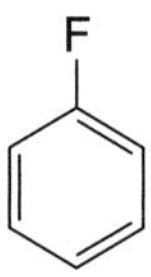

(b)
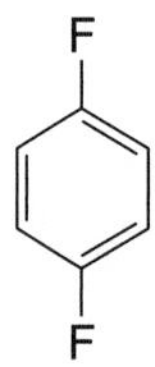

(c)
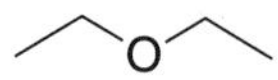

(d)
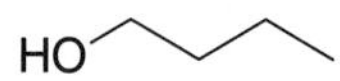

(e) NF_3

(f)
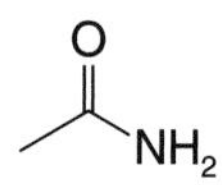

3.5 Determine all intermolecular interactions between each of the following pairs of species, and identify which is the most important.

a $CH_3O^{\ominus}$ with $K^{\oplus}$

b $CH_3O^{\ominus}$ with 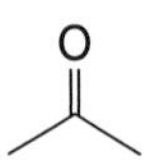

c Br_2 with 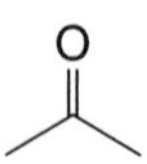

d CH_3CH_2OH with 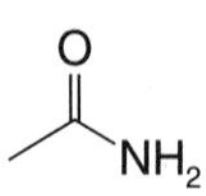

e CCl_4 with $CH_3(CH_2)_4CH_3$

f 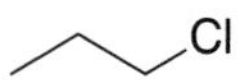 with 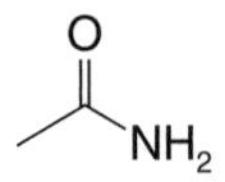

3.6 Circle all of the following species for which there exists a pair of *cis/trans* isomers.

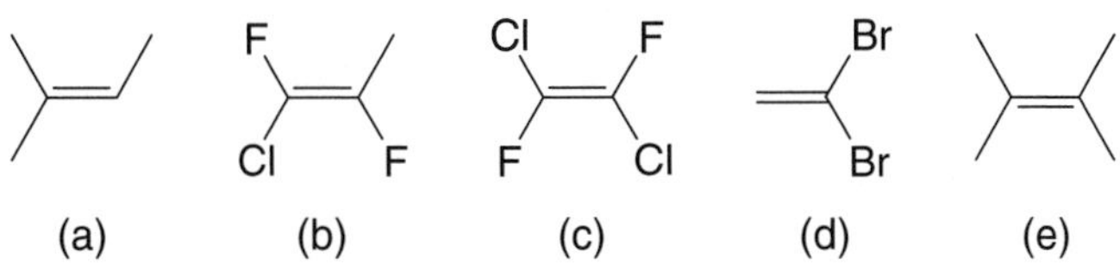

3.7 Rank the following in order of increasing boiling point.

a $CH_3CH_2CH_2OH$ **b** CH_3CO_2H **c** $(CH_3)_2C{=}O$
d $CH_3CH_2CH_2CH_3$ **e** CH_4 **f** $HO_2CCH_2CO_2H$
g $CH_3CH_2CH_2NH_2$

3.8 In general, an *enol* (recognized by a C=C—OH portion) is unstable in solution. Most enols spontaneously undergo a rearrangement to HC—C=O, called the keto form. The enol shown here, however, is more stable than usual. Explain.

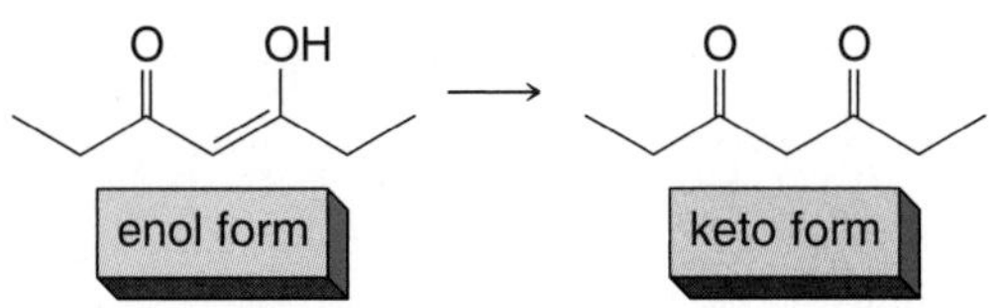

3.9 Rank the following in order of increasing solubility in acetone, $(CH_3)_2C{=}O$. (*Hint:* Look for pairs of similar molecules, and for each pair, determine which one is more soluble.)

a $CH_3CF_2CH_3$ **b** CO_2 **c** $HOCH_2CH_2CH_2OH$
d O_2 **e** $CH_3CHClCH_3$ **f** CS_2
g CH_3CHFCH_3 **h** $CH_3CH_2CH_2OH$

4 Isomerism

When you complete this chapter, you should be able to:

- Describe the specific structural relationships among constitutional isomers, stereoisomers, enantiomers, and diastereomers.
- Determine whether two molecules are isomers and, if they are, identify them as constitutional isomers, enantiomers, or diastereomers.
- Determine whether a molecule is chiral or achiral.
- Define *stereocenter* and locate all stereocenters in a given molecule.
- Predict whether the physical and chemical properties of a given pair of isomers will be the same or different.
- Calculate the index of hydrogen deficiency (or degree of unsaturation) of a compound, given only its molecular formula.
- Draw all constitutional isomers of a given molecular formula.
- Draw all stereoisomers of a given molecule.

Your Starting Point

Answer the following questions to assess your knowledge about isomerism.

1. What does it mean for molecules to be isomers of each other? ____________

2. What does *connectivity* refer to? ____________

3. How many different groups can be bonded to a tetrahedral atom? ____________

4. What determines whether an object's mirror image is identical to itself? ____________

5. Would you expect molecules that are mirror images of each other to have the same properties or different properties? ____________

6. Will compounds have the same properties if one is polar and the other is nonpolar? ____________

Answers: 1. Isomers are different molecules that have the same molecular formula. 2. Connectivity describes which atoms are bonded together and by what types of bonds. 3. A tetrahedral atom can be bonded to up to four different groups. 4. An object's mirror image is identical to itself if it has a plane of symmetry. Otherwise, the object and its mirror image are different. 5. They will have identical properties. 6. No. The two types of compounds will have different charge distributions. As we saw in Chapter 3, this gives rise to different physical properties, such as boiling point, melting point, and solubility.

4.1 Introduction

You have probably worked with, and become comfortable with, the notion of isomers and isomerism in general chemistry.

Two molecules are **isomers** of each other if they are *different molecules that have the same molecular formula*.

However, isomerism is one of the first concepts that many students struggle with in organic chemistry. Why?

Part of the answer is that there are several different types of isomers with which we must be concerned. That is, molecules that have the same formula can actually differ in any of several ways (discussed in detail later). If you encountered isomers in general chemistry, you probably dealt only with one main branch of isomers called *constitutional isomers,* which differ from each other in a specific way. In this chapter, we discuss constitutional isomers, but we will also discuss another main branch of isomers, called *stereoisomers,* which are further categorized as *enantiomers* or *diastereomers.*

Also, many students struggle with isomerism because of its connection to chemical reactivity. Two molecules that are isomers of each other may appear to be almost identical, but could behave quite differently. Alternatively, two molecules that appear to be structurally very different could have nearly identical chemical reactivity. We will have more to say about this in Section 4.5.

This chapter's first goal is to provide you with a solid understanding of the various ways in which two molecules can be isomers of each other. Simply keeping the definitions straight can often be quite difficult. The second goal is to teach you how to determine the precise relationship between any two molecules, using a very systematic method. Are they actually isomers of each other, or are they the same molecule drawn differently? If they are isomers, what type? Finally, we will examine the importance of isomerism in chemical reactions—something you will undoubtedly be held accountable for on your organic chemistry exams.

4.2 Isomers: A Relationship

The confusion with isomers and isomerism can begin with a lack of understanding of what it means to be an isomer. Many students can recite the definitions of the various isomer types quite well, but will try to look at a molecule and say, "That's an enantiomer" or "That's a diastereomer." Although we haven't yet covered the specific definition of *enantiomer* or *diastereomer,* you can be assured that such statements make no sense whatsoever. This is because

Any type of isomerism is a relationship between *two* molecules.

It does not describe a single molecule. Therefore, each of those statements is analogous to pointing at someone and saying "There's a cousin."

A statement using the word *cousin* that makes more sense is "Those two girls are cousins of each other" or "She is that man's cousin." Likewise, using the word *enantiomer, diastereomer,* or even *isomer* makes no sense unless you are talking about two or more molecules. It is okay to say, "Those two molecules are enantiomers of each other," or "This molecule is a diastereomer of that one." Keep this in mind as you go through these sections (in particular, the section on stereochemistry) and things will be less confusing.

4.3 Constitutional Isomerism

As mentioned briefly above, constitutional isomers are the first of two main branches of isomers—the other being stereoisomers. This is shown in the flowchart in Figure 4-1.

As we can see in the figure, in order to be constitutional isomers of each other, two molecules must first be isomers of each other; that is, they must be different molecules. Further,

Constitutional isomers must differ in their connectivity.

Connectivity describes which atoms are bonded together and by what types of bonds—single, double, or triple. Nothing less, nothing more. Connectivity specifically does not include anything about the three-dimensional shape. Therefore, the Lewis structure conveys all information about a molecule's connectivity. To better illustrate this point, we provide some examples.

Let's begin with a relatively simple example. In Figure 4-2, there are two Lewis structures of molecules with the formula C_3H_8. They are drawn differently, but their connectivity is identical. Notice that in Structure I there is a C atom labeled "*a*," which is singly bonded to three H atoms and to one C atom, and there is a corresponding C atom in Structure II, labeled "*z*." Notice, too, that in Structure I there is a C atom labeled "*b*," which is singly bonded to two H atoms and two C atoms, and there is a corresponding C atom, labeled "*y*," in Structure II. And finally, in

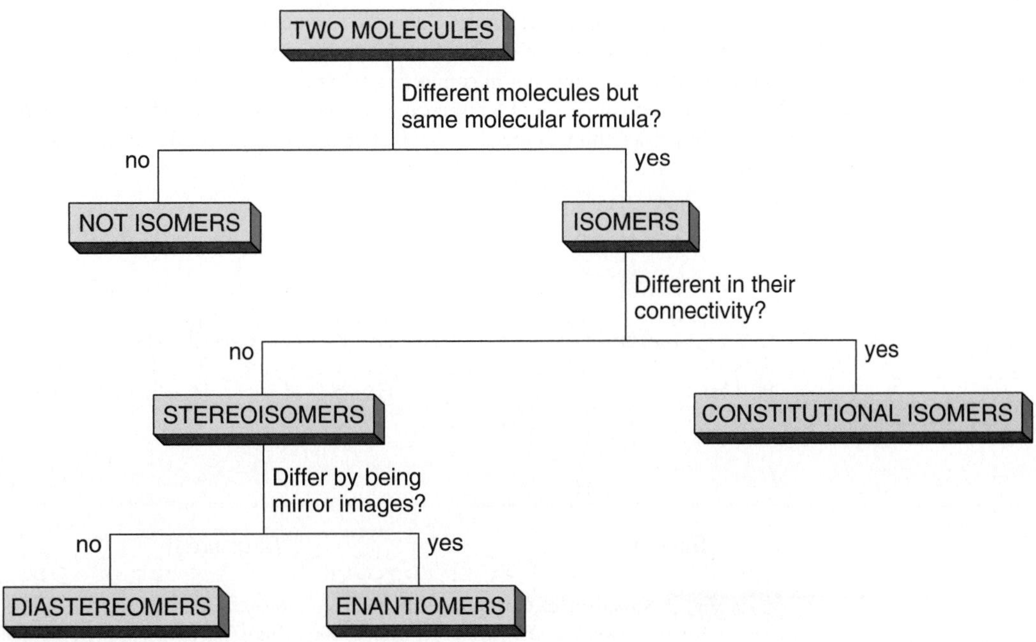

FIGURE 4-1 **Flowchart showing the relationships among various types of isomers.** The different types of isomers appear in capital letters in the blue boxes. The ways in which isomers are distinguished appear next to the vertical lines.

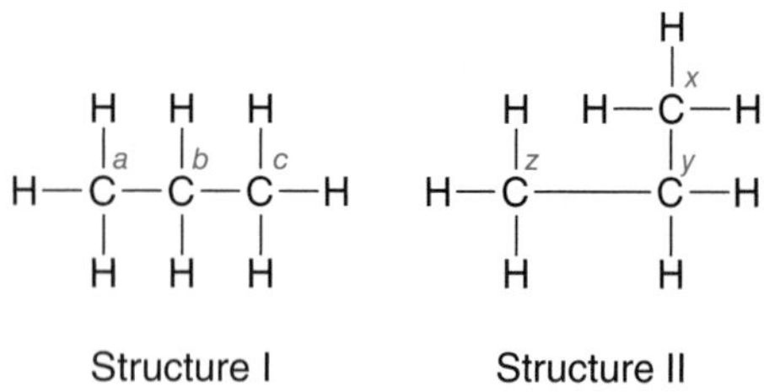

FIGURE 4-2 Structures of C_3H_8 with the same connectivity. Both Structure I and Structure II clearly have the same molecular formula of C_3H_8, but their Lewis structures are drawn differently. Nevertheless, they have the same connectivity. The corresponding C atoms are labeled "*a*," "*b*," and "*c*" in Structure I and "*z*," "*y*," and "*x*" in Structure II. Therefore, the molecules are *not* constitutional isomers of each other.

Structure I, there is a C atom labeled "*c*," which is singly bonded to three H atoms and to one C atom, and there is a corresponding C atom, labeled "*x*," in Structure II. Equally importantly, the corresponding C atoms in the two structures are bonded to each other in the same order: $C^a—C^b—C^c$ in Structure I and $C^z—C^y—C^x$ in Structure II. Therefore, even though the Lewis structures are drawn differently, they have identical connectivity. Consequently, they cannot be called constitutional isomers of each other.

✔ **QUICK CHECK**

There is another distinct way in which *x*, *y*, and *z* can be added to Structure II in Figure 4-2 to show that it has the same connectivity as Structure I. Do so by writing in the letters in Figure 4-2.

Answer: The top C atom is labeled "*z*," the leftmost C atom is labeled "*x*," and the bottom-right C atom remains "*y*."

Let's now look at another example. Figure 4-3 contains two Lewis structures, both with the same molecular formula of C_8H_{18}. In this case, the two structures are indeed constitutional isomers because they differ in connectivity. To prove this, you need find only one point of difference in the connectivity between the two molecules. Notice that, in Structure I in Figure 4-3, there is a C atom labeled "*a*" that is attached by single bonds to four other C atoms.

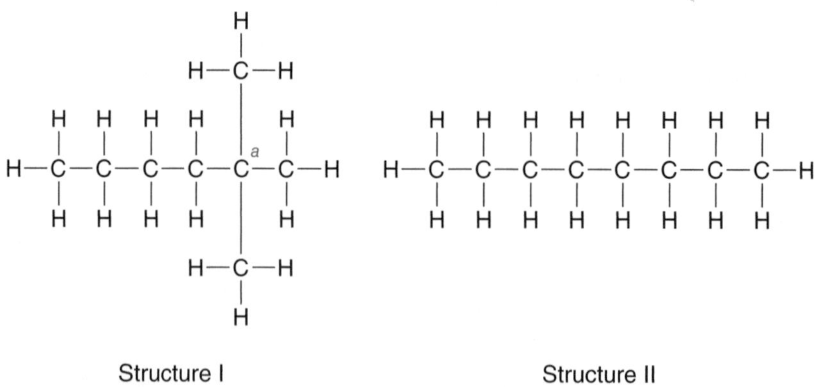

FIGURE 4-3 Structures of C_8H_{18} with different connectivity. Structure I and Structure II have the same molecular formula and differ in their connectivity. This is most easily seen by examining the C atom labeled "*a*," which is bonded to four other C atoms. There is no corresponding C atom in Structure II. Therefore, these must be constitutional isomers of each other.

In Structure II, there is no C atom like this. There are other points of difference, but we have already found our one.

TIME TO TRY

Another difference in connectivity between the structures in Figure 4-3 involves the number of CH_3 groups. How many such groups are there in Structure I? Structure II?

In yet another example, let's look at Figure 4-4, which again contains two Lewis structures having the molecular formula C_8H_{18}. Are these structures constitutional isomers of each other? Although at first glance it may appear as if they are, they are in fact not constitutional isomers. If you thought that they were, this is probably because you did not attempt the problem in a systematic fashion. Most likely, you just looked at the overall picture of each molecule, were convinced that they *looked* different, and concluded that they must be constitutional isomers. This will get you into trouble!

To approach a problem like this in a systematic fashion (which I encourage you to do *every* time), use the "two-finger method." Pick any atom in Structure I, and point to it with your left index finger. Let's choose the C atom labeled "*a*." It is bonded to three H atoms and to a C atom that is bonded to two more C atoms—one is part of a CH_3 group, and the other is part of a CH_2 group. The C atom labeled "*z*" in Structure II shares these same features. Keeping your left index finger where it is, use your right index finger to point to the C atom labeled "*z*." Now move your left index finger over to the next C atom, labeled "*b*," in Structure I. This is bonded to an H atom, a CH_3 group on the left, a CH_3 group above, and a CH_2 group on the right. Now move your right index finger over to the next C atom, labeled "*y*," in Structure II. Does the *y* carbon atom in Structure II share those same features with the *b* carbon atom in Structure I? Yes. So, thus far, we have not run into any points of difference. One more time, so you get

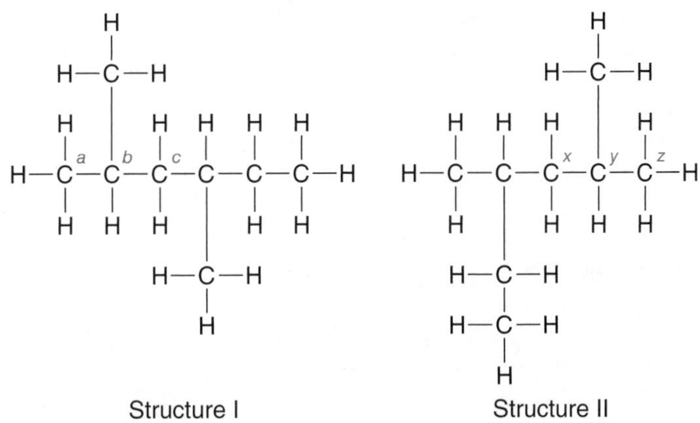

FIGURE 4-4 **Structures of C_8H_{18} with the same connectivity.** Structure I and Structure II have the same molecular formula of C_8H_{18}. They furthermore have exactly the same connectivity. Corresponding carbon atoms are *a-z*, *b-y*, and *c-x*. This correlation can be done for all C atoms in the two molecules, so these must *not* be constitutional isomers of each other.

used to it, move your left index finger over to the next C atom, labeled "*c*," in Structure I, and move your right index finger over to the next C atom, labeled "*x*," in Structure II. Do these two C atoms share the same connectivity features? Again, yes. Both of these C atoms are bonded to two H atoms and to two C atoms that are each bonded to a CH_3 group. If we continued like this all the way down the chain of C atoms, we would find that at each point the C atoms have identical connectivity. Therefore, there is no point of difference in connectivity between these two molecules.

✔ QUICK CHECK

Complete the labeling of the C atoms in Figure 4-4 to show that the two structures have the same connectivity throughout. The C atoms in Structure I should be labeled "*a*" through "*h*," and those in Structure II should be labeled "*s*" through "*z*."

Answer: In Structure I, *a*, *b*, and *c*, are already labeled in Figure 4-4. The next three C atoms to the right could be labeled "*d*," "*e*," and "*f*." The top C atom could be labeled "*g*," and the bottom C atom could be labeled "*h*." In Structure II, *z*, *y*, and *x*, are already labeled in Figure 4-4. The C atom to the left of the one labeled "*x*" would be labeled "*w*." The C atom below the one labeled "*w*" would be labeled "*v*," and the C atom below that would be labeled "*u*." In Structure I, the C atom labeled "*g*" is bonded directly to *b*, so in Structure II the C atom labeled "*t*" must be bonded directly to *y* and the topmost C atom must be *t*. Finally, because the C atom labeled "*h*" in Structure I is bonded directly to *d*, in Structure II the C atom labeled "*s*" must be bonded directly to *w*—that's the leftmost CH_3 group.

The reason this two-finger method is important is that it inherently checks to make sure that the corresponding atoms in the two structures are bonded together in the same order. In Figure 4-4, for example, notice that the three labeled C atoms in Structure I are bonded together in the order $C^a—C^b—C^c$ and the corresponding C atoms in Structure II are bonded together in the same order $C^z—C^y—C^x$. *If these orders were not the same, then the connectivities would be different, and the structures would be constitutional isomers of each other.*

Let's look at another example, which brings a multiple bond to the table. There are two Lewis structures in Figure 4-5, each of which has a double bond and each of which has the formula C_6H_{10}. As an added twist, each also has a five-carbon ring. As before, the safest way to test whether these are constitutional isomers is to use both index fingers. With your left index finger, pick any atom in Structure I—say, the C atom labeled "*a*," which is the C atom of the CH_3 group. It has three H atoms and one C atom, labeled "*b*," singly bonded to it. There is a C atom in Structure II, labeled "*z*," that seems to have the same connectivity. Put your right

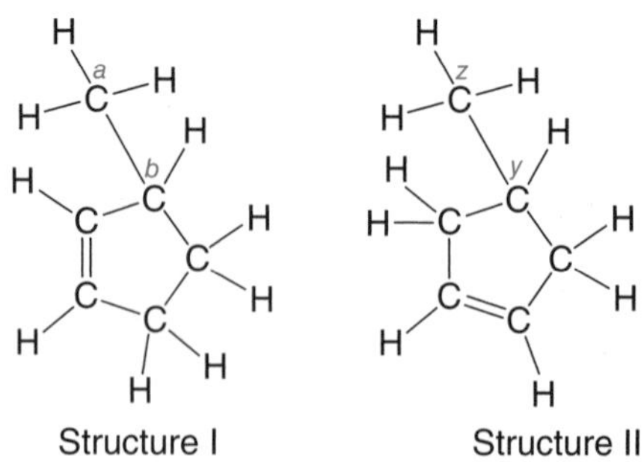

FIGURE 4-5 Structures of C_6H_{10} with different connectivity. Structure I and Structure II have the same molecular formula of C_6H_{10}. However, they have different connectivity. The C atom labeled "*b*" in Structure I is singly bonded to an H atom, a CH_3 group, a CH_2 group, and a doubly bonded C atom. No C atom in Structure II (including the one labeled "*y*") has the same connectivity. Therefore, these must be constitutional isomers of each other.

index finger on it. Now move your left index finger over to atom *b*, in Structure I, which is a C atom that is singly bonded to three C atoms and one H atom. Move your right index finger over to atom *y*, in Structure II, which is also a C atom that is singly bonded to three C atoms and one H atom. However, atom *b* and atom *y* are different C atoms—one of the C atoms to which *b* is bonded is doubly bonded to another C atom, whereas all of the C atoms to which *y* is bonded contain only single bonds. This is a subtle difference between the connectivities of atoms *b* and *y*, but one that clearly makes them different atoms. Therefore, since we have found our point of difference in connectivity, it must be that Structures I and II are constitutional isomers of each other.

✔ **QUICK CHECK**

Determine whether or not the following molecules are constitutional isomers.

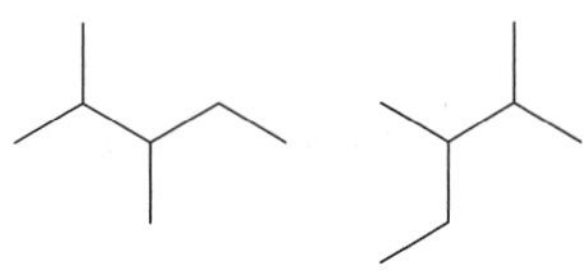

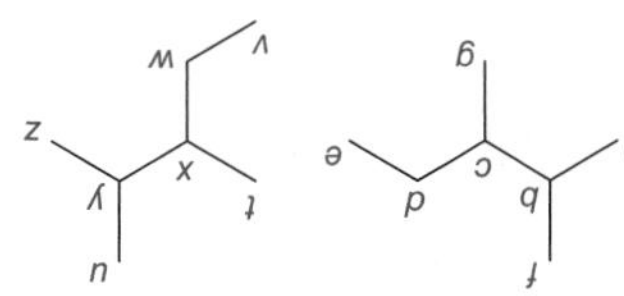

Answer: No, they are not constitutional isomers. Corresponding atoms are *a-z*, *b-y*, *c-x*, *d-w*, *e-v*, *f-u*, and *g-t*, as labeled below.

4.4 Stereoisomerism: Enantiomers and Diastereomers

Stereoisomerism is the second major branch of isomerism, which can be seen in Figure 4-1. Since there are only two branches, a pair of **stereoisomers** can be recognized as *two isomers that are NOT constitutional isomers*. In other words, stereoisomers must have the same molecular formula, and they must also have identical connectivity. To be isomers of each other, however, they must be different in some way. And if both molecules have the same connectivity (same atoms and same type of bonding), then the only way that the two molecules can be different is if the atoms are *arranged differently in space*. Thus,

Stereoisomers have the same connectivity, but differ in the way that their atoms are arranged in space.

Figure 4-1 indicates that there are two ways in which this can occur. One way results in *enantiomers*, and the other results in *diastereomers*.

Before we look at some examples of enantiomers and diastereomers, let's first define these two relationships. As we can see from the flowchart in Figure 4-1, both enantiomers and diastereomers are defined by comparing a molecule to its mirror image (we will practice drawing mirror images of molecules later).

Enantiomers (pronounced *en-ANT-ee-oh-mers*) are stereoisomers that are mirror images of each other.

Diastereomers (pronounced *die-uh-STER-ee-oh-mers*) are stereoisomers that are *not* mirror images of each other.

Because stereoisomers must be different molecules with the same connectivity, enantiomers can be defined more simply as nonsuperimposable mirror images. The term *nonsuperimposable* is just a fancy way of saying that the two molecules do not line up perfectly, atom for atom. So we can simply say that *enantiomers are different molecules that are mirror images of each other.*

WHY SHOULD I CARE?

With such relatively subtle differences between a pair of enantiomers or a pair of diastereomers, these types of isomers might seem unimportant. On the contrary, they are incredibly important, largely because of the types of molecules found in living organisms. For example, amino acids, which are the building blocks of proteins, have enantiomers. But only one of each pair of enantiomers is found exclusively in nature. As another example, each steroid typically has several diastereomers that are possible, but once again, in general only one of them is utilized in nature.

Let's now look at some examples to better see what gives rise to stereoisomerism and to understand how to determine whether given pairs of stereoisomers are enantiomers or diastereomers. The first example of a pair of stereoisomers is the most straightforward and involves a relationship that we discussed previously in Section 3.7—*cis/trans* isomerism. Let's specifically look at *cis*- and *trans*-1,2-dicholorethene, Cl—CH=CH—Cl (Figure 4-6). (Recall that the *trans* isomer is that in which the two non-H atoms are on opposite sides of the double bond and the *cis* isomer is that in which the two non-H atoms are on the same side.) Let's put these two structures through the flowchart in Figure 4-1.

First, we ask whether the structures are different molecules with the same formula. We quickly see that they do have the same molecular formula of $C_2H_2Cl_2$. Further, they differ by a 180° rotation about the central double bond, and as we learned in Section 3.4, double bonds cannot undergo such rotation.

Next, Figure 4-1 has us determine whether these isomers are stereoisomers by asking if they differ in their connectivity. We can do this in one of two ways. We have already used the first way—looking at the specific connectivity of each atom in the molecule (using both index fingers to keep track). The second way is to take advantage of the fact that a hypothetical rotation of 180° about the double bond would convert between the *cis* and the *trans* forms, but would not change the connectivity within the molecule—that's because, in doing so, no atoms are ever disconnected from each other. Therefore, no, these molecules do not differ in their connectivity, so they must be stereoisomers of each other.

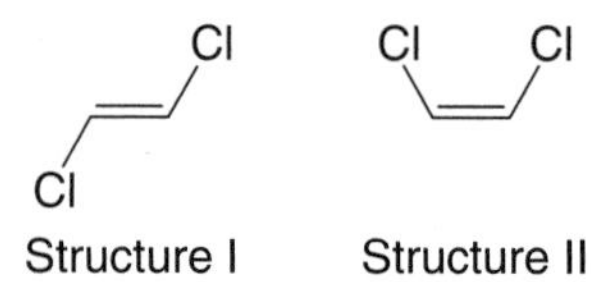

FIGURE 4-6 *trans* and *cis* isomers as diastereomers. Structure I is the *trans* isomer of 1,2-dichloroethene, and Structure II is the *cis* isomer. The two structures are different molecules with the same connectivity, but they are not mirror images, so they are diastereomers.

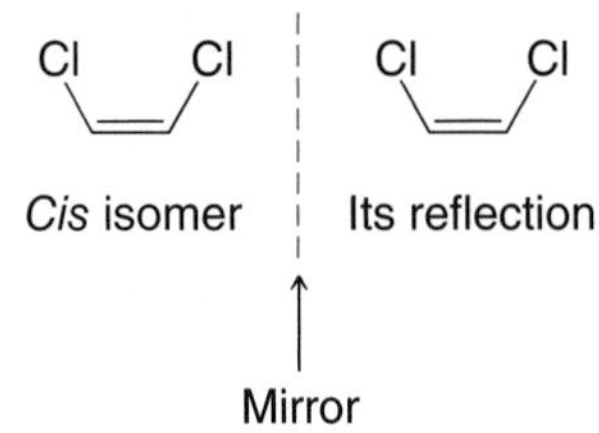

FIGURE 4-7 **A *cis* isomer and its mirror image.** The *cis* isomer of 1,2-dichloroethane is reflected through a mirror, indicated by the dashed line. The mirror image is identical to the original molecule, so *cis* and *trans* isomers cannot be mirror images of each other.

Finally, Figure 4-1 has us determine which kind of stereoisomers these molecules are by asking if they are mirror images. Although the answer to this question may be obvious, the *cis* isomer and its mirror image are shown in Figure 4-7. The mirror image of the *cis* isomer is also the *cis* isomer—that is, the mirror image is itself. Therefore, the *cis* and *trans* isomers cannot be mirror images of each other, which makes them diastereomers of each other.

TIME TO TRY

To show that *cis* and *trans* isomers are not mirror images of each other, we could have instead drawn the mirror image of the *trans* isomer. Draw that mirror image.

Given the importance of mirror images in stereoisomerism, we should practice drawing mirror images of molecules in general. In the previous example, drawing the mirror image of *cis*-1,2-dichloroethene was easy, but that is because the molecule is only two-dimensional and lies in a single plane. Most molecules are three-dimensional, so it is best to *use a systematic method to draw mirror images.*

STEPS TO DRAWING MIRROR IMAGES

1. Draw a dashed line next to the original molecule to represent the mirror. (Though it is not required, for consistency we will draw the line vertically to the right of the original molecule.)
2. Choose an atom whose mirror image you will draw, and draw its mirror image according to the following rules.
   - An atom and its mirror image are located on opposite sides of the mirror.
   - An atom and its mirror image are directly opposite each other and identical distances *along* the mirror.
   - An atom and its mirror image are identical distances *from* the mirror.
3. Repeat step 2 for another atom. If this atom is bonded to another atom, add the appropriate bond, paying attention to the following rules that pertain to dash-wedge notation.
   - A bond that is parallel to the plane of the paper (i.e., a solid line) in the original molecule must appear as a bond in the plane of the paper in the mirror image.
   - A bond that is a dash in the original molecule must appear as a dash in the mirror image.

(continued)

- A bond that is a wedge in the original molecule must appear as a wedge in the mirror image.

4. Continue with steps 2 and 3 until all atoms and all bonds have been drawn in the mirror image.

As an example of a three-dimensional molecule, let's choose CH_2FCl (Figure 4-8). Step 1 has us draw a dashed line adjacent to the original molecule in order to represent the mirror through which the molecule is to be reflected. This is shown in Figure 4-8a. (This mirror could be drawn anywhere, but in this case, it is drawn to the right of the original molecule.) For step 2, we must choose an atom to reflect through the mirror. Let's begin with the C atom. Having placed our mirror where it is, the rules say that the reflected C must be on the right side of the mirror and the same distances *along* the mirror and *from* the mirror as the original C. The result is shown in Figure 4-8b. For step 3, let's reflect the F atom through the mirror and connect the C—F bond. In the original molecule, that bond is a wedge, so the same must be true of the mirror image. The result is shown in Figure 4-8c. Step 4 has us continue with the reflections of the other atoms and other bonds until the entire mirror image is completed. Figure 4-8d is the result of having reflected the Cl atom through the mirror and connected the C—Cl bond. In this case, that bond is a dash in the original molecule, so the same must be

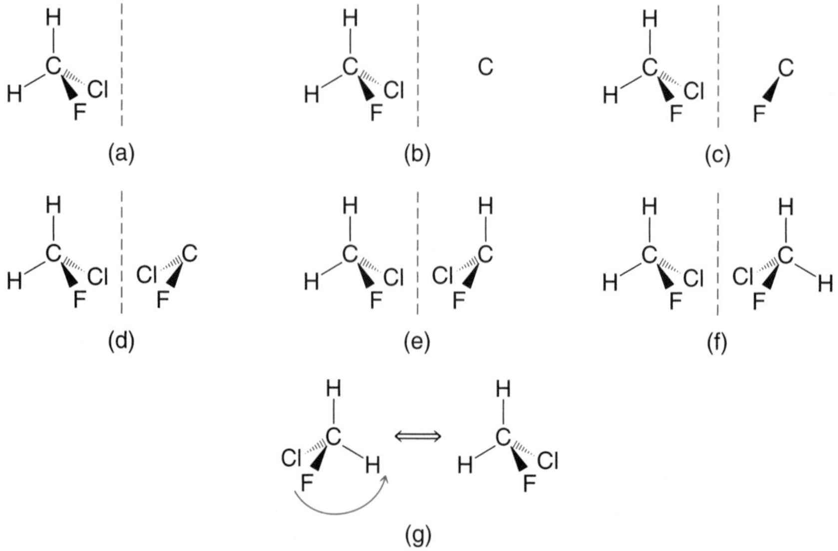

FIGURE 4-8 Drawing the mirror image of CH_2FCl. (a) The molecule to be reflected through the mirror (represented by the vertical dashed line). (b) The result of having reflected the C atom through the mirror. (c) The result of having reflected the F atom through the mirror. Note that the C—F bond is a wedge in both the mirror image and the original molecule. (d) The result of having reflected the Cl atom through the mirror. Note that the C—Cl bond is a dash in both the mirror image and the original molecule. (e) The result of having reflected one H atom through the mirror. Note that the C—H bond is in the plane of the paper in both the mirror image and the original molecule. (f) The result of having reflected the second H atom through the mirror, completing the mirror image. (g) Rotating the reflection, we can see that the mirror image is precisely the same as the original molecule.

true in the mirror image. And reflecting the two H atoms—shown in Figures 4-8e and 4-8f—completes the mirror image. Notice that the C—H bonds in the reflection are both in the plane of the paper, as they were in the original molecule.

✔ QUICK CHECK

Draw the mirror image of the following molecule.

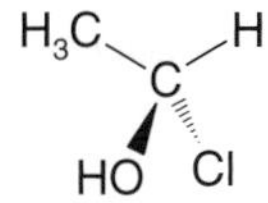

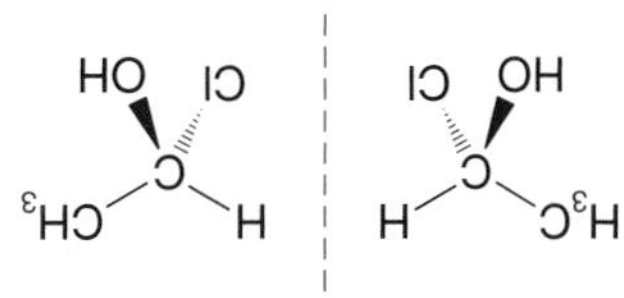

Answer:

What is the relationship between the original molecule and the mirror image in Figure 4-8? They have the same molecular formula and the same connectivity, since taking a reflection through a mirror does not change either of these. (If you don't believe this, then do the test with both index fingers!) Given that they are mirror images of each other, are they then enantiomers of each other? No. This is because the mirror image is superimposable on the original molecule—that is, the entire mirror image lines up perfectly with the original molecule, atom for atom. To see this, rotate the molecule's mirror image in space by roughly 120° about the axis that points in and out of the paper, as indicated in Figure 4-8g. The result is the original molecule. Therefore, the original molecule is exactly the same as its mirror image, just oriented differently in space.

TIME TO TRY

Using a molecular modeling kit, construct the original molecule and its mirror image exactly as shown in Figure 4-8f, paying attention to dash-wedge notation. Hold them up to each other, and orient them various ways in space until you can get every atom to line up perfectly.

LOOK OUT

If you know that two molecules are mirror images of each other, it is tempting to immediately conclude that they are enantiomers. However, this is not always true. The mirror image relationship is only half of the definition of enantiomers. The other half is that the molecules must be different—that is, nonsuperimposable. Keep in mind that every molecule will have a mirror image, but not every molecule will have a mirror image that is different from itself.

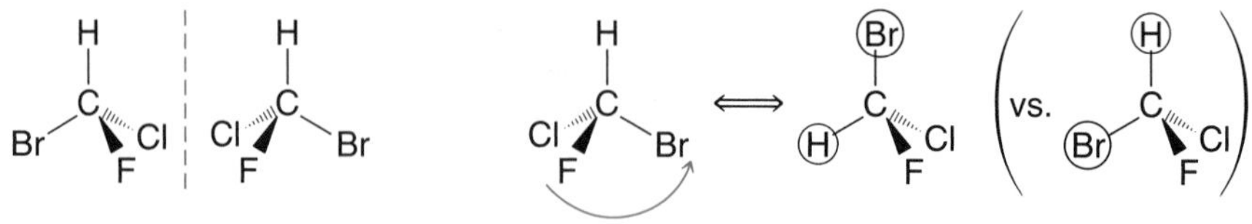

(a) (b)

FIGURE 4-9 Nonsuperimposable mirror images. (a) Three-dimensional representation of CHFClBr and its mirror image. (b) When the mirror image is rotated in space, three of the five atoms (C, F, and Cl) line up with atoms in the original molecule (shown in parentheses), but two of the atoms (H and Br) do not.

Finally, let's look at CHFClBr. It is shown with its mirror image in Figure 4-9 (convince yourself using the systematic method of reflecting one atom at a time). In this case, what is the relationship between the original molecule and its mirror image? You will find that no matter how you orient one molecule relative to the other, their atoms will not line up perfectly. The best that you can do is line up three of the five atoms, and the other two will not line up. This is depicted in Figure 4-9b, where C, F, and Cl line up perfectly in the original molecule and in the mirror image (after the reflection has been rotated in space), but H and Br do not. Therefore, the molecule and its mirror image are nonsuperimposable and thus are enantiomers of each other.

TIME TO TRY

Using a molecular modeling kit, construct the original molecule and its mirror image exactly as shown in Figure 4-9a, paying attention to dash-wedge notation. Hold them up to each other, and orient them in space every way you can imagine. Can you get all the atoms to line up perfectly?

4.4A CHIRALITY AND STEREOCENTERS

A term that goes hand in hand with stereoisomerism is *chirality*.

A molecule is **chiral** (pronounced *KAI-ruhl*) if it *has an enantiomer.*

A molecule is **achiral** (pronounced *EHY-kai-ruhl*) if it *has no enantiomer.*

This definition is straightforward enough, but remember that, as with all of the fundamental concepts covered in this book, you must be able to go beyond simple recitation of the definition. You must be able to apply it and therefore understand it well.

Before we go through some examples, let's look at the difference between the definition of *chiral* and the definition of *enantiomer*. Keeping these definitions straight can be frustrating, given that one is used in the definition of the other, but it is easy if you remember that *enantiomer* is a *relationship between two molecules. Chirality* is not—it is a word that *describes a single molecule.* A single molecule can be chiral or achiral, but a molecule can be an enantiomer only *of another molecule.*

So how do we determine whether a molecule is chiral or achiral? We start by taking a closer look at the original molecule in Figure 4-9 because it is chiral—its mirror image is different from itself (in fact, the original molecule and its mirror image are each chiral). This molecule is not special. As it turns out,

A molecule must be chiral if it has *exactly one* tetrahedral atom bonded to four different atoms or groups.

You can prove this by replacing the four atoms in Figure 4-9 with any four groups that are all different from each other—say, a CH_3 group, a CH_2CH_3 group, an OH group, and an NH_2 group. If you then use a molecular modeling kit to build this molecule and its mirror image, you will find that they are not superimposable. As in Figure 4-9, if you attempt to line up the molecule with its mirror image, the best you can do is line up the central C atoms and two of the remaining four groups. The other two groups will not line up.

It should be clear that a tetrahedral atom bonded to four different groups of atoms is important in stereoisomerism. Consequently, such atoms are given special designations.

A tetrahedral atom that is bonded to four different atoms or groups of atoms is called a **stereocenter** (sometimes called a **chiral center** or **chirality center**).

We can therefore reword the previous statement to say that *a molecule that contains exactly one stereocenter must be chiral.* (As we will see shortly, a molecule with two or more stereocenters can be chiral, but does not have to be. So be careful with this generalization!)

With this understanding, let's revisit the two molecules from Figures 4-7 and 4-8. Although we did not previously call explicit attention to it, we have already shown that both such molecules are in fact achiral. This was shown by having proven that each molecule's mirror image is the same as itself—that is, in each case, the two are superimposable. Thus, there is no way to produce a nonsuperimposable mirror image, meaning that neither molecule has an enantiomer.

What these molecules have in common is that they each have a **plane of symmetry,** or a **mirror plane.** That is, there is a way to divide each molecule in two, such that one half is the mirror image of the other half. For example, the molecule in Figure 4-7 lies entirely in a single plane—the plane of the paper. That plane is the molecule's plane of symmetry. Note that any atom that lies in the plane of symmetry can be cut in half and one half of each atom can be considered the mirror image of the other half. In other words, each atom that lies in a plane of symmetry is its own mirror image.

The plane of symmetry in the molecule in Figure 4-8, on the other hand, is that which contains the C, F, and Cl atoms and bisects the H—C—H angle. This is shown explicitly in Figure 4-10. The mirror images of the C, F, and Cl atoms are themselves, and the mirror image of one H atom is the other H atom.

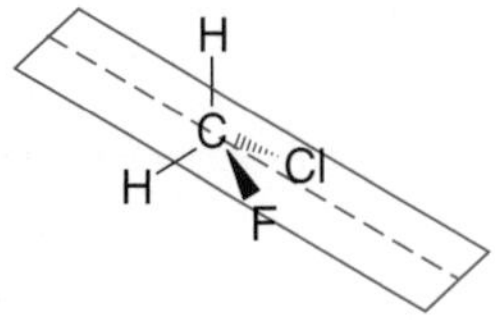

FIGURE 4-10 **The plane of symmetry in CH_2FCl.** The plane of symmetry contains C, F, and Cl and is perpendicular to the plane of the paper. The dashed line represents the line bisecting the H—C—H angle. Note that the H atoms are mirror images of each other. Each atom lying in the plane of symmetry is its own mirror image—one half of each of those atoms can be viewed as the mirror image of the other half.

The lesson to take away from this discussion about planes of symmetry is as follows:

Any molecule that has a plane of symmetry must be achiral.

Any molecule that has *no* plane of symmetry must be chiral.

REALITY CHECK

Chirality is something you may be considering for the first time here. But realize that it is not specific to molecules alone. Objects in the "real world" can also be chiral or achiral, depending on whether they have a plane of symmetry. For example, your left and right hands are chiral—one is the mirror image of the other, and they are certainly not the same. (If you are right-handed, have you ever tried to golf with left-handed clubs?) This is consistent with the fact that neither has a plane of symmetry. By contrast, a coffee mug is achiral—its mirror image is exactly the same as itself—consistent with the fact that it does have a plane of symmetry that splits the handle in half.

QUICK CHECK

Determine whether each of the following everyday objects is chiral or achiral.

1. Gloves.
2. Baseball bat.
3. Screw.
4. Nail.
5. Fan.
6. Chair.

Answers: 1. Chiral. 2. Achiral. 3. Chiral. 4. Achiral. 5. Chiral. 6. Achiral.

As you can probably gather, spotting a stereocenter is important. However, students often have difficulty recognizing stereocenters in relatively large, complex molecules. Consider the molecule in Figure 4-11, for example. It is tempting to say that this molecule has no stereocenters, but it in fact has one—the C atom indicated. The reason it is tempting to say

FIGURE 4-11 Spotting a stereocenter in a relatively complex molecule. This molecule has one stereocenter, which is the C atom indicated. The four groups attached to that C atom are circled—H, CH_3, CH_2CH_3, and $CH_2CH_2CH_3$—and are all different from one another.

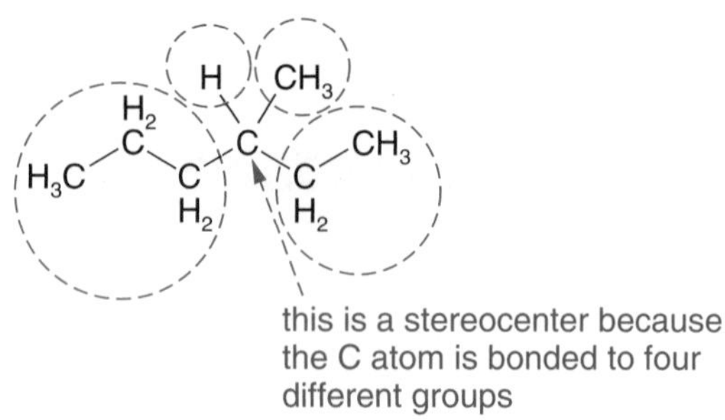
H
CH3
H2
C
C
CH3
H3C
C
C
H2
H2
this is a stereocenter because the C atom is bonded to four different groups

that this C atom is not a stereocenter is that it is bonded to two CH_2 groups. Thus, the C atom of interest is bonded to only three different groups—H, CH_3, and CH_2—which would be inconsistent with the definition of a stereocenter. However, this is not the intended use of the word *groups*. What is intended, instead, is that we view each group as *the entire portion of the molecule connected to that atom*. With this understanding, we can see that the stereocenter in Figure 4-11 is indeed bonded to four different groups—H, CH_3, CH_2CH_3, and $CH_2CH_2CH_3$, as indicated in the figure.

Rings can also make it difficult for some students to spot a stereocenter. For example, the molecule in Figure 4-12 has one stereocenter, as indicated. To be a stereocenter, that C atom must be bonded to four different groups. Clearly, the C atom is bonded to at least three different groups—H, CH_3, and two additional groups that are part of the ring. But are those groups that are part of the ring different from each other? To answer this question, we split the ring in half in such a way as to also split the atom of interest in half—this is indicated in the figure. As we can see, in this case, such a split yields two halves of the ring that are not the same. This allows us to say that the groups are indeed different and that there are in fact four different groups attached to the C atom of interest.

✔ QUICK CHECK

Identify any stereocenters in the following molecule.

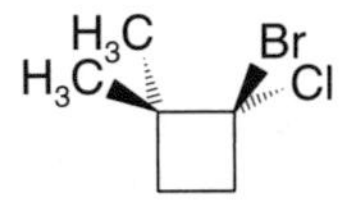

Answer: There is one stereocenter—the C atom to which the Br and Cl atoms are bonded. The four different groups are Br, Cl, and the two different halves of the ring.

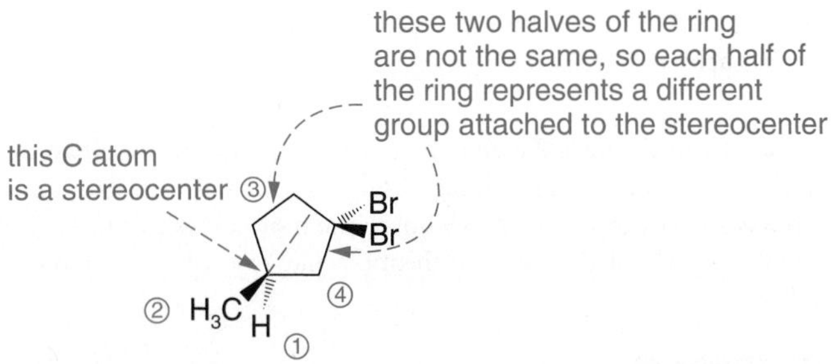

FIGURE 4-12 **Spotting a stereocenter in a ring.** The C atom indicated is a stereocenter because it has four different attached groups, which are specified. Those groups are H, CH_3, and the two halves of the ring. The two halves of the ring are different from each other because one half has two Br atoms, whereas the other doesn't.

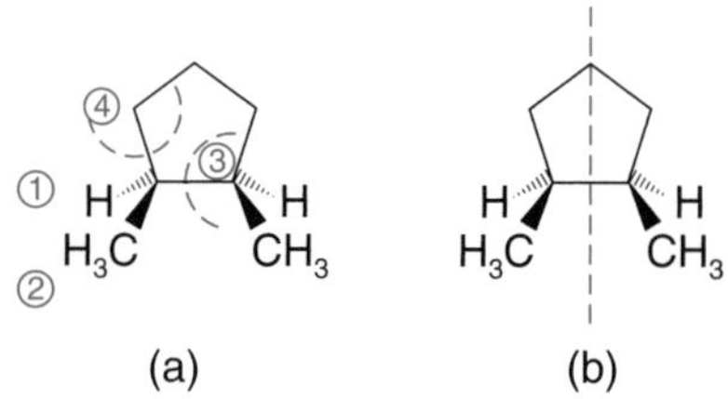

FIGURE 4-13 An achiral molecule with two stereocenters. (a) The bottom-left C atom of the ring has four different groups attached, including two different groups that are part of the ring, so it is a stereocenter. So is the bottom-right C of the ring. Thus, the molecule has two stereocenters in all. (b) The molecule has a plane of symmetry, so overall it is achiral.

Let's now step it up a bit and examine molecules that have more than one stereocenter. Figure 4-13a, for example, shows a molecule that contains two stereocenters—each C of the ring that is bonded to a CH_3 group. As indicated in Figure 4-13a, the bottom-left C atom of the ring has four different groups—H, CH_3, and the two halves of the ring—so that C atom is a stereocenter. The same is true of the bottom-right C atom of the ring.

TIME TO TRY

In Figure 4-13a, label the four different groups to which the bottom-right C of the ring is attached.

Is that molecule chiral or achiral? There are two ways to determine this. The foolproof way (which you can *never* practice enough) is to draw the reflection alongside the original molecule, build both of those molecules using a molecular modeling kit, and then try to line them up—if all the atoms line up perfectly, then the original molecule is achiral, but if they don't, it is chiral.

The second way is to apply the plane of symmetry test. As shown in Figure 4-13b, the molecule does have a plane of symmetry, which is depicted as a dashed line. Therefore, that molecule must be achiral. As such, it is also said to be *meso*.

A molecule is **meso** (pronounced MEE-zoh) if it contains at least two stereocenters, but overall is achiral.

A similar molecule is shown in Figure 4-14; it, too, possesses two stereocenters. If you build that molecule and its mirror image using a molecular modeling kit, you will find that there is no way to perfectly line the two up, so the molecule is chiral. Alternatively, take the time to search for a plane of symmetry in the molecule. You will not find one.

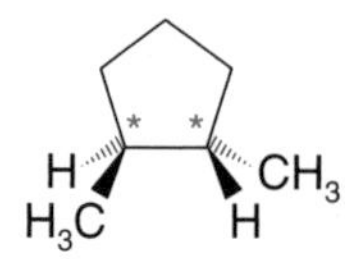

FIGURE 4-14 A chiral molecule with two stereocenters. The two stereocenters are the C atoms indicated by *. The molecule does not have a plane of symmetry. If you build this molecule and its mirror image using a molecular modeling kit, you will find that they do not line up atom for atom and are therefore different molecules.

TIME TO TRY

Draw the molecule in Figure 4-14 along with its mirror image. Using a molecular modeling kit, construct both of these molecules. Hold the two molecules up to each other, and orient them in every way imaginable. Is there any orientation in which the two molecules line up perfectly, atom for atom?

The previous two examples provide a very important lesson.

A molecule with more than one stereocenter can be chiral, or it can be achiral.

What about the relationship between the two molecules in Figures 4-13 and 4-14? These molecules each have the same molecular formula and the same connectivity. Furthermore, they are *different* molecules. In one molecule, both CH_3 groups are on the same side of the ring (i.e., both are pointing toward you), and in the other molecule, they are on opposite sides (if you cannot immediately see that these molecules are different, build each molecule with the kit, and try to line them up). Consequently, these molecules are stereoisomers of each other, meaning that they must be either enantiomers or diastereomers of each other. To test which one, we must determine whether they are mirror images of each other or not. It turns out that they are not mirror images of each other (again, if you cannot immediately see this, go through the complete exercise of drawing each molecule's mirror image), which makes them diastereomers.

You have just seen a second example of a pair of diastereomers. The first was a pair of planar molecules containing no stereocenters (*cis/trans* isomers). This one is three-dimensional, containing more than one stereocenter. The lesson to learn from this is as follows:

Molecules that have more than one stereocenter can have potential enantiomers as well as potential diastereomers.

LOOK OUT

One thing about stereoisomerism that tends to trip up students relates to the definition of a stereocenter. Even though there is no mention of a C atom in the definition, students tend to assume that in order to be a stereocenter the tetrahedral atom must be C. However, this is not true. The N atom is another stereocenter that you can encounter in organic chemistry. It can be bonded to four different groups, making it a tetrahedral atom. It just turns out that an N atom with four different groups also bears a + charge. A second thing that often trips up students is the assumption that, since the existence of stereocenters within a molecule has a tendency to make a molecule chiral, then a molecule that has no stereocenters must be achiral. This is not true. There is a problem at the end of the chapter that gets at the heart of this idea.

✔ **QUICK CHECK**

Determine whether each molecule is chiral or achiral. If the molecule is achiral, can you find at least one plane of symmetry?

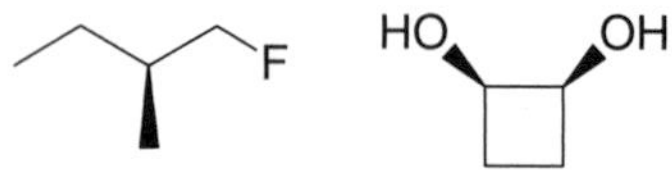

Answer: The first molecule is chiral, but the second is achiral. The plane of symmetry in the second molecule splits the ring in half vertically.

4.5 Physical and Chemical Behavior of Isomers

In the opening section of this chapter, we mentioned that in some cases isomers can behave quite similarly and in other cases they can behave quite differently. Much of this has to do with the specific type of isomerism that two molecules share. For example,

Two molecules that are constitutional isomers of each other must have different chemical properties (the product formed, reaction rate, equilibrium constant, etc.) and physical properties (boiling point, melting point, water solubility, etc.).

The reason is that, by definition, constitutional isomers must have different connectivity, which, in turn, means that they must have different bonding in some way. For example, it may be that one of two constitutional isomers has a C=O bond, whereas the second molecule has a C=C bond. It may be that one of the molecules contains a ring in the structure, while the other does not.

Just because constitutional isomers must have different properties does not necessarily mean that their properties must be drastically different.

How differently constitutional isomers behave largely depends on how different their connectivities are.

Figure 4-15a shows two constitutional isomers of C_5H_{10}. Both structures have one C=C double bond, and the rest are C—H and C—C single bonds. The structural difference between them is not great—the double bonds are simply found at different locations within the molecules. Not surprisingly, these two molecules behave quite similarly in terms of both their physical and their chemical characteristics. On the other hand, Figure 4-15b shows two constitutional isomers

(a) (b)

FIGURE 4-15 Properties of constitutional isomers. (a) Constitutional isomers of C_5H_{10}. The only difference is in the placement of the double bond along the carbon chain, suggesting that the two molecules should have quite similar (though not identical) chemical behavior. (b) Constitutional isomers of C_3H_5NO. They are constructed from different types of bonds, so they have very different chemical behavior.

that have quite different physical and chemical behavior. This should not be a surprise, given their more substantial differences in connectivity—for example, one has a $C{\equiv}N$ triple bond, whereas the other has nothing of the sort.

In contrast to constitutional isomers,

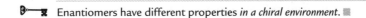

Two molecules that are enantiomers of each other have identical properties *in an achiral environment*.

An **achiral environment** simply means that nothing present is chiral except the enantiomers we are speaking of.

Enantiomers have identical properties because, even though their atoms are arranged differently in space, they are mirror images of each other—what takes place on one side of a mirror is identical to what takes place on the other.

PICTURE THIS

To better understand the fact that enantiomers have identical properties in an achiral environment, let's use an analogy of feet and socks. Your left and right feet are chiral—one is the mirror image of the other, and they are not superimposable. A sock, however, is achiral because it has a plane of symmetry (running down the length of it). From experience, we know that the same sock will fit equally well on either foot. That is, each of those enantiomers "interacts" in exactly the same way with the achiral sock. Both enantiomers behave the same in the achiral environment of a sock.

On the other hand,

Enantiomers have different properties *in a chiral environment*.

As its name suggests, a **chiral environment** means that something present is chiral other than the enantiomers we are speaking of. A solvent, for example, could be chiral. Alternatively, a catalyst could be chiral. Rest assured, though, that almost every situation you will encounter in your yearlong organic chemistry course will involve an achiral environment.

PICTURE THIS

We can better understand the different properties of enantiomers in a chiral environment by considering the "interaction" between feet and shoes. We already know that your left and right feet are chiral. For the same reason, left and right shoes are chiral—they are enantiomers of each other. We know that a left shoe does not fit on a right foot, and vice versa. That is, each of those enantiomers behaves differently in the chiral environment we are calling a shoe.

REALITY CHECK

Although your organic chemistry course will deal almost exclusively with achiral environments, the fact that enantiomers have different properties in a chiral environment is extremely important to consider in biochemical processes. One instance that made international news in the early 1960s involved *thalidomide*, which was prescribed as an antinausea medication to pregnant women. Thalidomide is a chiral compound and was sold as an equal mixture of its two enantiomers. It was later determined that, whereas one enantiomer was effective in combating nausea and was relatively safe, the other enantiomer caused birth defects—specifically, babies were born with deformed and/or missing limbs. From this, it is clear that the body is a chiral environment. As you continue in your studies, you will see that such a chiral environment is provided specifically by certain biochemical compounds, such as proteins, carbohydrates, and lipids.

Finally, unlike enantiomers,

Diastereomers must have different properties.

This is readily seen with the *cis* and *trans* isomers of 1,2-dichloroethene, Cl—CH=CH—Cl (see again Figure 4-6). Notice that there is no net dipole in the isomer with both Cl atoms on opposite sides of the double bond, whereas there is a considerable net dipole with both Cl atoms on the same side of the double bond. As we saw in Chapter 3, this translates into significant differences in intermolecular interactions, which lead to different physical properties such as boiling point, melting point, and solubility. As we will be able to see later in this book, such differences in charge distribution throughout the molecule also translate into differences in chemical reactivity.

Unlike the case with the *cis* and *trans* isomers of Cl—CH=CH—Cl, it is not always easy to see precisely why diastereomers should have different properties. For example, the pair of diastereomers we saw previously in Figures 4-13 and 4-14 are structurally very similar and would be expected to have nearly the same properties. However, no matter how similar diastereomers appear, their behavior must differ in some way, given that they are not identical and are not mirror images of each other.

4.6 Application: Index of Hydrogen Deficiency (Degree of Unsaturation)

A molecule's *index of hydrogen deficiency* (IHD), also known as its *degree of unsaturation*, is quite useful as an aid in determining the structure of an unknown compound. Knowing nothing more than the molecular formula, you can determine how many double bonds, triple bonds, or rings can possibly exist. Having such information can drastically reduce the number of candidates for the molecular structure. Combining the IHD with other information, such as the information that might be obtained from spectroscopy experiments (whose discussion will be left to your traditional organic chemistry course), usually can lead to clear identification of the compound.

The IHD for a given molecule can be defined as follows:

A molecule's **index of hydrogen deficiency (IHD)**, or **degree of unsaturation**, is the number of H_2 molecules absent from an analogous completely "saturated" molecule.

This requires us to know what it means to be completely saturated.

A molecule is said to be **saturated** (with H atoms) if it *has the maximum possible number of H atoms* for the given non-H atoms.

For example, a molecule of ethane, $H_3C{-}CH_3$, is a saturated molecule because it is not possible for a molecule with two C atoms to possess any more than six H atoms. Each C atom is singly bonded to a total of four atoms (which is the maximum possible, given the octet rule), three of which are H atoms. By contrast, a molecule of ethene, $H_2C{=}CH_2$, has two fewer H atoms, or one fewer H_2 molecule, than the completely saturated two-carbon molecule. Therefore, its IHD is 1. Another H_2 molecule can be removed, yielding $HC{\equiv}CH$; such a molecule therefore has an IHD of 2.

✔ QUICK CHECK

How many H atoms are there in a saturated molecule with three C atoms?

Answer: The three C atoms must be bonded together using two bonds, C—C—C. That leaves eight bonds that can be used to attach H, so a saturated compound will have eight H atoms.

The above examples can be generalized as follows.

Each double bond increases a molecule's IHD by 1.

Each triple bond increases a molecule's IHD by 2.

To demonstrate this, we can examine $CH_3CH{=}CHCH{=}CHC{\equiv}CH$. There are two double bonds, each contributing 1 to the IHD, and there is one triple bond, contributing 2 to the IHD, suggesting that the total IHD is 4. This can be verified by comparing this seven-carbon molecule to a completely saturated seven-carbon molecule, such as $CH_3(CH_2)_5CH_3$, which contains 16 H atoms. The difference is eight H atoms, or four H_2 molecules, consistent with an IHD of 4.

There is one more structural feature that contributes to an IHD.

Each ring increases a molecule's IHD by 1.

For example, the cyclic seven-carbon molecule cycloheptane (Figure 4-16a) has a molecular formula of C_7H_{14}. Compared to the completely saturated seven-carbon molecule previously mentioned, which has a formula of C_7H_{16}, there is a single H_2 molecule that is missing, for an IHD of 1.

In a molecule that contains double bonds, triple bonds, and rings, determining the total IHD simply involves adding the IHDs contributed by each. For example, benzene, which is a cyclic six-carbon molecule that contains three double bonds (Figure 4-16b), has a total IHD of 4.

FIGURE 4-16 **Index of hydrogen deficiency in cyclic compounds.** (a) Structure of cycloheptane, C_7H_{14}. The ring contributes an IHD of 1 to the molecule. (b) Structure of benzene, C_6H_6. The ring contributes an IHD of 1, and so does each of the three double bonds. Therefore, the total IHD is 4.

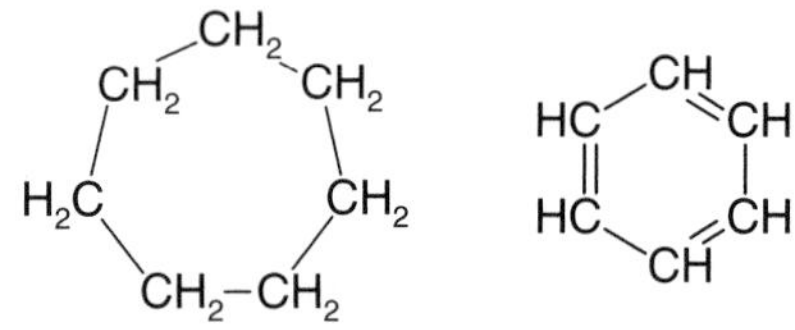

(a) (b)

The same result can be obtained by comparing the formula of benzene, C_6H_6, to the analogous completely saturated (open-chain) molecule, which would have a formula of C_6H_{14}. Benzene is therefore missing eight H atoms, for an IHD of 4.

✔ **QUICK CHECK**

What is the IHD for each of the following compounds?

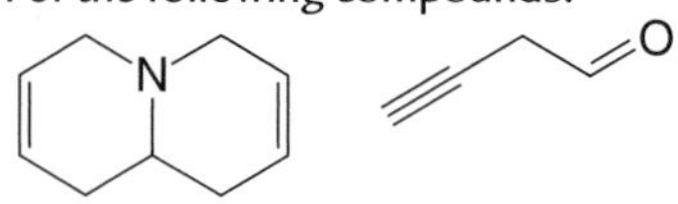

Answer: The first compound has two rings and two double bonds, for a total IHD of 4. The second compound has one triple bond and one double bond, for a total IHD of 3.

The IHD is most useful when given nothing more than the molecular formula of a molecule. From the molecular formula, you can determine the IHD, and with the IHD, you can subsequently determine the number of double bonds, triple bonds, or rings—or the combination thereof—that could be present. This works because *constitutional isomers have the same IHD as one another*. So, if you are given a molecular formula of C_7H_6, then by comparing it to the completely saturated molecule of C_7H_{16}, you can quickly deduce that the IHD is 5. A molecule that contains five double bonds is therefore consistent with the formula C_7H_6. So is a molecule that contains three double bonds and one triple bond as well as a molecule that contains two rings, a double bond, and a triple bond. All of these examples are constitutional isomers of one another.

The situation becomes slightly more tricky when the molecular formula you are given contains atoms other than just C and H. It may contain O, N, F, or Cl, for example. However, a systematic approach (outlined on the next page) can make the solution very straightforward.

With this, let's determine the IHD of $C_7H_7NO_2$. Step 1 is to draw a Lewis structure of a completely saturated molecule with seven Cs, one N, and two Os. The completely saturated molecule should contain all single bonds and no rings. If we connect all of the non-H atoms together with single bonds (Figure 4-17a) and then add enough H atoms to complete the octets, we might come up with the structure shown in Figure 4-17b, which has the formula $C_7H_{17}NO_2$. (Whether or not such a molecule exists does not matter. All we care about is that it is completely saturated.) Thus, taking care of step 2, we can see that the completely saturated molecule has 17 H atoms. For step 3, we subtract 7 (the number of H atoms in the formula given to us) from 17 (the number of H atoms in the saturated molecule) to determine that the formula we have been given is missing 10 H atoms. Step 4 has us divide this number by two to obtain 5 H_2 molecules missing, for an IHD of 5.

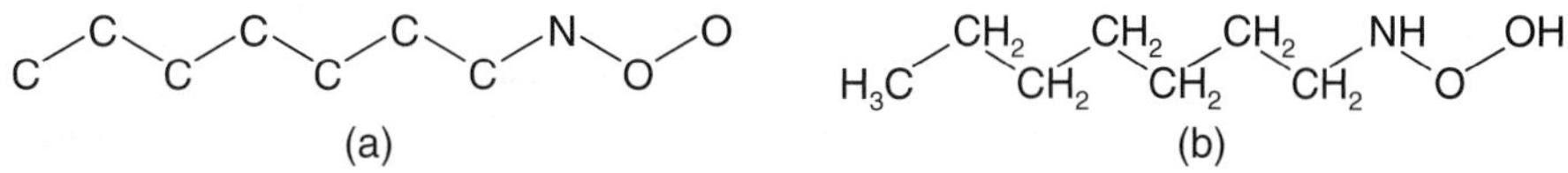

FIGURE 4-17 **Constructing a saturated molecule.** (a) A possible arrangement of the non-H atoms in a fully saturated molecule containing seven C atoms, one N atom, and two O atoms. (b) The same arrangement with enough H atoms added to give each non-H atom its octet, yielding a formula of $C_7H_{17}NO_2$.

STEPS FOR DETERMINING A MOLECULE'S IHD FROM ITS FORMULA

1. Draw an analogous completely saturated molecule.
   a. The saturated molecule should have the same number of each type of non-H atom as the formula you are given.
   b. Do not include any double bonds, triple bonds, or rings in the saturated molecule.
2. Count the number of H atoms in the completely saturated molecule you just drew.
3. Subtract the number of H atoms in the formula you are given from the number of H atoms in the completely saturated molecule. This gives you the number of *H atoms* missing from the formula you are given.
4. Divide that number by 2 to yield the number of *H_2 molecules* missing from the formula you are given. This is the IHD of the molecule.

✔ QUICK CHECK

What is the IHD for a compound with the formula $C_4H_5O_3NCl_2$?

Answer: One possible analogous completely saturated compound is $H_3C—CH_2—CH_2—CH_2—O—O—O—NCl_2$. It has nine H atoms, so the formula we have been given is missing two H_2 molecules. This gives it an IHD of 2.

This method can work with ions as well. Let's determine the IHD of $C_3H_3O_2^-$. In coming up with a completely saturated species, we can choose to put the negative charge on either a C atom or an O atom—the formula will be the same. Let's put it on an O atom, which means that this O atom will have three lone pairs of electrons and one covalent bond. If we connect the three C atoms in a row, followed by the two O atoms, we end up with the following as a completely saturated species: $CH_3CH_2CH_2OO^-$ (the O atoms have only single bonds). Its molecular formula is $C_3H_7O_2^-$, so the molecular formula we are examining, $C_3H_3O_2^-$, is short four H atoms, for an IHD of 2.

LOOK OUT

Many organic chemistry textbooks present formulas or equations that can be used to determine the number of H atoms in a completely saturated molecule containing *n* carbon atoms. Whereas these equations can be quite convenient, each formula

(continued)

represents another thing to memorize. Trouble arises because the formula is different depending on whether the molecule has only Cs and Hs or whether it contains Fs, Ns, Os, and so on. For example, a completely saturated molecule that contains *n* Cs will possess $2n+2$ H atoms. For each F atom the molecule contains, there will be one less H atom in the analogous completely saturated molecule. For each N atom, there will be one additional H atom. And each O atom does not change the number of H atoms in the analogous completely saturated compound. Furthermore, working with charged species requires additional adjustments to those formulas!

Needless to say, memorizing these formulas will serve to confuse you more than help you. Instead, I believe the best method for determining the formula for the completely saturated molecule is that which was used earlier for $C_7H_7NO_2$—a method that is simple to apply and involves no memorization. This method is independent of the types of bonds contained in the molecule and can even work with charged species.

4.7 Application: Draw All Constitutional Isomers of...

A common homework or exam problem you may encounter begins: "Draw all constitutional isomers that have the molecular formula. . . ." We devote time to working this type of problem not merely to teach you how to tackle it, but also because the ability to work these problems comfortably demonstrates an in-depth understanding of constitutional isomerism; in addition, it provides your brain with an exercise in problem-solving skills—an absolute necessity for organic chemistry.

Drawing all isomers can be quite challenging, but the following steps can help.

STEPS FOR DRAWING ALL ISOMERS HAVING A PARTICULAR FORMULA

1. Determine the formula's IHD in order to know the possible combinations of double bonds, triple bonds, and rings required in each isomer you draw.
2. Draw all possible **backbones** that are unique in their *connectivity*.
   a. Each backbone includes atoms that form two or more bonds, such as C, N, and O, because they allow molecular chains to be extended through them; backbones do not include H atoms or halogen atoms (F, Cl, Br, and I), which participate in only one bond and therefore represent the termination of a chain.
   b. Leave out any double bonds or triple bonds; these will be added later.
   c. Do include rings. The number of rings must not exceed the IHD computed from step 1; each backbone, however, can have fewer rings than the IHD.
3. For each backbone generated in step 2, add double bonds and/or triple bonds in order to achieve the total IHD calculated in step 1.
   a. Try to add these double/triple bonds at various locations so as to generate as many unique connectivities as possible.
4. For each structure generated in step 3, add halogen atoms at various locations so as to generate as many unique connectivities as possible.
5. Add H atoms to complete each atom's octet.

Let's jump right in with a problem in which we are asked to draw all constitutional isomers of $C_4H_8F_2$. Step 1 is to calculate the IHD in order to determine the possible number of double bonds, triple bonds, and rings. As we saw before, this entails drawing *any* completely saturated molecule using all of the non-H atoms—in this case, four C atoms and two F atoms—and comparing it to the molecular formula we have been given. In this case, it turns out that $C_4H_8F_2$ is already completely saturated. Therefore, the IHD is 0, and any molecule we come up with must contain only single bonds and no rings. Knowing this dramatically decreases the possibilities!

TIME TO TRY

Verify that $C_4H_8F_2$ is completely saturated by drawing a chain of four C atoms and by adding two F atoms and as many H atoms as possible.

Step 2 has us draw backbones with various connectivities—in this case, ones that contain four C atoms and only single bonds. One arrangement is that in which all of the C atoms form a linear chain—that is, $C^1—C^2—C^3—C^4$ (the C atoms are labeled to help us keep them straight), as shown in Figure 4-18a. Another arrangement of the C atoms (in the hope of producing a backbone with different connectivity) is that in which C^4 is disconnected from C^3, leaving $C^1—C^2—C^3$, and is reconnected to one of the other two Cs (either C^1 or C^2). If C^4 is connected to C^1, this yields $C^4—C^1—C^2—C^3$, as shown in Figure 4-18b. Although the atom labels are in a different order, this connectivity is exactly the same as the original connectivity of $C^1—C^2—C^3—C^4$, so it does not count as being unique. If, on the other hand, C^4 is connected to C^2, the result is a T-shaped backbone, the connectivity of which is unique from the linear connectivity (Figure 4-18c). The T-shaped backbone is unique in that it contains a C atom that is bonded to three other C atoms—something that is not present in the linear arrangement.

Step 3 deals with double bonds and triple bonds. Because the IHD of our formula is 0, we skip this step. Step 4 deals with halogen atoms, of which there are two—the two F atoms. In order to arrive at as many different connectivities for the backbone as possible, let's add these F atoms one at a time, first producing various connectivities of C_4F.

Working with the linear backbone from Figure 4-18a, realize that there are two distinct C atoms—C^1 has exactly the same connectivity as C^4, and C^3 has exactly the same connectivity as C^2. Therefore, only two unique connectivities of C_4F can be produced from the linear carbon backbone, shown in Figure 4-19a and 4-19b.

Working with the T-shaped backbone from Figure 4-18c, notice once again that there are only two distinct C atoms—the central C atom is the only one that is bonded to three other

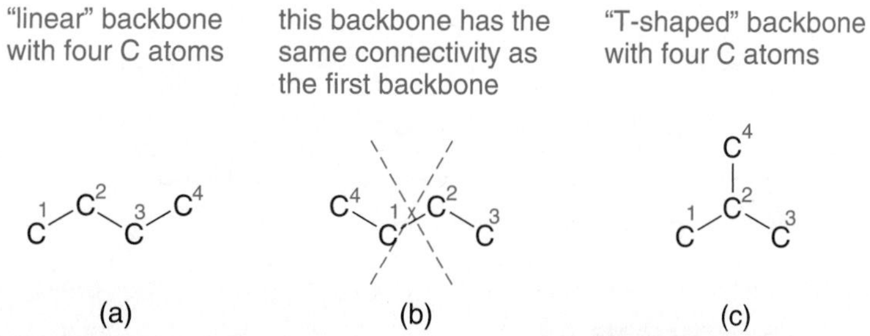

FIGURE 4-18 **Different possible backbones in which four carbon atoms are singly bonded together.** The connectivities in (a) and (b) are the same, despite different atom labels, making that in (b) redundant with that in (a). The backbone in (c) has a different connectivity from that in (a).

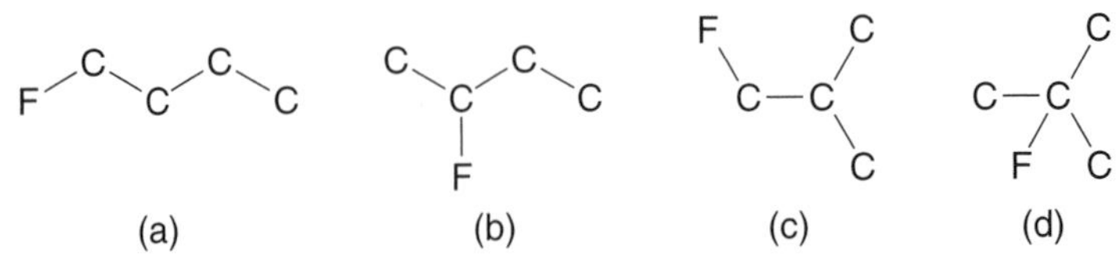

FIGURE 4-19 Different connectivities of C_4F. The connectivities in (a) and (b) are the result of adding an F atom to the linear carbon backbone. Those in (c) and (d) are the result of adding an F atom to the T-shaped carbon backbone.

C atoms, and the three remaining C atoms all share the same connectivity. Therefore, as shown in Figures 4-19c and 4-19d, two unique connectivities of C_4F can be produced from the T-shaped carbon backbone. In all, then, there are four unique connectivities with the formula C_4F.

Let's now complete step 4 by adding the second F atom to the four connectivities in Figure 4-19. To do this systematically, it helps to temporarily label the F atom added first as "F^1" and the one we are about to add as "F^2." With this in mind, let's add F^2 to the structure from Figure 4-19a. In that structure, notice that all four C atoms have different connectivity, so adding F^2 to any of those C atoms produces a unique connectivity of C_4F_2. These are shown in Figures 4-20a through 4-20d.

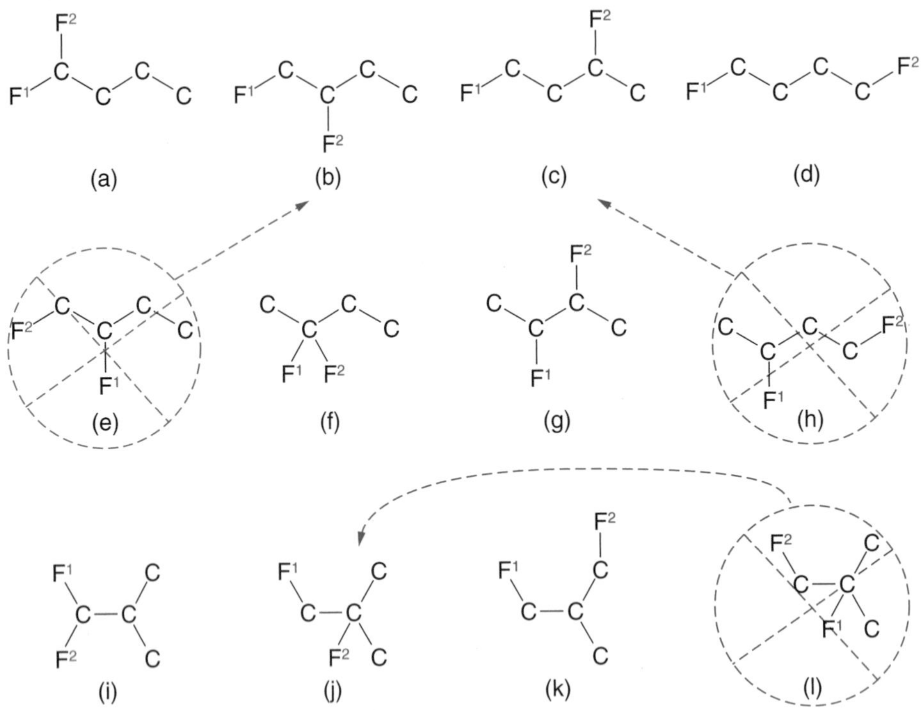

FIGURE 4-20 Unique connectivities of C_4F_2. The structures in (a) through (d) are produced from the C_4F structure in Figure 4-19a. Those in (e) through (h) are produced from Figure 4-19b. Those in (i) through (k) are produced from Figure 4-19c. That in (l) is produced from Figure 4-19d. Redundant connectivities are crossed out.

We repeat this process of adding F^2 to the C_4F structure from Figure 4-19b, yielding the four structures in Figures 4-20e through 4-20h. At this point, however, realize that we have produced two redundancies. Even though the F atoms are labeled differently, the connectivity of the structure in Figure 4-20e is the same as that in Figure 4-20b. Likewise, the connectivity of the structure in Figure 4-20h is the same as that in Figure 4-20c. Therefore, of the first eight structures of C_4F_2, only six have unique connectivities.

Let's continue now by adding F^2 to the C_4F structure from Figure 4-19c. This yields the three connectivities of C_4F_2 shown in Figures 4-20i through 4-20k. Finally, F^2 can be added to the C_4F structure from Figure 4-18d to yield just one unique connectivity, shown in Figure 4-20l. However, as indicated, the structure of C_4F_2 in Figure 4-20l has the same connectivity as that in Figure 4-20j, so it is redundant. In all, then, there are nine total structures of C_4F_2 with unique connectivities.

Finally, for step 5, we add enough H atoms to complete each atom's octet. In each case, this will require adding eight total H atoms to yield the formula of $C_4H_8F_2$. For example, in Figure 4-20a, one H is added to the first C atom, two to the second C atom, two to the third C atom, and three to the fourth C atom, resulting in $CHF_2CH_2CH_2CH_3$. Doing the same to each of the unique connectivities in Figure 4-20 yields nine total constitutional isomers of $C_4H_8F_2$.

Let's perform this exercise one more time with a different formula—C_3H_6O. For step 1, we must compare the formula we have been given with a completely saturated molecule with three C atoms and one O atom. One possibility is $CH_3CH_2CH_2OH$, which has the formula C_3H_8O. Consequently, the formula we have been given has an IHD of 1.

Step 2 has us draw all of the backbones of C_3O without any multiple bonds. As shown in Figure 4-21, there are three unique connectivities with one ring—two of them have a three-membered ring (Figures 4-21a and 4-21b), and one has a four-membered ring (Figure 4-21c). There are also three different connectivities of C_3O without a ring: C—C—C—O (Figures 4-21d through 4-21f), C—C—O—C (Figure 4-21g), and C—C(O)—C (Figures 4-21h and 4-21i). (Notice that there is no T-shaped backbone with O at the center because that would require the O atom to be bonded to three C atoms.)

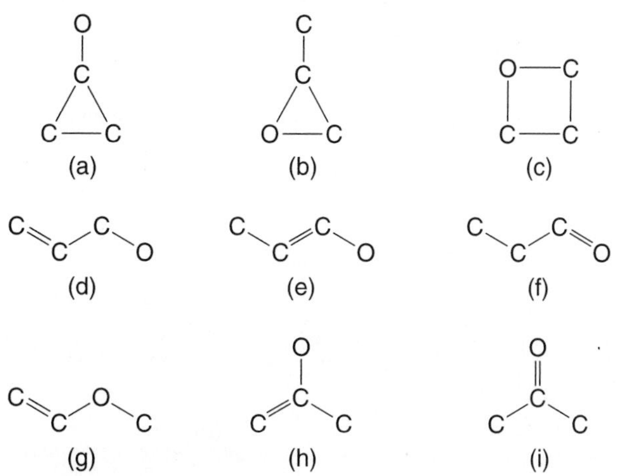

FIGURE 4-21 **Backbones of C_3O with an IHD of 1.** Backbones (a) through (c) have one ring, so they satisfy the requirement that IHD = 1. Backbones (d) through (i) have no rings, so in order to achieve an IHD of 1 they must have one double bond.

Step 3 has us add multiple bonds at various locations in each backbone to achieve the required IHD of 1. The backbones with a ring already have an IHD of 1, so no double or triple bonds can be added. Each backbone without a ring, however, must have one double bond. As shown in Figures 4-21d through 4-21f, a double bond can be added to the C—C—C—O backbone in three different places, each yielding a unique connectivity. By contrast, a double bond can be added to only one location in the C—C—O—C backbone, which, as shown in Figure 4-21g, is between the two C atoms; if the double bond were to involve an O atom, three bonds to O would be required. Finally, a double bond can be added to the C—C(O)—C backbone to yield two unique connectivities, as shown in Figures 4-21h and 4-21i.

At this point, we would normally proceed to step 4, which has us deal with halogen atoms. In this case, because the formula has no halogen atoms, we can go directly to step 5, which has us simply add the necessary H atoms to complete each atom's octet. For example, for the structure in Figure 4-21a, one H atom should be added to the O atom, one H atom should be added to the C atom below the O atom, and two H atoms should be added to each of the two bottom C atoms. That's a total of six H atoms, yielding the formula C_3H_6O, which is the same as the one we were originally given.

4.8 Application: Draw All Stereoisomers of...

In this section, we cover how to draw all stereoisomers of a given formula. The reasons for doing so are similar to the ones for practicing how to draw all constitutional isomers of a given formula, as we did in the previous section. First, you will likely be expected to tackle such problems in your yearlong organic chemistry course. Second, doing so forces you to understand more deeply the relationships among stereoisomers.

As we learned in Section 4.4A, stereoisomers often (but are not required to) contain at least one stereocenter—that is, a tetrahedral atom that is bonded to four different groups. These stereocenters allow stereoisomers to be different, while the connectivity remains the same. To begin to understand why, let's reexamine Figure 4-9, shown again as Figure 4-22.

We previously established that in this case the molecule and its mirror image are different molecules, making them enantiomers of each other. As we can see in Figure 4-22b, the reason is that the groups that are attached to the stereocenter are arranged differently in space. The stereocenter and two of the attached groups (in this case, F and Cl) line up perfectly, but the remaining two attached groups (H and Br) do not. In other words, the stereocenter in each enantiomer has a different **configuration.**

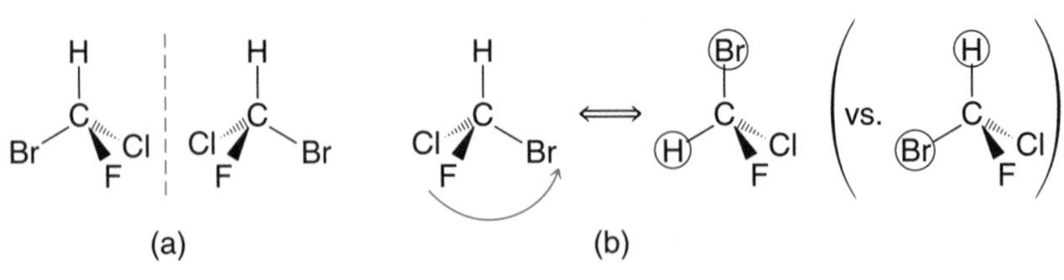

FIGURE 4-22 **This is the same as Figure 4-9.** (a) Three-dimensional representation of CHFClBr and its mirror image. (b) The mirror image is rotated in space, so that three of the five atoms (C, F, and Cl) line up with atoms in the original molecule (shown in parentheses), but two atoms (H and Br) do not.

With this example, we can see specifically how the configurations about a stereocenter are related.

The configuration about a stereocenter will be changed upon

- Taking its mirror image.
- Exchanging any two groups attached to the stereocenter.

Furthermore, we are given insight into how many configurations are possible about a given stereocenter.

For any stereocenter, there are exactly two configurations that are possible.

The reason, quite simply, is that for any molecule there can be only one mirror image.

Because there are only two stereochemical configurations possible, chemists have devised a way to specify exactly which configuration a given stereocenter has and have designated such configurations as either *R* or *S*. We will not discuss that topic here—it will be left for your yearlong organic chemistry course. However, what we can say here about such designations is that whenever a mirror image is taken or whenever we exchange two groups attached to a stereocenter, the configuration is changed from *R* to *S*, or vice versa. Another way of saying this is that the configurations are *reversed*.

With this in mind, the general strategy for drawing all stereoisomers becomes relatively straightforward.

STEPS FOR DRAWING ALL STEREOISOMERS OF A GIVEN COMPOUND

1. Draw the first stereoisomer by specifying the dash-wedge notation for every stereocenter in the molecule. (It does not matter what the dash-wedge notation is for the initial stereoisomer.)
2. Redraw the molecule with the configuration reversed at the first stereocenter, leaving all the other stereocenters alone. This can be done by exchanging any two groups on that stereocenter.
3. If there is a second stereocenter,
   a. Redraw the molecule with the original configuration at the first stereocenter, but with the configuration reversed at the second stereocenter.
   b. Redraw the molecule again, but with the configuration at each of the first two stereocenters reversed from the original configuration.
4. If there is a third stereocenter,
   a. Redraw the molecule with the original configurations at each of the first two stereocenters, but with the configuration reversed at the third stereocenter.
   b. Keeping the reversed configuration at the third stereocenter, repeat steps 2 and 3.
5. If there are additional stereocenters, repeat the pattern.
6. Cross out any redundant stereoisomers.

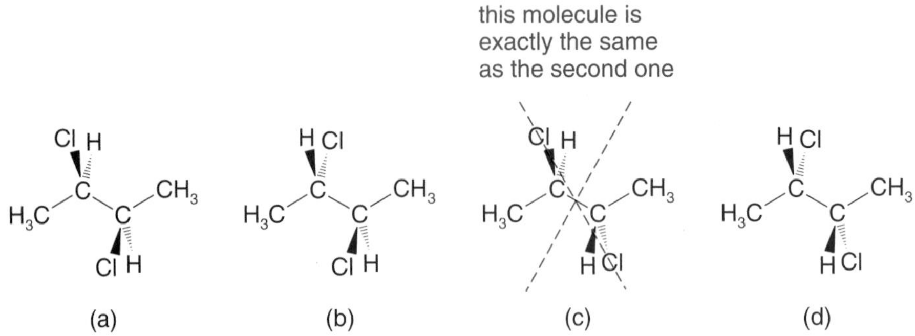
this molecule is
exactly the same
as the second one
(a)
(b)
(c)
(d)

FIGURE 4-23 **Drawing all stereoisomers of $CH_3CHClCHClCH_3$.** (a) The dash-wedge notation at each stereocenter is specified. (b) The configuration of the stereocenter at the left is reversed, while that at the right remains unchanged. (c) The configuration of the stereocenter at the left is the same as in the original molecule, but that at the right is reversed. (d) Both configurations are reversed from the original molecule. Molecule (c) is crossed out because it is identical to (b).

Let's apply these rules to drawing all stereoisomers of $CH_3CHClCHClCH_3$. We see that there are two stereocenters—each C atom that is bonded to a Cl atom—so for step 1 we need to specify the configuration by supplying the dash-wedge notation at those carbons. This is shown in Figure 4-23a. For step 2, we can reverse the configuration at the stereocenter at the left by exchanging H with Cl. This is shown in Figure 4-23b. Figure 4-23c is what we obtain from step 3a—the original configuration is drawn for the stereocenter at the left, but the reverse configuration is drawn for the stereocenter at the right. Figure 4-23d is what is obtained from step 3b—both configurations are reversed from Figure 4-23a. As a final step, we check for redundant molecules. In this case, the molecules in Figures 4-23b and 4-23c are identical—this can be seen by flipping the molecule over and rotating about the central C—C bond.

TIME TO TRY

Use a molecular modeling kit to construct the structures shown in Figures 4-23b and 4-23c. Orient the two structures in various ways, including rotation about the central C—C bond, until all of the atoms line up perfectly.

✔ QUICK CHECK

What is the relationship among all the molecules in Figure 4-23?

Answer: Structures (a) and (b) are diastereomers. Structures (b) and (d) are diastereomers. Structures (a) and (d) are enantiomers.

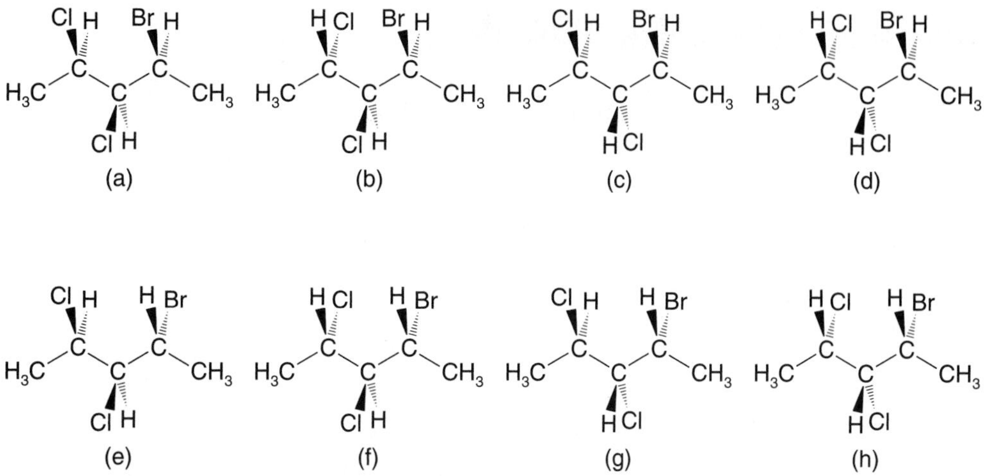

FIGURE 4-24 Drawing all stereoisomers of $CH_3CHClCHClCHBrCH_3$. (a) through (d) are the results of steps 1 through 3, analogous to what is shown in Figure 4-23. (e) through (h) are the results of step 4. Each structure is identical to the one above it except that the configuration is reversed at the stereocenter that has the Br atom.

As a second example, let's draw all stereoisomers of $CH_3CHClCHClCHBrCH_3$. The first three steps are shown in Figure 4-24a through 4-24d. For all four of these structures, notice that the configuration of the stereocenter with the Br atom remains unchanged. To complete step 4, we then reverse the configuration at that stereocenter, draw the first two stereocenters with their original configurations, and repeat steps 1–3. This is shown in Figures 4-24e through 4-24h. Finally, we must check for redundant molecules. In this case, they are all different from one another, so there are eight total stereoisomers.

From the protocol that we are using to draw all stereoisomers of a compound, you should be able to see that the number of structures that are generated is doubled with each additional stereocenter. In other words,

If there are *n* stereocenters in a molecule, there will be at most 2^n total stereoisomers.

This statement says "at most" because some of the structures that are generated could be redundant. This was shown in Figure 4-23 and will be the case whenever a structure that is generated has a plane of symmetry, as was the case for the structure in Figure 4-23b.

✔ QUICK CHECK

If a molecule has four stereocenters, what is the maximum number of stereoisomers possible?

Answer: The maximum possible number is $2^n = 2^4 = 16$.

WHAT DID YOU LEARN?

4.1 Identify the specific type of relationship between each of the following pairs of molecules (i.e., *same molecules, constitutional isomers, enantiomers, diastereomers,* or *unrelated*).

(a) 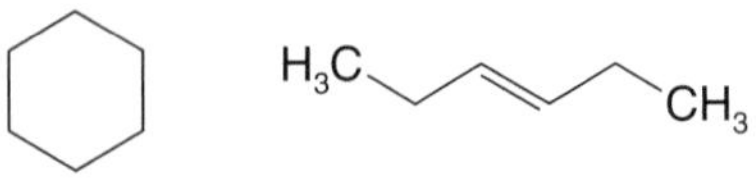

(b) 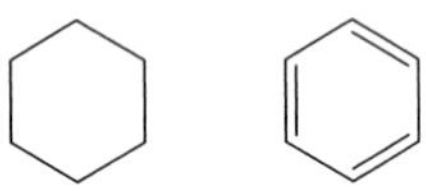

(c) 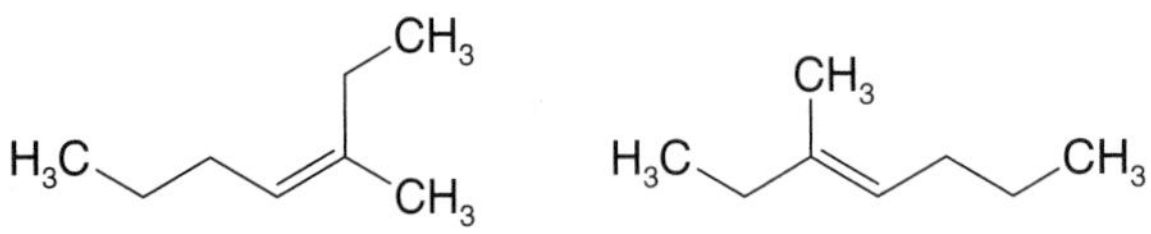

(d) 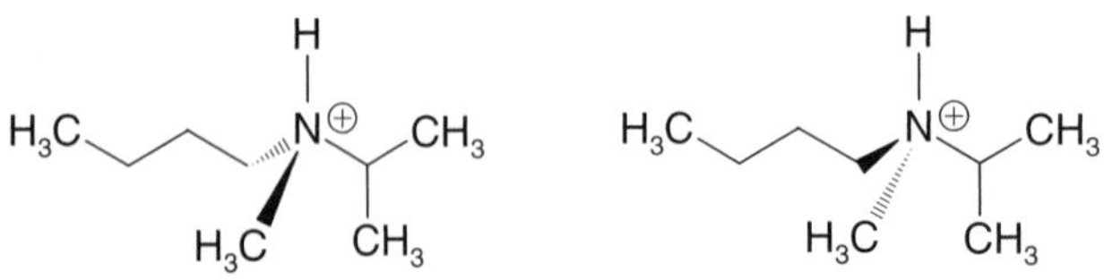

(e) 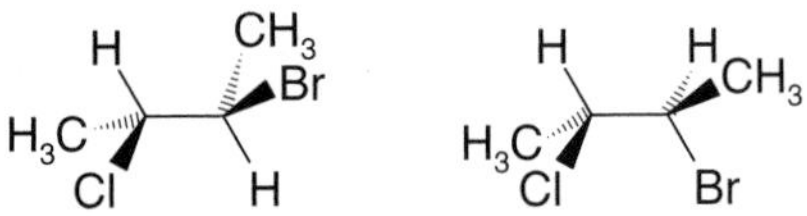

(f) 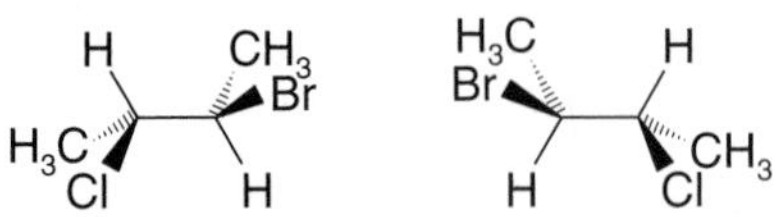

(g) 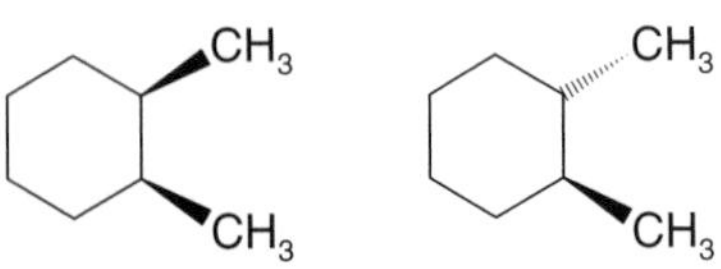

4.2 Determine which pairs of species in the previous problem will have identical physical and chemical properties.

4.3 Which of the following molecules are chiral?

(a)

(b)

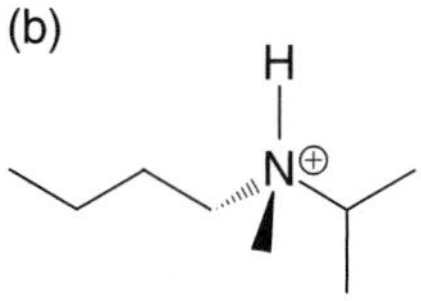

(c)

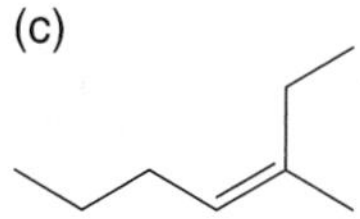

(d)

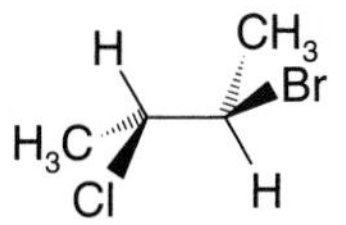

(e)

(f)

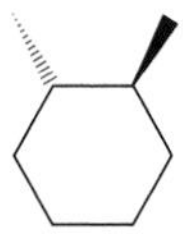

(g)

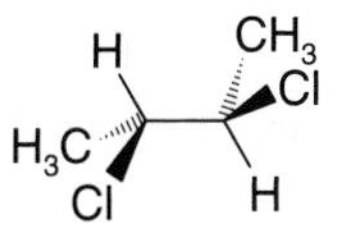

(h)

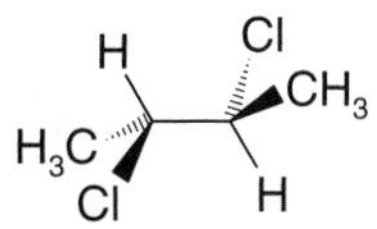

4.4 Calculate the index of hydrogen deficiency for each of the following molecular formulas:

a C_6H_6 **b** $C_6H_5NO_2$ **c** $C_8H_{13}NOF_2$
d $C_4H_{12}Si$ **e** $C_6H_5O^-$ **f** $C_4H_6O_3S$

4.5 Draw all constitutional isomers that have the molecular formula $C_3H_6OF_2$ in which the O is bonded to only one C atom. (There are 14 isomers.)

4.6 Draw all constitutional isomers that have the molecular formula C_4H_6. (There are nine isomers.)

4.7 Draw all stereoisomers of each of the following compounds:

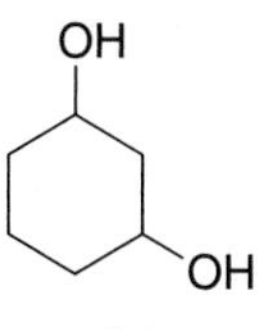

(a)

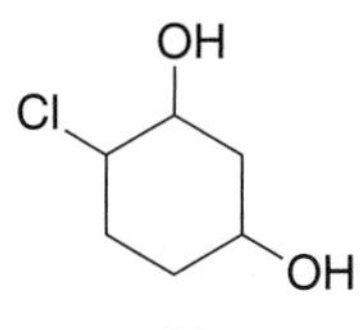

(b)

5 Reaction Mechanisms 1: Elementary Steps

When you complete this chapter, you should be able to:

- Distinguish among an overall reaction, an elementary step, and a reaction mechanism.
- Identify an elementary step as one of the following:
  - Coordination
  - Heterolysis
  - Proton transfer
  - Bimolecular nucleophilic substitution (S_N2)
  - Nucleophilic addition
  - Nucleophilic elimination
  - Electrophilic addition
  - Electrophilic elimination
  - Bimolecular elimination (E2)
- Draw the curved arrow notation for any of the above elementary steps.
- Describe each of the above elementary steps in terms of the flow of electrons from an electron-rich site to an electron-poor site.
- Simplify the structures of metal-containing reactant compounds as they pertain to identifying electron-rich and electron-poor sites.
- Recognize when a mixture of stereoisomers is produced from an elementary step and predict whether the mixture will have equal or unequal amounts of those stereoisomers.
- Explain why S_N2 steps are stereospecific and predict which stereoisomer is produced, based on the particular structure of the substrate.

Your Starting Point

Answer the following questions to assess your knowledge about reaction mechanisms.

1. What is the definition of a chemical reaction? ______________________

2. How do an overall reaction, an elementary step, and a reaction mechanism differ? ______________________

3. Are opposite charges attracted or repelled? ______________________

4. Is a species stabilized or destabilized with additional bonds? ______________________

5. Is a species stabilized or destabilized when an atom gains an octet? ______________________

6. What is a reaction coordinate? What is a transition state? ______________________

7. What is a Brønsted acid? A Brønsted base? ______________________

8. What types of ions typically behave as spectator ions? ______________________

9. How many configurations are possible for a stereocenter? How are they related? ______________________

Answers: 1. A chemical reaction is a process that causes one species to be converted into another, typically by the formation and breaking of covalent bonds. 2. An overall reaction is the process that converts the compounds added to a reaction mixture into compounds that can be isolated; an elementary step is a reaction that takes place in a single event; a reaction mechanism is the sequence of elementary steps that shows how overall reactants become overall products. 3. Opposite charge attract; like charges repel. 4. Additional bonds stabilize a species. 5. A species is stabilized when an atom gains an octet. 6. A reaction coordinate describes the geometry changes that occur to species as they are converted from reactants to products; a transition state is the structure that appears at an energy maximum along the reaction coordinate. 7. A Brønsted acid is a proton donor; a Brønsted base is a proton acceptor. 8. Metal cations from Group 1A typically behave as spectator ions. 9. There are two configurations for any stereocenter. They are related by the exchange of any two attached groups or by reflection through a mirror.

5.1 Introduction

To this point in the book, we have focused primarily on concepts that pertain to a molecule's structure. Chapter 2 presented aspects of bonding, Chapter 3 discussed issues surrounding molecular geometry, and Chapter 4 dealt with isomerism. In this chapter, we turn our attention to chemical reactions.

A **chemical reaction** is the conversion of one substance—a reactant—into another—the products—typically through the breaking and/or formation of covalent bonds.

When we watch such a reaction take place in the laboratory, it may seem to proceed with a smooth disappearance of the **overall reactants** (the substances you physically add in order to initiate the reaction) and the smooth appearance of the **overall products** (the substances you can isolate when the reaction is over). However, if we could see the reaction take place on the molecular level, this is not what we would observe. Rather, such reactions take place in distinct *elementary steps.*

An **elementary step** is a reaction that occurs in a single event (almost instantaneously), each time transforming a reactant molecule into a product molecule.

Thus, an elementary step cannot be broken down into simpler reactions.

Depending on the nature of the reaction, the overall reactants may be converted into the overall products in a single elementary step (i.e., Overall Reactants $\rightarrow$ Overall Products), or the conversion may require several such steps (e.g., Overall Reactants $\rightarrow$ A $\rightarrow$ B $\rightarrow$ C $\rightarrow$ Overall Products). In other words, the reactions may have different *mechanisms.*

A **mechanism** is the precise sequence of elementary steps that results in the conversion of the overall reactants into the overall products.

WHY SHOULD I CARE?

As we mentioned in Chapter 1, there are several reasons why it is important to be able to work comfortably with reaction mechanisms. One is that, if you *understand* reaction mechanisms—that is, *why* they occur the way they do—then you will remember reactions much longer than if you try to memorize them (e.g., using flash-card techniques). Another reason is that, if you know reaction mechanisms, and not just the reactions themselves (i.e., memorizing which reactants form which products), then you will begin to see the many similarities between reactions that otherwise seem completely different. Still another reason is that, if you understand why each elementary step takes place, then you can begin to actually predict mechanisms of reactions that you have never seen before!

Given the importance of reaction mechanisms, we have actually devoted two chapters to the topic—this chapter and Chapter 7. This chapter deals primarily with beginning concepts pertaining to individual elementary steps, whereas Chapter 7 focuses on topics pertaining to mechanisms composed of multiple steps.

More specifically, this chapter will give you the tools necessary to recognize and understand the most common elementary steps in organic reaction mechanisms. These elementary steps—10 in all—comprise the vast majority of the reactions you will encounter in your organic chemistry course. You will see how each step is described using *curved arrow notation,* and you will also learn about some of the factors that drive each step. In other words, you will begin to be able to speak the language of reaction mechanisms.

5.2 Bond Formation (Coordination) and Bond Breaking (Heterolysis)

We begin with two of the most straightforward elementary steps—simple bond formation and bond breaking—which we call *coordination* and *heterolysis*, respectively.

In a **coordination step,** a lone pair of electrons from one atom is used to form a new *single bond* to another atom, without any bonds breaking.

An example is shown in Equation 5-1.

coordination step

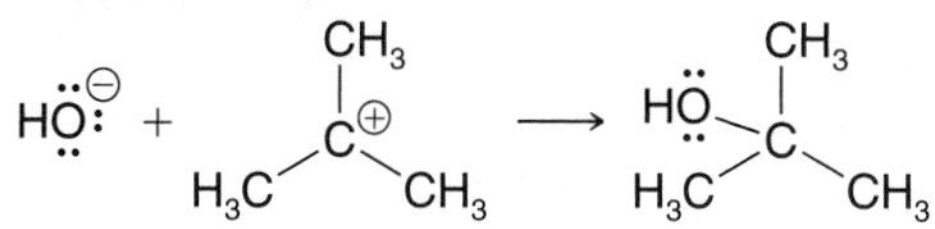

(5-1)

You will notice that in order for a coordination step to occur, without violating the octet rule in the product, the atom to which the new bond forms must initially lack an octet.

QUICK CHECK

In Equation 5-1, identify the atom lacking an octet, the new bond formed, and the pair of electrons used to form that bond.

Answer: The C atom with the positive charge initially lacks an octet in the reactants; a new O—C bond is formed; a lone pair of electrons from O is used to form that bond.

A heterolysis step is essentially the opposite of coordination.

In a **heterolysis step** (also called **heterolytic bond dissociation**), a single bond between two atoms is broken, and the pair of electrons from that bond ends up on just one of those two atoms.

An example is shown in Equation 5-2.

heterolysis step

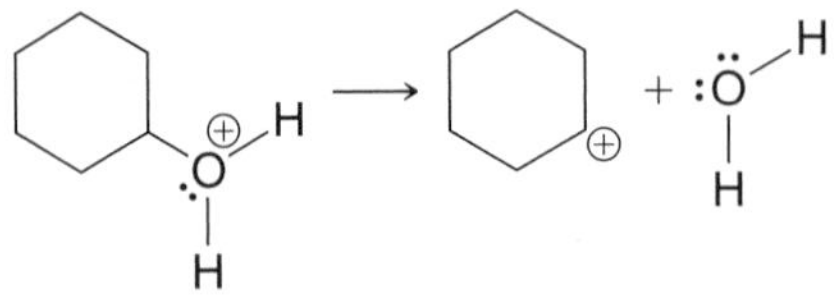

(5-2)

✔ QUICK CHECK

In Equation 5-2, identify the bond that is broken, where the two electrons from that bond end up, and the atom in the products that lacks an octet.

Answer: The C—O bond is broken; the two electrons from that bond end up as a lone pair on O; the positively charged C atom in the products lacks an octet.

Having seen examples of coordination and heterolysis steps, let's spend some time examining certain aspects of elementary steps, which will help us better work with and understand such steps. This is done in Sections 5.2A through 5.2C.

5.2A FREE-ENERGY DIAGRAMS, REACTION COORDINATES, AND TRANSITION STATES

We have learned that coordination and heterolysis are two types of elementary steps, meaning that they each take place in a single event, almost instantaneously. What is key here is the word *almost*. The transformation of a reactant molecule into a product molecule does in fact take a small amount of time, and that transformation involves a continuous change in molecular structure. With this in mind, chemists can specify the extent to which a reactant molecule has transformed into a product molecule, using what is called a *free-energy diagram*.

In a **free-energy diagram** for a particular reaction, *Gibbs free energy* (symbolized G) is plotted on the *y*-axis, and the *reaction coordinate* is plotted on the *x*-axis.

Gibbs free energy is something you will have encountered in general chemistry, and it has a specific definition. For this course, however, the most important thing to know about it is how it relates to stability.

The lower (more negative) the Gibbs free energy, the greater the stability.

The higher (more positive) the Gibbs free energy, the less the stability.

The **reaction coordinate**, on the other hand, tracks changes in *geometry* that take place throughout the course of a reaction.

The farther to the right in a free-energy diagram, the more the species involved in the reaction look like the products.

The farther to the left, the more the species involved look like the reactants.

For example, for the coordination step in Equation 5-1, an increase in the reaction coordinate corresponds to the shortening of the C—O distance and to a decrease in both the negative charge on O and the positive charge on C.

Figure 5-1 is the free-energy diagram of the coordination step in Equation 5-1. As you can see, the reactants appear at the far left, and the products appear at the far right. Notice that the reactants and products differ in energy—that difference is denoted as ΔG_{rxn}.

Notice also that, as the reaction coordinate progresses to the right, free energy changes smoothly—first it increases, and after passing through the *transition state*, it then decreases until the products are reached. In other words,

The transition state of an elementary step appears at an *energy maximum* along the reaction coordinate between the reactants and the products. The structure that corresponds to the transition state is typically enclosed in square brackets and has a double-dagger symbol in the upper-right corner outside the brackets.

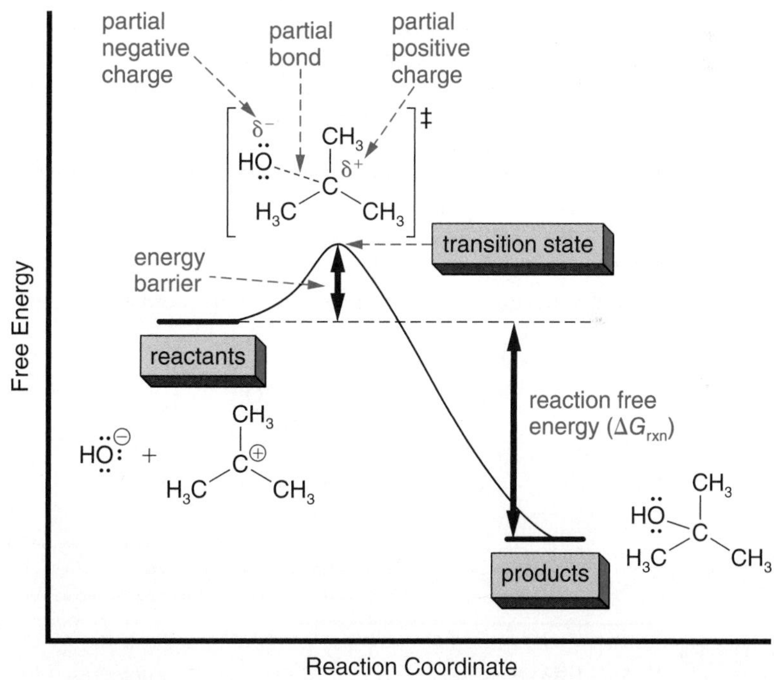

FIGURE 5-1 **Free-energy diagram of the coordination step in equation 5-1.** The products have a lower free energy—and therefore greater stability—than the reactants, giving the reaction a negative value for ΔG_{rxn}. The transition state is the structure that appears at the energy maximum between the reactants and the products and is something of a hybrid of the reactants and the products. The energy barrier of the reaction is the energy difference between the reactants and the transition state.

Because the transition state is located between the reactants and the products along the reaction coordinate (which corresponds to changes in geometry), we can further say that the structure of a transition state is something of a hybrid of the reactants and the products. More specifically,

A bond that is either formed or broken during the course of an elementary step appears as a partial bond in the transition state.

A formal charge that appears or disappears during the course of an elementary step appears as a partial charge in the transition state.

In Figure 5-1, for example, notice that during the course of the elementary step, an O—C bond forms, a negative charge disappears from O, and a positive charge disappears from C. Therefore, in the transition state, there is a partial O—C bond, a partial negative charge on O, and a partial positive charge on C.

One major feature of an energy diagram is the **energy barrier,** also called the **free energy of activation.** It is identified as the difference in energy between the reactants and the transition state. The significance of the energy barrier is its impact on the *rate* of an elementary step.

The larger the energy barrier of an elementary step, the slower the reaction rate.

This should make sense because the larger that barrier is, the more energy the reactants are required to expend in order to get to the products.

Now let's take a look at the free-energy diagram for the heterolysis step from Equation 5-2, shown in Figure 5-2. Once again, the reactants are on the far left, the products are on the far right, and the energy difference between the two is ΔG_{rxn}. Furthermore, the transition state is located at the energy maximum between the reactants and the products, and the energy barrier is the difference in energy upon going from the reactants to the transition state.

QUICK CHECK

Draw the structure of the transition state for the coordination step in Equation 5-2.

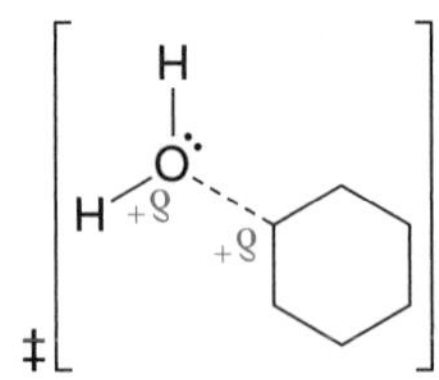

Answer: The C—O bond is broken during the course of the elementary step, so in the transition state it is a partial bond. Also, the positive charge is on an O atom in the reactants, but on a C atom in the products, so each atom receives a partial positive charge in the transition state.

QUICK CHECK

Which elementary step will proceed faster—the coordination step in Equation 5-1 or the heterolysis step in Equation 5-2?

Answer: The coordination step in Equation 5-1 will proceed faster because, as we can see in Figures 5-1 and 5-2, it has a smaller energy barrier.

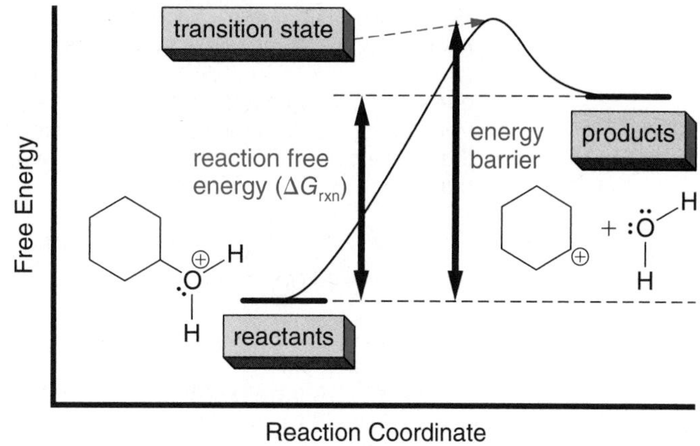

FIGURE 5-2 Free-energy diagram of the coordination step in equation 5-2. The products have a higher free energy—and therefore less stability—than the reactants, giving the reaction a positive value for ΔG_{rxn}. The transition state is the structure that appears at the energy maximum between the reactants and the products, and is something of a hybrid of the reactants and the products. The energy barrier of the reaction is the energy difference between the reactants and the transition state.

5.2B CURVED ARROW NOTATION

In the examples in Equations 5-1 and 5-2, only the reactants and the products are shown. However, in order to better understand elementary steps, it is important to trace the movement, or flow, of valence electrons. This is done through what is called **curved arrow notation,** also called **pushing electrons.** There are three basic rules of curved arrow notation.

BASIC RULES OF CURVED ARROW NOTATION

- A curved arrow represents the flow of *valence electrons*, not the movement of atoms.

- Each double-barbed curved arrow (↷) represents the movement of two valence electrons.

- To show bond formation, the head of the arrow points to the region where the bond is formed, and to show bond breaking, the tail of the arrow originates from the center of a bond.

At this point, you should ask yourself: Why show the movement of valence electrons instead of atoms? The answer is rather simple. The movement of valence electrons corresponds to the formation and/or breaking of bonds, and that is what *defines* a chemical reaction; the movement of atoms does not.

Realize, however, that during the course of an elementary step, atoms do need to move. This is taken care of in curved arrow notation with the following rule.

- When an atom needs to move in an elementary step, the atom is understood to follow its own electrons.

Let's now see how curved arrow notation is used to describe the elementary steps shown earlier. The reaction shown previously in Equation 5-1, for example, requires only one curved arrow because only two electrons are flowing—the lone pair on the O atom. A bond forms to the C atom lacking an octet, so a curved arrow is drawn from a lone pair on the O atom to that C atom, as shown in Equation 5-3.

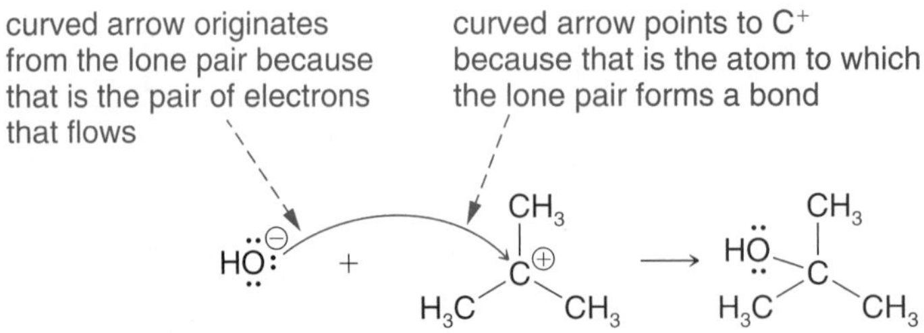
curved arrow originates from the lone pair because that is the pair of electrons that flows
curved arrow points to C+ because that is the atom to which the lone pair forms a bond
HO
CH3
H3C
CH3
HO
CH3
H3C
CH3

(5-3)

Notice that this is an example where atoms have to move. Namely, the O and C atoms must come close together in order for the O—C bond to form. As mentioned above, this is taken care of by the understanding that the O atom (and everything it is attached to) follows the lone pair of electrons that forms a bond. In other words, in this case, the curved arrow not only indicates the formation of a new single bond, but also implies that the HO^- species approaches to within a bonding distance of C.

To show how curved arrow notation is used to indicate bond breaking, let's examine the heterolysis reaction previously shown in Equation 5-2. Once again, this reaction calls for only one curved arrow because there are two total electrons that flow—the pair of electrons from the C—O bond. More specifically, the C—O bond breaks, so the curved arrow must originate from the center of that bond. And because the two electrons end up as a lone pair on the O atom, the curved arrow must point to O, as illustrated in Equation 5-4.

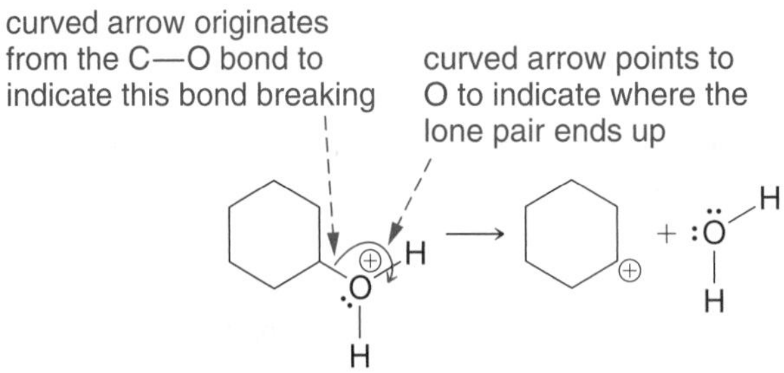
curved arrow originates from the C—O bond to indicate this bond breaking
curved arrow points to O to indicate where the lone pair ends up
H
O
H
O
H
H

(5-4)

5.2C ELECTRON-RICH TO ELECTRON-POOR

More than simply accounting for changes in bonding in an elementary step, curved arrow notation gives us a *feel* for how electrons tend to flow.

Electrons tend to flow from an electron-rich site to an electron-poor site.

This means that, when drawing a curved arrow, the tail of the arrow tends to originate from an electron-rich site and should point toward an electron-poor site. (We will elaborate on how to identify such sites shortly.)

The reason for this rule should make sense, as it is due to a simple law of physics with which you are probably familiar: *Opposite charges attract; like charges repel.* At an electron-rich site, there is an excess of negative charge, so the electrons at that site are not terribly happy. Those negatively charged electrons will be repelled from the electron-rich site and will be drawn toward a positively charged site—that is, to a site that is deficient of electrons, or electron-poor.

To apply the notion of *electron-rich to electron-poor* to a coordination step, let's examine, once more, the curved arrow notation for the reaction shown previously in Equation 5-3, which is shown again in Equation 5-5. Notice that the O atom from HO^- bears a negative charge, which is indicative of an electron-rich site, and the C^+ atom bears a positive charge, which is indicative of an electron-poor site. The C^+ atom is further identified as electron-poor by the fact that it lacks an octet. Thus, the curved arrow is drawn from the electron-rich O atom to the electron-poor C atom.

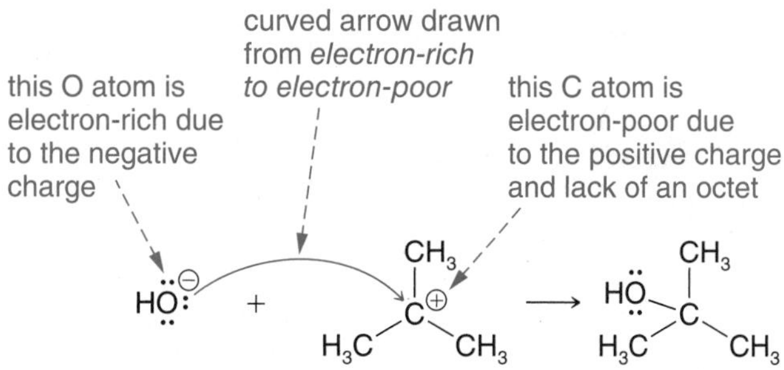
curved arrow drawn
from *electron-rich*
to electron-poor
this O atom is
electron-rich due
to the negative
charge
this C atom is
electron-poor due
to the positive charge
and lack of an octet
HO
+
CH3
C
H3C
CH3
HO
CH3
C
H3C
CH3

(5-5)

Similarly, we can apply the notion of *electron-rich to electron-poor* to the heterolysis step in Equation 5-4, as shown in Equation 5-6. The bonding region is a site that is relatively electron-rich because two electrons are effectively confined there. By contrast, the positively charged O atom is electron-poor. Thus, the curved arrow is drawn from the electron-rich bond to the electron-poor O atom.

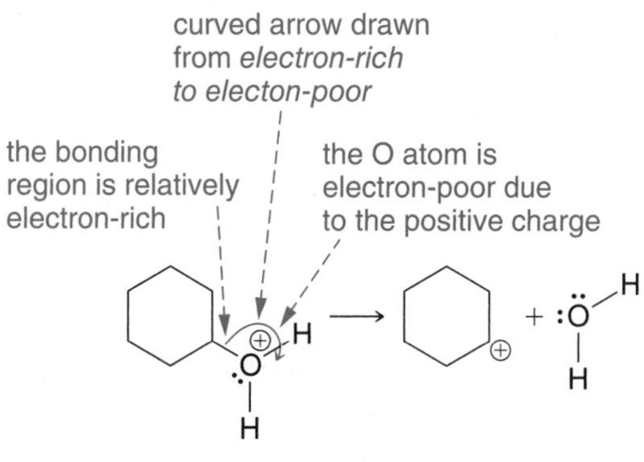
curved arrow drawn
from *electron-rich*
to electon-poor
the bonding
region is relatively
electron-rich
the O atom is
electron-poor due
to the positive charge
O
H
H
+ :O
H
H

(5-6)

5.3 Proton Transfers

In the previous section, we learned about two of the most fundamental elementary steps—coordination and heterolysis. In a coordination step, a covalent bond is formed, and in a heterolysis step, one is broken. Some elementary steps, however, involve both the formation and the breaking of a covalent bond simultaneously. One such step is a *proton transfer*.

In a **proton transfer step,** an H atom simultaneously undergoes the breaking of a bond to one atom and the formation of a bond to another.

An example is shown in Equation 5-7, with the curved arrow notation included.

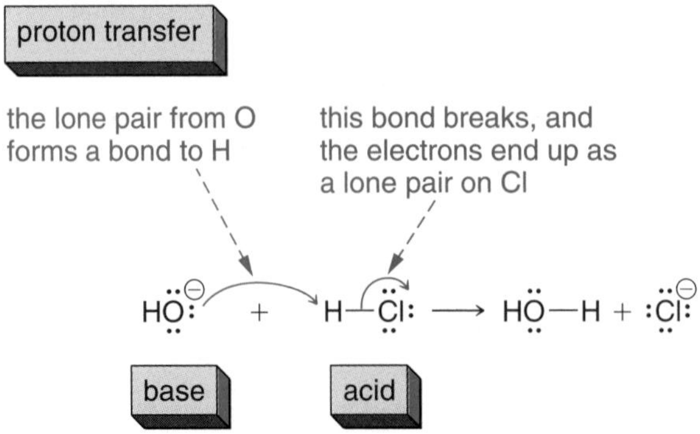
proton transfer
the lone pair from O forms a bond to H
this bond breaks, and the electrons end up as a lone pair on Cl
base
acid

(5-7)

In Equation 5-7, a lone pair from O forms a bond to H, and when the H—Cl bond breaks, the electrons end up as a lone pair on Cl.

You can see why such a reaction is called a *proton transfer.* Notice that H leaves the H—Cl molecule, but doesn't take any electrons with it. Thus, we can think of it leaving as H^+—a species with one proton at the nucleus, but no electrons. However, realize that H^+ is never actually formed because the would-be H^+ simultaneously gains a bond to O.

Viewing these steps in this way, you should be able to see that proton transfer steps are also **Brønsted acid–base reactions.** Recall from general chemistry that a **Brønsted acid** is a proton (i.e., H^+) donor and a **Brønsted base** is a proton acceptor. Thus, in Equation 5-7, H—Cl is the Brønsted acid and HO^- is the Brønsted base.

✔ QUICK CHECK

In the proton transfer step $H_3O^+ + H_2N^- \rightarrow H_2O + NH_3$, which species is the Brønsted acid and which is the Brønsted base?

Answer: H_3O^+ is the Brønsted acid and H_2N^- is the Brønsted base.

Very importantly, realize that the curved arrow notation for the proton transfer step in Equation 5-7 follows the *electron-rich to electron-poor* guideline we learned about in Section 5.2C. Notice that the O atom is electron-rich due to its negative charge. By contrast, the H atom is electron-poor due to the partial positive charge it has—the Cl atom's electronegativity creates a dipole along the H—Cl bond, which points toward Cl and leaves a partial positive charge on H. Thus, the curved arrow that is drawn from O to H represents the flow of electrons from electron-rich to electron-poor. The second curved arrow represents the breaking of the H—Cl bond, which must take place in order to avoid two bonds to H.

TIME TO TRY

In Equation 5-7, circle and label the electron-rich site and the electron-poor site. Also, circle and label the curved arrow that represents the flow of electrons from electron-rich to electron-poor.

5.4 Bimolecular Nucleophilic Substitution (S_N2)

In the previous section, we saw that a proton transfer step involves the formation of a single bond and the breaking of another at the same time. Another elementary step in which this occurs is called **bimolecular nucleophilic substitution,** abbreviated **S_N2**. An example is shown in Equation 5-8. In this case, a lone pair of electrons from O forms a bond to C, and at the same time, the C—Cl bond breaks, with the pair of electrons ending up on Cl.

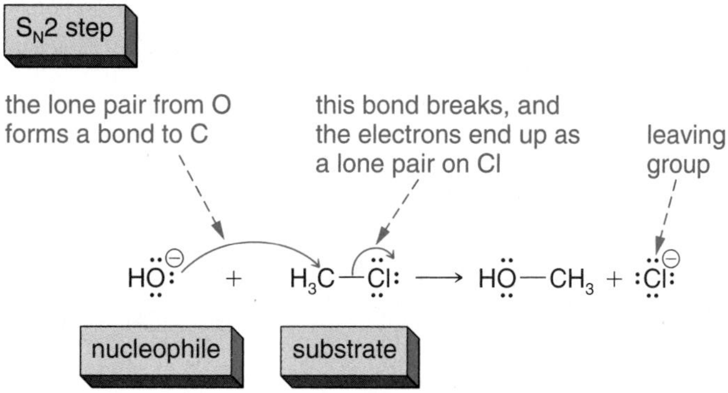

S$_N$2 step
the lone pair from O forms a bond to C
this bond breaks, and the electrons end up as a lone pair on Cl
leaving group
HO:⁻ + H$_3$C—Cl: ⟶ HO—CH$_3$ + :Cl:⁻
nucleophile
substrate

(5-8)

Despite the similarities in the curved arrow notation between proton transfer and S_N2 steps, the species that are involved receive different labels. Whereas in a proton transfer step the reactants are called a *base* and an *acid*, in an S_N2 step they are called a *nucleophile* and a *substrate.*

A **nucleophile** is a species in which an atom has a full or partial negative charge, as well as a lone pair of electrons, and tends to form a single bond to a non-H atom.

The word *nucleophile* derives from Latin and means "nucleus loving." This is because a nucleophile, having an excess of negative charge, tends to form a bond to an atom with an excess of positive charge—the same type of charge that a nucleus has. Notice in Equation 5-8, for example, that in the HO^- nucleophile the O atom has both a full negative charge and a lone pair of electrons.

A **substrate** is a species in which a non-H atom is singly bonded to a *leaving group*.

A leaving group, quite simply, is the species that departs from the substrate. As indicated in Equation 5-8, for example, the leaving group is the Cl^- anion. In general,

A **leaving group** is a species that is relatively stable in the form in which it leaves from the substrate and includes neutral molecules and halide anions (Cl^-, Br^-, I^-) as well as other species.

Because leaving groups frequently include negative charges, this idea will become clearer in Chapter 6, where we discuss charge stability to some depth.

With what we have discussed so far, you can see why this step is called *bimolecular nucleophilic substitution.* It is **bimolecular** because it involves two reactant molecules in a single elementary step. It is nucleophilic because it involves a nucleophile as one of the reactants.

And it is a substitution reaction because, in the substrate, one portion—the leaving group—is replaced by another—a nucleophile.

The *electron-rich to electron-poor* nature of an S_N2 step is very similar to what we saw with proton transfers. In Equation 5-8, the nucleophile is relatively electron-rich, whereas the substrate is relatively electron-poor. Thus, the curved arrow from the nucleophile to the substrate represents the flow of electrons from electron-rich to electron-poor. The second curved arrow in each case, representing the breaking of a bond, is required in order to prevent exceeding an octet. These ideas are summarized in Equation 5-9.

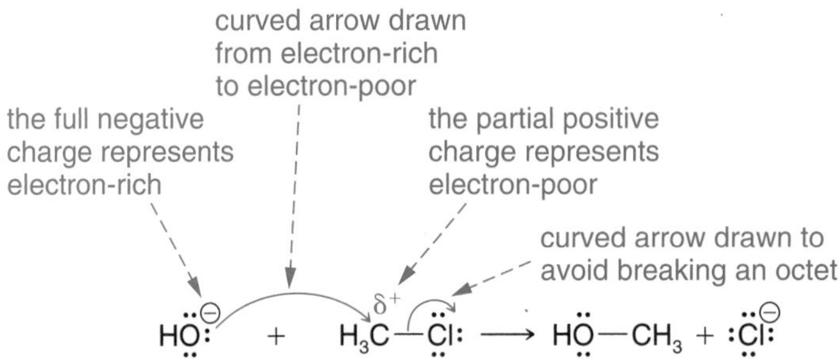
curved arrow drawn from electron-rich to electron-poor
the full negative charge represents electron-rich
the partial positive charge represents electron-poor
curved arrow drawn to avoid breaking an octet
HO: + H3C—Cl: ⟶ HO—CH3 + :Cl:

(5-9)

✔ **QUICK CHECK**

For the following S_N2 reaction, identify the pertinent electron-rich and electron-poor sites, supply the curved arrow notation, and draw the products.

$Cl^- + H_3C{-}Br \rightarrow$

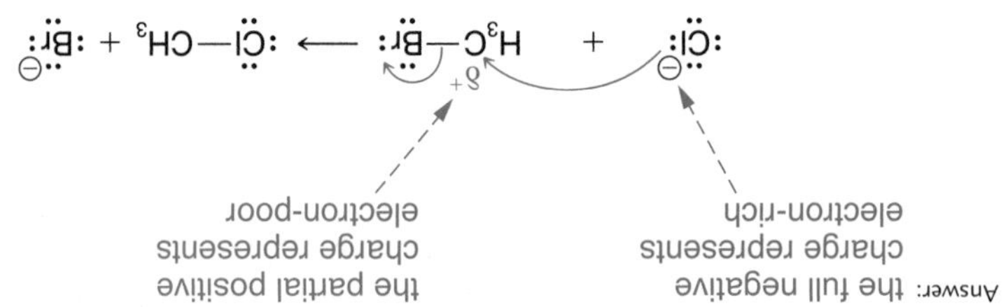

5.5 Nucleophilic Addition and Elimination

In the previous section, we examined the S_N2 step, in which a nucleophile attacks an atom bonded to a suitable leaving group. A nucleophile can also attack an atom involved in a *polar multiple bond*—a double or triple bond having a significant dipole moment. An example is shown in Equation 5-10. In such reactions, a lone pair from the nucleophile forms a bond to the less electronegative atom of a polar multiple bond—a C atom in this case. At the same time, one of the bonds of the multiple bond breaks, and that pair of electrons ends up on the more electronegative atom of the multiple bond—an O atom in this case.

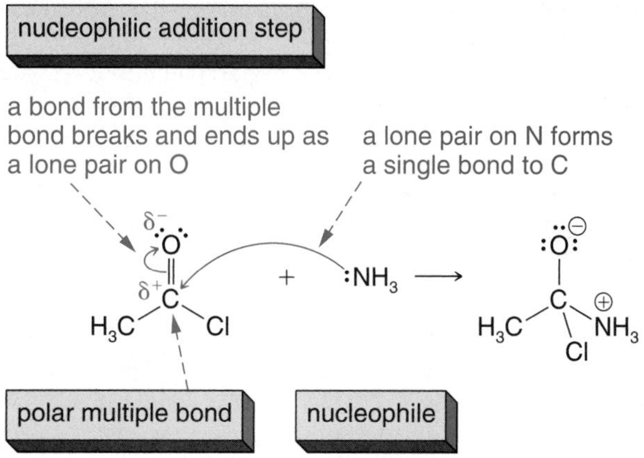
nucleophilic addition step
a bond from the multiple bond breaks and ends up as a lone pair on O
a lone pair on N forms a single bond to C
polar multiple bond
nucleophile

(5-10)

Not surprisingly, such an elementary step is called a **nucleophilic addition step.** The nucleophile is said to undergo *addition* to the polar multiple bond.

✔ QUICK CHECK

Consider the following nucleophilic addition step. Label the nucleophile and the polar multiple bond. Assuming that the nucleophile forms a bond to the C atom of the multiple bond, draw the product.

$HO^- + CH_3C{\equiv}N \rightarrow$

Answer: HO^- is the nucleophile, and the C≡N bond is the polar multiple bond. A lone pair of electrons from O forms a bond to the C atom of the C≡N bond, and a pair of electrons from the triple bond ends up as an additional lone pair on N, giving N a −1 formal charge.

$HO^- + CH_3C{\equiv}N \rightarrow CH_3C(OH){=}N^-$

The *electron-rich to electron-poor* nature of a nucleophilic addition step is fairly straightforward. The nucleophile, which has an excess of negative charge, is relatively electron-rich. Conversely, the electronegative atom of the polar multiple bond is relatively electron-poor. In this case, this is evidenced by a partial positive charge there. Thus, the curved arrow drawn from the nucleophile to the polar multiple bond represents the flow of electrons from an electron-rich site to an electron-poor site.

TIME TO TRY

In Equation 5-10, label the pertinent electron-rich and electron-poor sites, and also label the curved arrow that represents the flow of electrons from electron-rich to electron-poor.

Notice that once the curved arrow representing *electron-rich to electron-poor* is drawn, a second curved arrow must be drawn in order to avoid having five bonds to C. In Equation 5-10, that second curved arrow represents the breaking of a bond from the C=O bond.

Why doesn't the C—C or C—Cl bond break instead? There are multiple reasons for this, but one of the major reasons has to do with the *strength* of each type of bond: The second and third bonds of a multiple bond are weaker than an analogous single bond. (The specific reasons for this will be discussed in your yearlong organic chemistry course.) Therefore, it is generally easier to break one bond from a double or triple bond than it is to break a single bond. With this in mind, we arrive at the following rule:

When the formation of a bond to a particular atom forces another bond on that atom to be broken, you should in general choose to break a bond from a multiple bond instead of a single bond.

The reverse of nucleophilic addition is also a common elementary step in organic chemistry, an example of which is shown in Equation 5-11. In this step, a lone pair of electrons from an electronegative atom, such as O or N, forms an additional bond to an adjacent C atom to which it is initially bonded. Simultaneously, a *leaving group* must depart from that C in order to avoid having five bonds to C. Notice that in the products the leaving group has the characteristics of a nucleophile, possessing an atom that has a lone pair of electrons and a partial or full negative charge. Therefore, this step can be called a **nucleophilic elimination step.**

nucleophilic elimination step

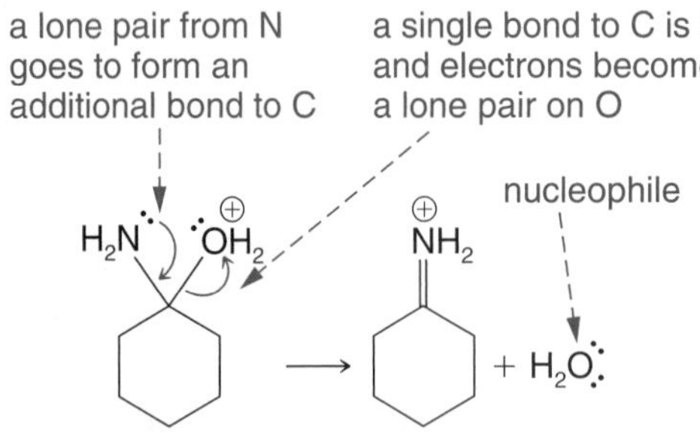
a lone pair from N
goes to form an
additional bond to C
a single bond to C is broken
and electrons become
a lone pair on O
nucleophile
H2N
OH2
NH2
+ H2O

(5-11)

In a nucleophilic elimination step, the electronegative atom bonded to C is relatively electron-rich—it has a full or partial negative charge. The leaving group, on the other hand, is relatively electron-poor. Thus, with the combination of the two curved arrows, there is a flow of electrons from an electron-rich site to an electron-poor site. These ideas are shown in Equation 5-12.

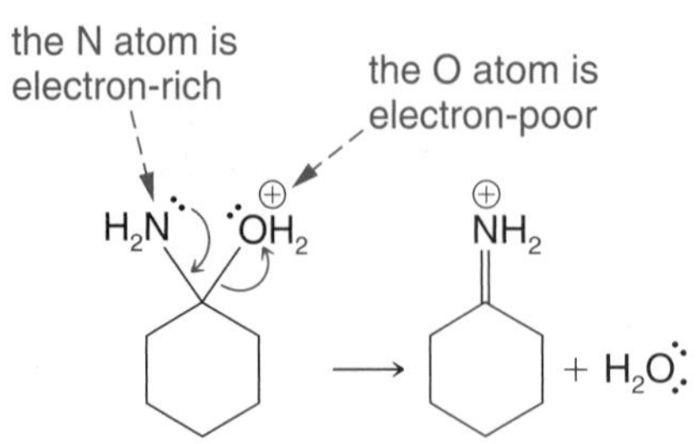
the N atom is
electron-rich
the O atom is
electron-poor
H2N
OH2
NH2
+ H2O

(5-12)

5.6 Electrophilic Addition and Elimination

When a species containing a nonpolar multiple bond (generally as part of a C=C double bond or C≡C triple bond) approaches a strongly electron-deficient species, called an **electrophile,** an atom of the multiple bond tends to form a bond to the electrophile. This is called an **electrophilic addition step,** examples of which are shown in Equations 5-13 and 5-14.

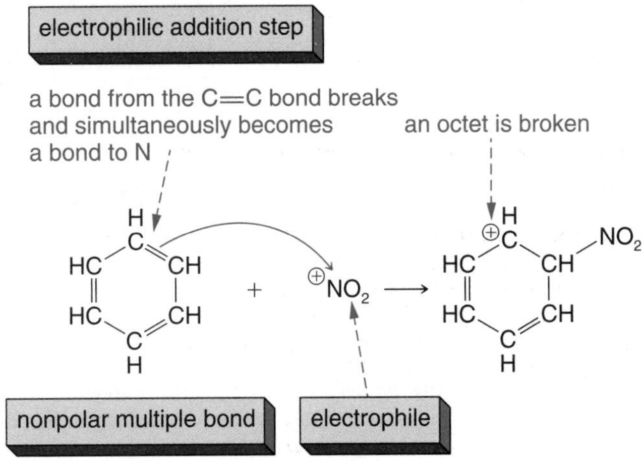
electrophilic addition step
a bond from the C=C bond breaks and simultaneously becomes a bond to N
an octet is broken
nonpolar multiple bond
electrophile

(5-13)

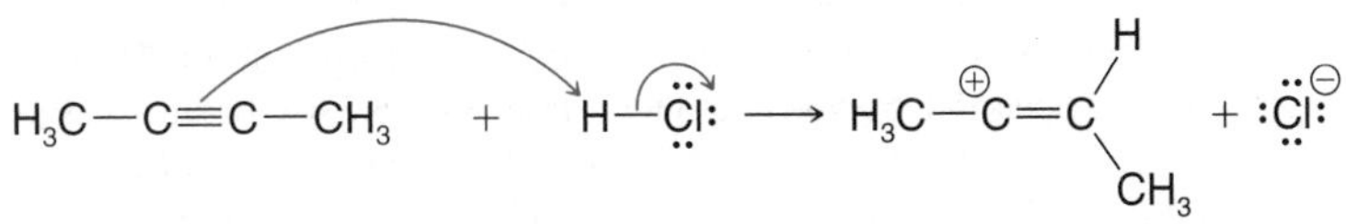

(5-14)

In Equation 5-13, a pair of electrons from a C=C double bond becomes a new bond to the N atom of the electrophile. In Equation 5-14, a pair of electrons from the C≡C triple bond becomes a new bond to H. Simultaneously, the H—Cl bond breaks (in order to avoid two bonds to H), and the electrons from that bond become a lone pair on Cl. Importantly, notice that each electrophilic addition step produces a C atom that lacks an octet.

TIME TO TRY

In Equation 5-14, identify the nonpolar multiple bond, the electrophile, and the atom that loses an octet during the course of the reaction.

The *electron-rich to electron-poor* nature of electrophilic addition is straightforward. In an electrophilic addition step, the C=C double bond or C≡C triple bond is electron-rich, due to the confinement of electrons to the region between two atoms. In contrast, the electrophile is electron-poor—in Equation 5-13, for example, the electrophile has a full positive charge on N, and in Equation 5-14, the electrophile has a partial positive charge on H. Therefore, the flow of electrons from electron-rich to electron-poor is achieved by the curved arrow drawn from the multiple bond to the electrophile. These ideas are summarized Equation 5-15.

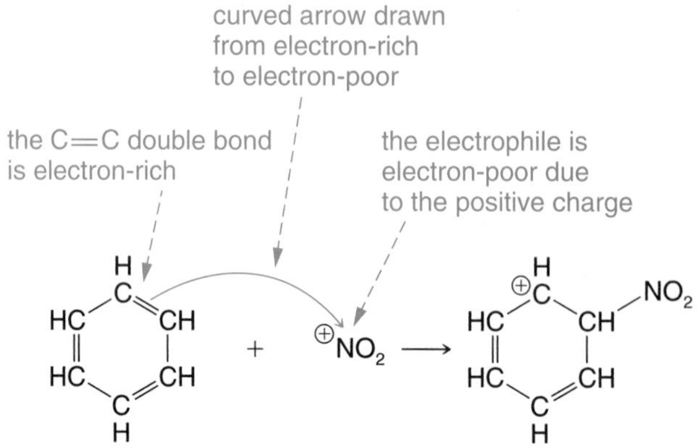
curved arrow drawn
from electron-rich
to electron-poor
the C=C double bond
is electron-rich
the electrophile is
electron-poor due
to the positive charge
HC
CH
HC
CH
C
H
+
NO2
NO2

(5-15)

TIME TO TRY

In Equation 5-14, identify the pertinent electron-rich and electron-poor sites, and indicate which curved arrow represents the flow of electrons from electron-rich to electron-poor.

The reverse of electrophilic addition is also a common elementary step in organic chemistry, an example of which is shown in Equation 5-16. In such a step, the electrophile that is eliminated is one that is attached to an atom that is *adjacent* to an atom lacking an octet. In Equation 5-16, the electrophile that is eliminated is a proton (H^+), which is attached to a C atom that is adjacent to a C^+ atom (the atom that initially lacks an octet).

electrophilic elimination step

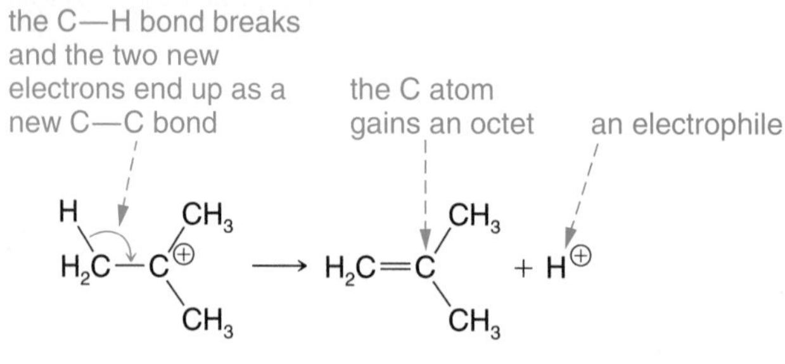
the C—H bond breaks
and the two new
electrons end up as a
new C—C bond
the C atom
gains an octet
an electrophile
H
CH3
H2C
C
CH3
H2C=C
CH3
CH3
+ H

(5-16)

When considering the *electron-rich to electron-poor* nature of an electrophilic elimination step, it is clear that the C atom with the +1 formal charge is electron-poor. The C—H bond, on the other hand, is electron-rich, due to the confinement of electrons to the region between two atoms. Therefore, the flow of electrons from electron-rich to electron-poor is achieved by the single arrow that is shown in the reaction.

TIME TO TRY

Label the pertinent electron-rich and electron-poor sites in the electrophilic elimination step in Equation 5-16.

5.7 Carbocation Rearrangements

We have seen carbocations (species with a C^+ atom) appear as reactants in a couple of the elementary steps we have examined thus far. We saw in Section 5.2 that carbocations can form a bond to an electron-rich species in a coordination step. And we saw in Section 5.6 that a carbocation can eliminate an H^+ electrophile. Carbocations can also undergo *rearrangements.*

Equations 5-17 and 5-18 provide examples of the two most common types of carbocation rearrangements, called a **1,2-hydride shift** and a **1,2-methyl shift,** respectively. The "1,2" is simply a way of specifying that the indicated species shifts to a neighboring atom. In a 1,2-hydride shift, a hydride anion (an H atom with a pair of electrons, H^-) migrates to the adjacent C^+ atom, and in a 1,2-methyl shift, a methyl (CH_3) group migrates.

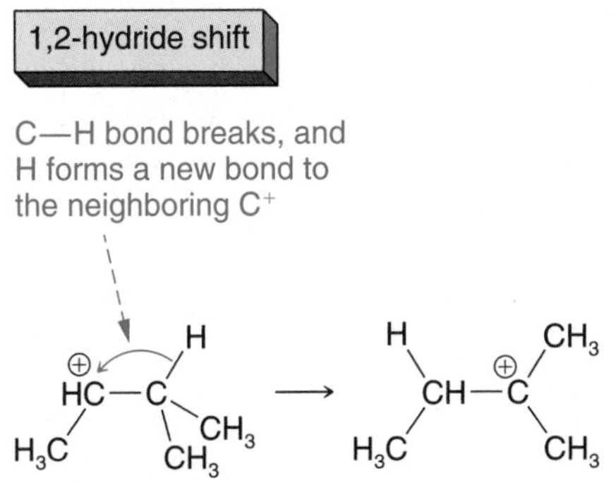
1,2-hydride shift
C—H bond breaks, and
H forms a new bond to
the neighboring C+

(5-17)

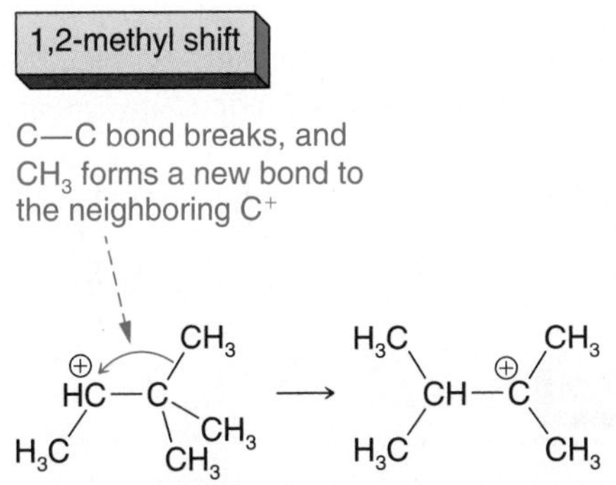
1,2-methyl shift
C—C bond breaks, and
CH3 forms a new bond to
the neighboring C+

(5-18)

The curved arrow notation is essentially the same for the two carbocation rearrangements. A single curved arrow originates from the middle of a bond connecting either H or CH_3 to the C atom that is adjacent to the C^+ atom and points toward the C^+ atom. Thus, the bond connecting H or CH_3 to the uncharged C atom breaks, and simultaneously those electrons become a bond to the C atom that initially has the positive charge.

It should make sense that such steps are called carbocation rearrangements. Notice that the C atom bearing the positive charge in the products is not the same as the one in the reactants—the positive charge has moved to the atom that is initially bonded to the migrating group.

The *electron-rich to electron-poor* nature of a carbocation rearrangement is fairly straightforward. The positively charged C atom is electron-poor, owing to the full positive charge and the fact that it lacks an octet. The bond connecting the migrating group to the uncharged C atom is a region that is electron-rich. Therefore, the single curved arrow that appears represents the flow of electrons from an electron-rich site to an electron-poor site.

TIME TO TRY

Label the pertinent electron-rich and electron-poor sites in Equation 5-18.

5.8 Bimolecular Elimination (E2)

All of the elementary steps we have discussed so far involve the flow of either two electrons or four electrons, requiring either one or two curved arrows, respectively. A **bimolecular elimination (E2) step,** however, involves the flow of six electrons, which requires three curved arrows. An example is provided in Equation 5-19.

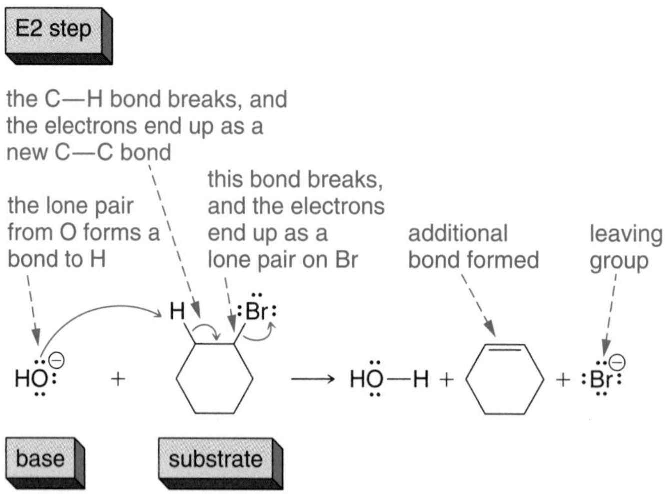
E2 step
the C—H bond breaks, and the electrons end up as a new C—C bond
the lone pair from O forms a bond to H
this bond breaks, and the electrons end up as a lone pair on Br
additional bond formed
leaving group
H
:Br:
HO:
HO—H +
+ :Br:
base
substrate

(5-19)

In this particular example, a lone pair from the negatively charged O atom forms a bond to H, the H atom's bond to C breaks and the two electrons become an additional carbon–carbon bond, and the C—Br bond breaks, with the pair of electrons ending up as a lone pair on Br. And because this is an elementary step, *all of this happens simultaneously!*

Equation 5-19 has some familiar species involved. One reactant, HO^-, behaves as a *base*, given that it gains an H^+ ion during the course of the reaction. The other reactant is a *substrate*, similar to what we saw in an S_N2 step (Section 5.4), given the fact that a leaving group is attached by a single bond to C—in this case, the leaving group departs as Br^-.

With all of this in mind, you can see why it is called *bimolecular elimination*. It is called bimolecular because, just as we saw in an S_N2 step, there are two reactant molecules. And it is called elimination because two portions are removed from the substrate—a proton (H^+) and a leaving group. In the specific case of Equation 5-19, we would say that the net result of the reaction is the elimination of H—Br from the substrate.

It is important to realize that not every type of substrate can participate in such an E2 step.

The only substrates that can participate in an E2 step are those in which an H atom and a leaving group are attached to adjacent atoms.

This is essential if a new multiple bond is to be formed.

What about the *electron-rich to electron-poor* nature of an E2 step? Clearly, the base is electron-rich—in the case of Equation 5-19, the base carries a full negative charge. Similar to what we saw with S_N2 steps, the C atom bonded to the leaving group is electron-poor, given that it has a partial positive charge. But a curved arrow is not drawn directly from the base to the electron-poor atom—that would have described an S_N2 step instead. Rather, the flow of electrons from electron-rich to electron-poor is achieved in two curved arrows. The curved arrow from the base to the H atom represents the flow of electrons from the electron-rich site, and the curved arrow from the H—C bond to the C atom bonded to the leaving group represents the flow of electrons to the electron-poor site.

TIME TO TRY

In Equation 5-19, label the pertinent electron-rich and electron-poor sites, and circle the two curved arrows that together represent the flow of electrons from electron-rich to electron-poor.

5.9 Application: Simplifying Assumptions about Electron-Rich and Electron-Poor Sites

Almost all of the reactions shown in this chapter involve at least one ion as a reactant. Consider the simple proton transfer step shown previously in Equation 5-7: $HO^- + H—Cl \rightarrow H_2O + Cl^-$. The equation seems to suggest that we can add HO^- to HCl to initiate the reaction, but doing so is impossible, for the following reason.

Ions do not exist in pure form in the solid or liquid phase—there must be an equal amount of positive and negative charge present in order to produce a neutral compound.

We can, however, add a *source* of HO^-, such as NaOH; NaOH is an ionic compound, so in solution it dissolves as Na^+ and HO^- (recall our discussion in Section 3.9).

This may initially seem to complicate the picture because Na^+ can be considered electron-poor. We can thus envision Na^+ reacting with an electron-rich site of another species. However, as you may recall from general chemistry, Na^+ is relatively inert in solution and tends not to react—it behaves as a **spectator ion.** Although we will not discuss the specific reasons why, this is generally true of Group 1A metal cations—Li^+, Na^+, and K^+. Consequently,

When we envision the flow of electrons from an electron-rich site to an electron-poor site, we can disregard Group 1A metal cations.

Metals are also found in **organometallic compounds,** in which *a metal atom is bonded directly to a carbon atom.* Unlike in NaOH, the bond to the metal atom in an organometallic compound is covalent, so organometallic compounds do not dissociate into ions in solution. However, because metal atoms have such low electronegativity, carbon–metal bonds are in general highly polar—the C atom is more electronegative, so it receives a partial negative charge, and the metal atom has a partial positive charge. Thus, in such a carbon–metal bond, the C atom is electron-rich and the metal atom is electron-poor. This leads to a very useful simplifying assumption.

In a covalent bond involving a metal atom and a nonmetal atom, we can treat the metal atom as a spectator ion, generated by giving both electrons of the bond to the nonmetal atom.

Consider, for example, an **alkyllithium compound** such as CH_3CH_2Li, in which Li is directly bonded to C. As shown in Figure 5-3, we can treat Li as an Li^+ spectator ion, generated by giving both electrons of the C—Li bond to C. The resulting reactive species is $CH_3CH_2^-$, which is highly electron-rich at the C atom. Similarly, in a **Grignard reagent** such as C_6H_5MgBr, in which Mg is bonded directly to C, we can treat the MgBr portion as a spectator, $Mg^{2+}Br^-$, generated by giving both electrons of the C—Mg bond to C. That gives us $C_6H_5^-$ as the reactive species, in which a negative charge appears on C.

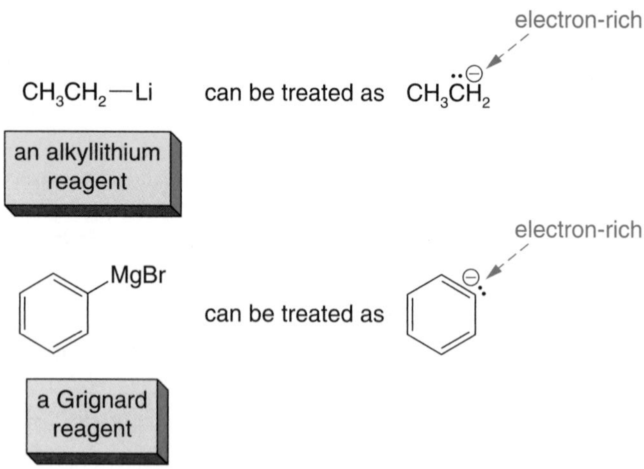

FIGURE 5-3 Simplifying assumptions for organometallic reagents. (*top*) In an alkyllithium reagent, the Li atom can be treated as an Li^+ spectator ion, generated by giving both electrons of the C—Li bond to the C atom. Thus, the C atom becomes negatively charged and is electron-rich. (*bottom*) In a Grignard reagent, the MgBr portion can be treated as a spectator $Mg^{2+}Br^-$ species, generated by giving both electrons of the C—Mg bond to the C atom. Thus, the C atom becomes negatively charged and is electron-rich.

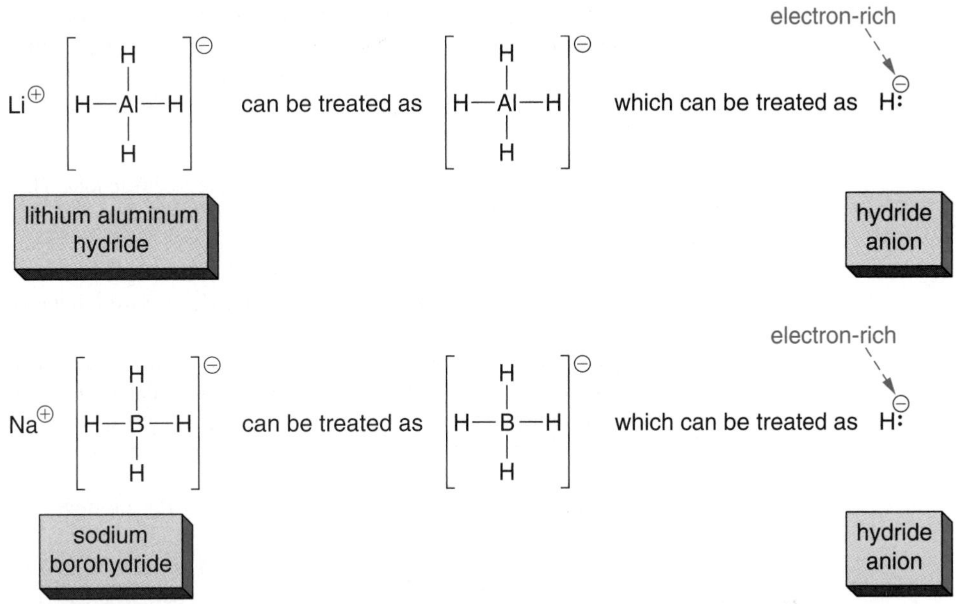

FIGURE 5-4 **Simplifying assumptions for hydride reagents.** (*top*) In lithium aluminum hydride, Li^+ is a spectator ion and can be ignored. In the remaining AlH_4^- species, we can treat the Al atom as an Al^{3+} spectator ion, generated by giving both electrons of each Al—H bond to the H atom. Thus, each H atom becomes negatively charged and is electron-rich. (*bottom*) In sodium borohydride, Na^+ is a spectator ion and can be ignored. The remaining BH_4^- species can be treated simply as H^-, which is electron-rich.

As shown in Figure 5-4, this simplifying assumption also works for **hydride reagents,** such as **lithium aluminum hydride,** $LiAlH_4$. $LiAlH_4$ is an ionic compound that consists of Li^+ and AlH_4^- ions. Clearly, then, the Li^+ portion can be ignored, since it is a true spectator ion. The remaining AlH_4^- ion consists of polar covalent Al—H bonds. The simplifying assumption above allows us to treat the Al atom as an Al^{3+} spectator ion, generated by giving the pair of electrons from each Al—H bond to H. Thus, each H atom can be treated as H^-, called a **hydride anion**. By analogy, **sodium borohydride,** $NaBH_4$, can be treated simply as H^-, given that B and Al are in the same column of the periodic table and thus have similar behavior.

✔ QUICK CHECK

How would you simplify each of the following compounds?

1. $LiN(CH_3)_2$
2. $KOCH_2CH_3$
3. $CH_3C{\equiv}CNa$
4. NaH

Answers: 1. $^-N(CH_3)_2$. 2. $^-OCH_2CH_3$. 3. $CH_3C{\equiv}C^-$. 4. H^-.

These simplifying assumptions for organometallic and hydride reagents come in handy when trying to predict the products of reactions such as the one in Equation 5-20.

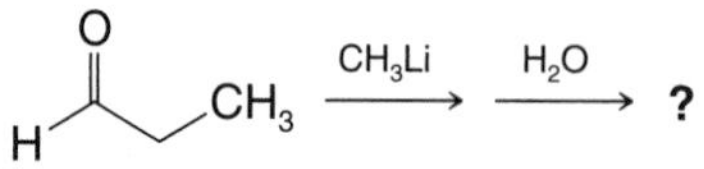

(5-20)

This notation actually represents a sequence of two successive reactions. The first has us add CH_3Li to the compound at the very left, and after that reaction has come to completion, we add water.

To tackle this problem, we begin by simplifying CH_3Li to CH_3^-, which has an electron-rich C atom. The C atom of the C=O bond has a partial positive charge, so it is electron-poor. As shown in Figure 5-5, we draw a curved arrow from electron-rich to electron-poor, representing the formation of a new C—C bond. Recognize that this now represents the beginning of a nucleophilic addition step (Section 5.5), where CH_3^- acts as the nucleophile and the C=O bond is the polar multiple bond. To complete such a step, a curved arrow is drawn from the center of the C=O bond toward the O atom, representing a conversion of a pair of bonding electrons to a lone pair.

Next, water is added. In the product of the first reaction, the negatively charged O atom is electron-rich, and an H atom of H_2O is electron-poor. Again, a curved arrow is drawn from electron-rich to electron-poor. In order to avoid two bonds to H, an H—O bond of the water molecule must break, with the electrons ending up as a lone pair on O. This is a proton transfer step (Section 5.3).

TIME TO TRY

In each step of Figure 5-5, label the pertinent electron-rich and electron-poor sites, and identify the curved arrow that represents the flow of electrons from electron-rich to electron-poor. In addition, write the name of each type of elementary step below the corresponding reaction arrow.

In a similar fashion, we can predict the products of the reaction in Equation 5-21.

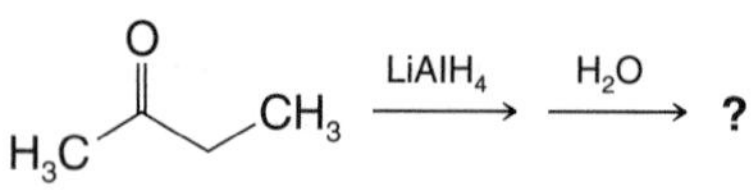

(5-21)

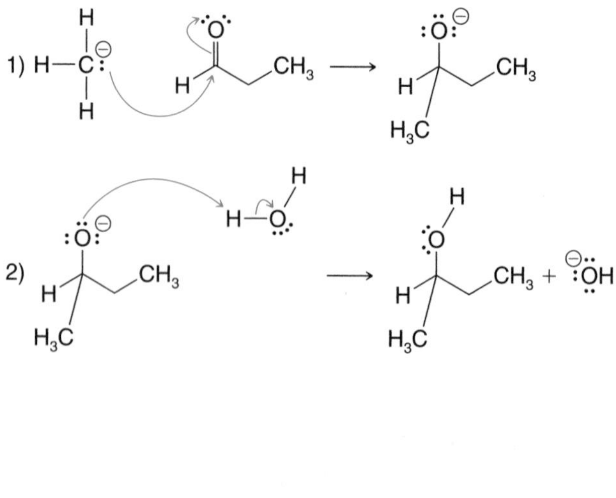

FIGURE 5-5 Mechanism for the sequence of reactions in equation 5-20. In step 1 (top reaction), the alkyllithium reagent is treated as CH_3^-, which is electron-rich. It forms a bond to the electron-poor C atom of the C=O double bond, forcing a bond from C=O to break, with the electrons becoming a lone pair on O. This is a nucleophilic addition step. In step 2 (bottom reaction), the electron-rich O atom forms a bond to the electron-poor H atom, forcing the H—O bond of water to break, with the electrons becoming a lone pair on O. This is a proton transfer step.

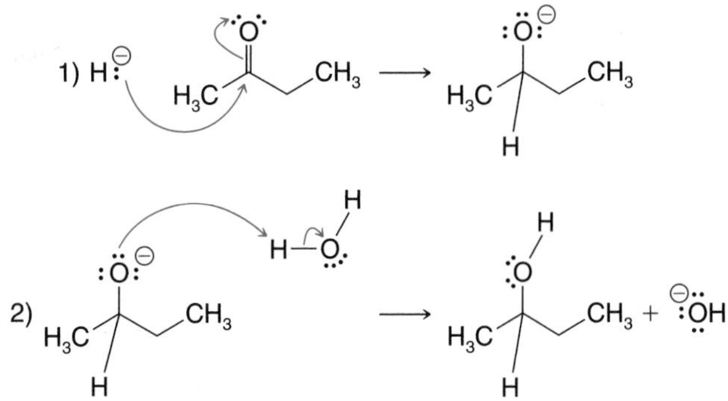

FIGURE 5-6 Mechanism for the sequence of reactions in equation 5-21. In step 1 (top reaction), $LiAlH_4$ is treated as H^-, which is electron-rich. It forms a bond to the electron-poor C atom of the C═O double bond, forcing a bond from C═O to break, with the electrons becoming a lone pair on O. This is a nucleophilic addition step. In step 2 (bottom reaction), the electron-rich O atom forms a bond to the electron-poor H atom, forcing the H—O bond of water to break, with the electrons becoming a lone pair on O. This is a proton transfer step.

Again, the notation indicates that there are two successive reactions. First, $LiAlH_4$ is added to the reactant at the far left. Once that reaction has come to completion, water is added.

The complete mechanism is shown in Figure 5-6. We begin by applying the simplifying assumption, which allows us to treat $LiAlH_4$ as simply H^-, an electron-rich species. The C atom of the C═O bond has a significant partial positive charge and is therefore electron-poor. Thus, a curved arrow is drawn from electron-rich to electron-poor, signifying the formation of a new H—C bond. This initiates a nucleophilic addition step, and in order to avoid five bonds to C, a bond of the C═O double bond breaks, and those electrons end up as an additional lone pair on O.

The addition of water initiates the second reaction. The negatively charge O atom is electron-rich, and an H atom of water is electron-poor. A curved arrow is drawn from electron-rich to electron-poor, signifying the formation of a new O—H bond. In order to avoid two bonds to H, an H—O bond of water must break, with the electrons ending up as a lone pair on O. As in Figure 5-5, this is a proton transfer step.

TIME TO TRY

In each step of Figure 5-6, label the pertinent electron-rich and electron-poor sites, and identify the curved arrow that represents the flow of electrons from electron-rich to electron-poor. In addition, write the name of each type of elementary step below the corresponding reaction arrow.

✔ QUICK CHECK

Using the appropriate simplifying assumptions, draw the complete mechanism of the following sequence of reactions.

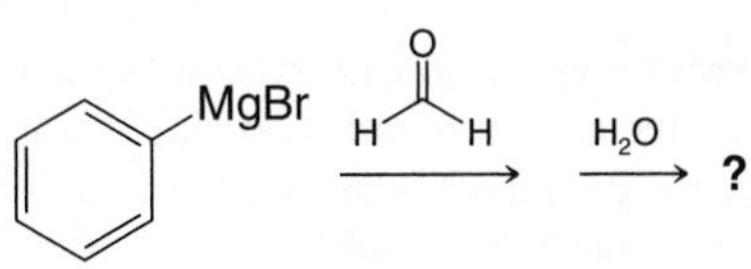

Answer: The C_6H_5MgBr species can be simplified to $C_6H_5^-$, as previously shown in Figure 5-3, so it has an electron-rich C atom. The C═O bond has an electron-poor C atom, so the flow of electrons from electron-rich to electron-poor initiates a nucleophilic addition step, as shown below. When water is added, the flow of electrons from electron-rich to electron-poor initiates a proton transfer step.

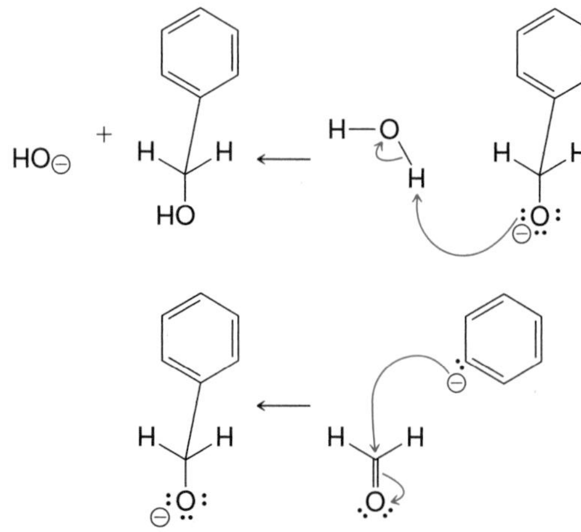

5.10 Application: Stereochemistry of Reactions and the Production of a New Stereocenter

In Section 4.5, we saw that stereoisomers can have substantially different properties. For this reason, it is important to know whether stereoisomers can be produced from a reaction, and, if so, what their relative amounts are. In other words, we must know the **stereochemistry** of the various reactions.

One of the most common ways in which stereochemistry comes into play in a reaction is when a *new* stereocenter is produced. This is possible whenever a reaction converts a planar atom into a tetrahedral one. If the newly produced tetrahedral atom has four *different* groups, it will have become a stereocenter. An example can be seen in the nucleophilic addition step shown in the first reaction of Figure 5-6.

TIME TO TRY

On the reactant side of the nucleophilic addition step in Figure 5-6, circle and identify the atom that *becomes* a new stereocenter.

Recall from Section 4.8 that for any stereocenter there are two possible configurations. We should therefore ask, In this reaction, is one of those configurations produced exclusively, or is a mixture produced? To answer this question, let's examine more carefully the dynamics of this reaction, shown in Figure 5-7.

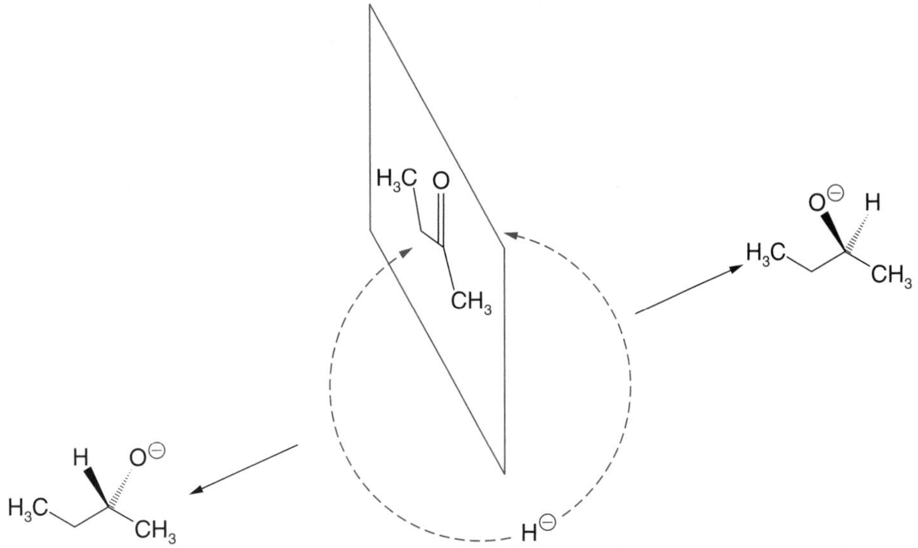

FIGURE 5-7 **Formation of the stereocenter in figure 5-6.** Nucleophilic addition of H^- to a C=O bond to form a new stereocenter. The plane shown is the plane of the initial trigonal planar C atom. If H^- approaches the C atom from the left side of the plane, the bond forms on the left to produce the species at the lower left. If H^- approaches the C atom from the right side of the plane, the bond forms on the right to produce the species at the upper right. Notice that the two product species differ only by the exchange of two groups on the new sterocenter (the O and H atoms), so the configuations are different—one is *R* and one is *S*.

As we can see, the geometry about the C atom is planar, and the H^- ion can approach the C atom from either side of that plane. If H^- approaches from the left side of the plane (as it appears in the figure), then the new C—H bond forms on the left, and the result is a wedge. By the same token, if H^- approaches from the right side of the plane, the new C—H bond forms to the right, and the result is a dash. Because a reaction involves a very large number of reactant molecules, a certain percentage will proceed with one approach, and the remaining percentage will proceed with the other approach. In other words, the reaction will produce a mixture of the two products. Importantly, these products differ by the exchange of two groups on the new stereocenter—in this case, the O and H atoms. Thus, those stereocenters have opposite configurations—one is *R* and the other is *S*.

This specific case involving nucleophilic addition can be generalized to other elementary steps as well.

Whenever a stereocenter is generated from an atom that is initially trigonal planar, both types of stereochemical configurations can be produced, resulting in a mixture of stereoisomers.

The question then becomes, What are the relative proportions of those stereoisomers in the product mixture? To answer that question, we can take advantage of what we already know about the properties of various stereoisomers, discussed previously in Section 4.5. Namely, *enantiomers must have identical properties*, whereas *diastereomers must have different properties.* The specific property of interest here is the amount produced in a chemical reaction. Therefore, we can say the following.

If a reaction produces a mixture of *enantiomers*, they must be produced in exactly equal amounts.

If a reaction produces a mixture of *diastereomers*, they must be produced in unequal amounts.

When two enantiomers are present in equal amounts, we say that there is a **racemic mixture** (pronounced *ruh-SEE-mik*) of the enantiomers. Looking back at Figure 5-7, for example, we can see that the two products are enantiomers. Therefore, we know that they must be produced in equal amounts.

TIME TO TRY

Using a molecular modeling kit, construct both product species in Figure 5-7, and show that they are nonsuperimposable mirror images.

Suppose, now, that the nucleophilic addition step in Figure 5-7 involves a molecule that initially contains a stereocenter, as shown in Figure 5-8. (The stereocenter is marked with *.) Just as we saw previously, a new stereocenter is generated from a trigonal planar C, so the result must be a mixture of the two configurations. This time, however, the stereoisomers that are produced are diastereomers of each other, so we know that they must be produced in *unequal* amounts.

TIME TO TRY

Using a molecular modeling kit, construct both product species in Figure 5-8, and convince yourself that they are not mirror images.

LOOK OUT

When diastereomers are produced, such as in Equation 5-8, we know that they must be produced in unequal amounts. That is all we can predict. We cannot predict the relative amounts; they might be 99% to 1%, or they might be 51% to 49%. Another thing that we cannot predict is which diastereomer will be produced in the greater amount. For example, it might be tempting to say that the diastereomer on the left of Figure 5-8 will be the favored product because it is the result of the nucleophile attacking on the opposite side of the relatively

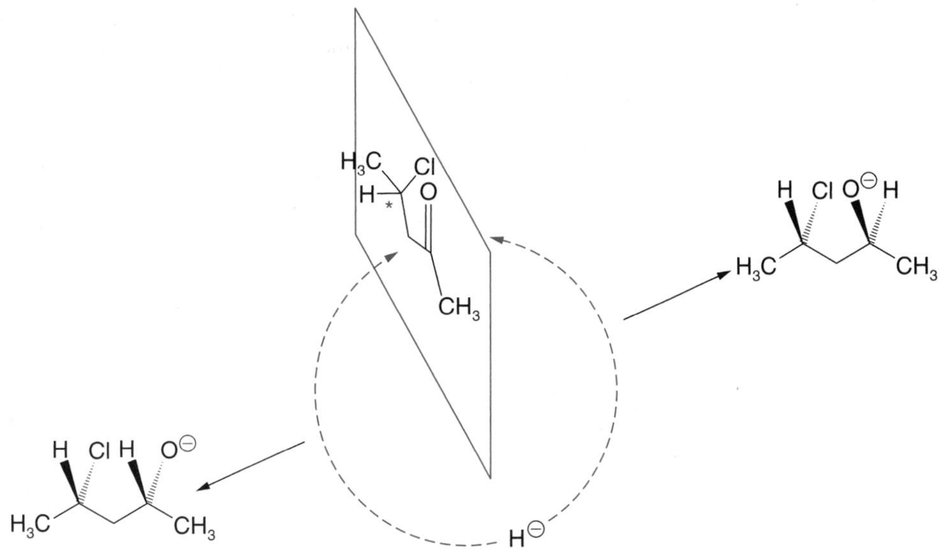

FIGURE 5-8 **Formation of a stereocenter involving a chiral reactant.** This illustrates the same reaction as in Figure 5-7, but the molecule containing the C=O bond has a stereocenter (marked with *), making that molecule chiral. As a result, the two products are diastereomers, which will be produced in *unequal* amounts.

large Cl atom. But there are many subtle things that control a reaction like this, which are not accounted for in such an analysis. The take-home message is that you cannot reliably predict which of two diastereomers is the major product unless you know more about the details of the specific reaction.

5.11 Application: Stereospecificity of S_N2 Steps

The previous section introduced us to the concept of stereochemistry of a reaction. There we learned that, if a stereocenter is produced from a trigonal planar atom, then the product will be a mixture of both the *R* and *S* configurations. Furthermore, we learned that the mixture will be equal—a *racemic mixture*—if the products are enantiomers, and it will be unequal if they are diastereomers.

An S_N2 step, however, involves the simultaneous formation and breaking of single bonds on a tetrahedral C. Therefore, a stereocenter can be produced from an atom that is initially a stereocenter, and as such, the rules in the previous section do not apply. Rather, for an S_N2 step, when two stereocenters are possible for the C atom involved in the reaction, only one configuration is produced. These reactions are said to be **stereospecific.**

The particular stereoisomer that is produced is governed by the following rule.

In an S_N2 step, the nucleophile attacks and forms a bond to the substrate on the side opposite the leaving group.

This is exemplified in Equation 5-22. As we can see, the stereocenter in the substrate—the C atom bonded to the leaving group—is directly involved in the reaction, in that during the course of the reaction one of its bonds is broken and another is formed. Therefore, as shown, we can envision two possible stereoisomers as products, which differ by the configuration of the stereocenter. However, the one that is exclusively produced is the one in which HO^- (the nucleophile) attacks from the side opposite the Br (the leaving group). Because the initial C—Br bond is pointing toward us, the HO^- nucleophile attacks the substrate from behind the plane of the paper, leading to a new C—OH bond that remains behind the plane of the paper.

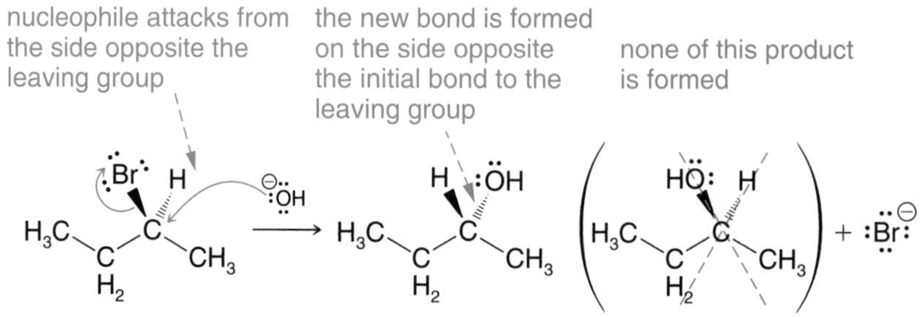
nucleophile attacks from the side opposite the leaving group
the new bond is formed on the side opposite the initial bond to the leaving group
none of this product is formed

(5-22)

The reasons for this rule can be seen in Figure 5-9. If the nucleophile were to attack the substrate from the side on which the leaving group is attached, as shown in Figure 5-9a, it would encounter two problems. One is charge repulsion. Notice that the leaving group has a partial negative charge, and a nucleophile, by definition, must also have a partial or full negative charge. Thus, the nucleophile and leaving group prefer to be as far away from each other as possible (reminiscent of VSEPR theory).

The second reason has to do with the fact that leaving groups take up space. If a nucleophile were to attack the substrate from the side on which the leaving group is attached, it would physically crash into the leaving group. In other words, we say that the nucleophile would encounter **steric hindrance** in such an approach.

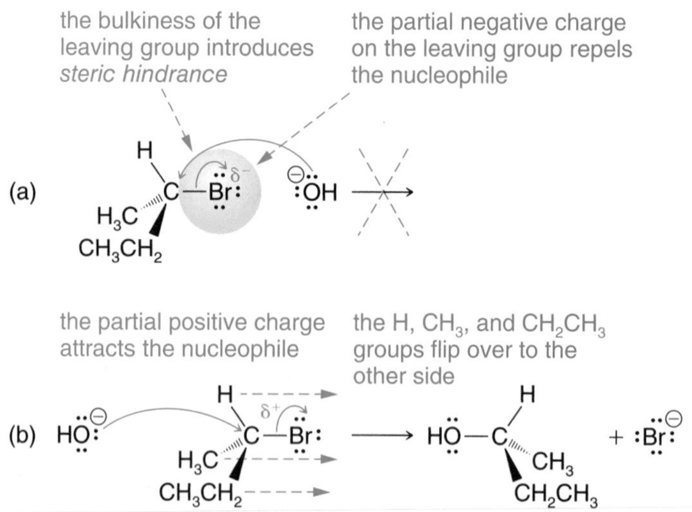
the bulkiness of the leaving group introduces steric hindrance
the partial negative charge on the leaving group repels the nucleophile
(a)
the partial positive charge attracts the nucleophile
the H, CH3, and CH2CH3 groups flip over to the other side
(b)

FIGURE 5-9 **Stereospecificity of the S_N2 step.** (a) Nucleophilic attack does *not* take place on the side of the substrate on which the leaving group is attached, due to charge repulsion and steric hindrance. (b) Nucleophilic attack is favored on the side opposite the leaving group. The partial positive charge on the C atom helps attract the nucleophile, and steric hindrance by the leaving group is not a problem. The dashed arrows indicate that the C atom bonded to the leaving group undergoes Walden inversion, whereby the H, CH_3, and CH_2CH_3 groups flip over to the other side.

By contrast, the nucleophile does not encounter these problems if it attacks from the side opposite the leaving group, as shown in Figure 5-9b. Notice that, due to the bond dipole, the C atom attached to the leaving group has a partial positive charge. As a result, when the nucleophile attacks from that side, it experiences charge *attraction* instead of charge repulsion. Furthermore, because the leaving group departs from the side opposite the nucleophile's attack, the two species do not crash into each other. Thus, the nucleophile does not experience severe steric hindrance.

With the understanding that the nucleophile forms a bond to C on the side opposite the leaving group, realize that the other three groups on C—in this case, H, CH_3, and CH_2CH_3—cannot remain frozen in place during the course of the reaction. If they did not move, then in the product of Figure 5-9b there would be four groups attached to the left side of the C atom, which would be a geometry that is *not* tetrahedral. Instead, as indicated, those three groups must flip over to the other side—the side from which the leaving group departed. This motion of those three groups is called **Walden inversion.**

PICTURE THIS

When considering Walden inversion in an S_N2 reaction, it helps to think of an umbrella inverting in a windstorm, as shown below. In this analogy, the C atom and its H, CH_3, and CH_2CH_3 groups make up the umbrella.

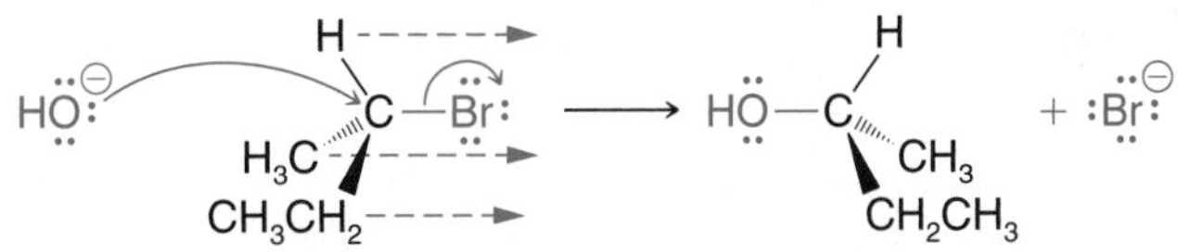

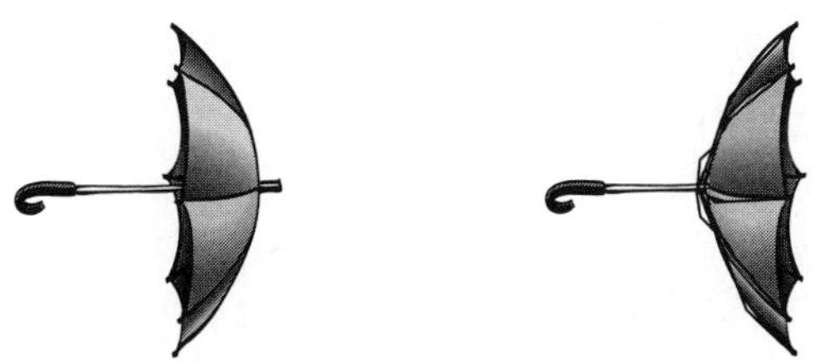

WHAT HAVE YOU LEARNED?

5.1 For each of the following elementary steps, the reactants and products are shown, but the curved arrow notation is not. Supply the curved arrow notation, and provide the name of each elementary step. (You might need to add lone pairs of electrons, covalent bonds, or H atoms in order to show enough detail.)

(a)

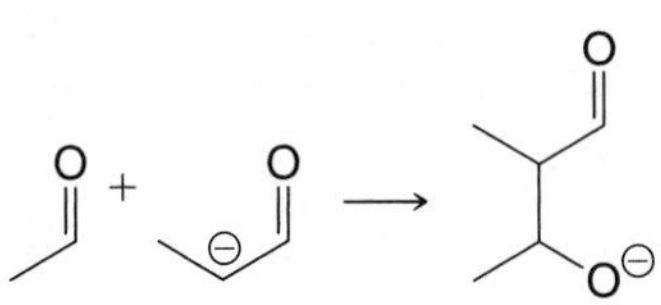

(b) $Cl^{\ominus} + AlCl_3 \longrightarrow {}^{\ominus}AlCl_4$

(c) 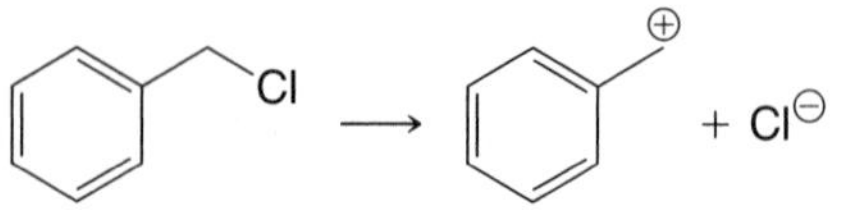

(d) $H_3C^{\ominus} + H_2O \longrightarrow CH_4 + HO^{\ominus}$

(e) 

(f) 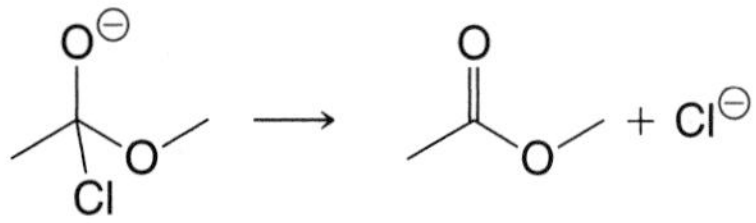

(g) 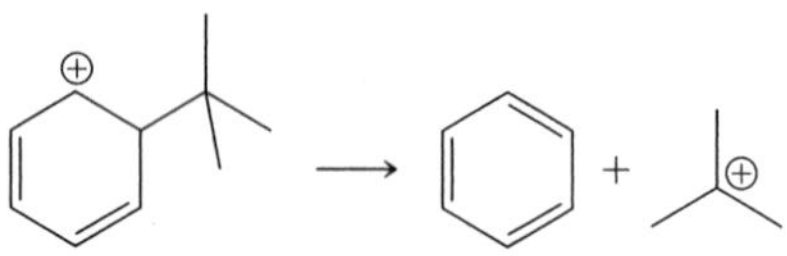

5.2 For each of the following elementary steps, the curved arrow notation is shown, but the products are not. Draw each set of products, and provide the name of each elementary step. Label the important electron-rich and electron-poor sites, and identify which curved arrows represent the flow of electrons from electron-rich to electron-poor.

(a) 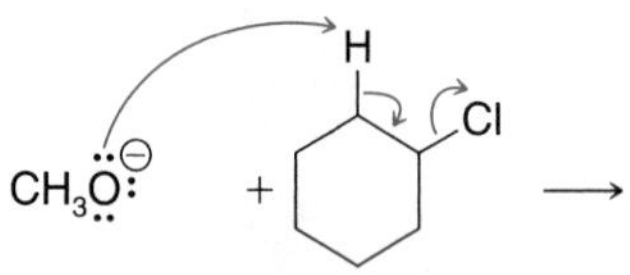

(b) 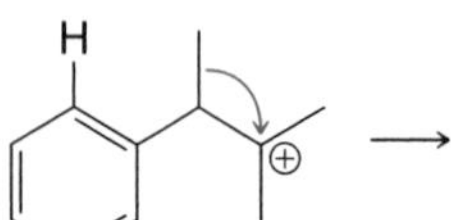

(c) 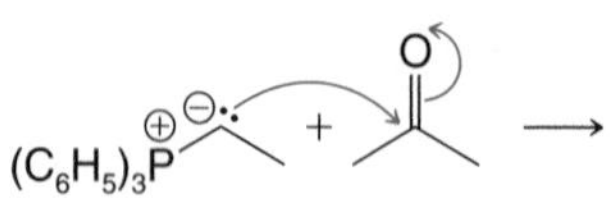

(d) 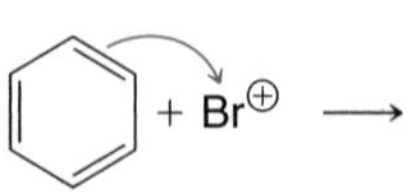

(e) 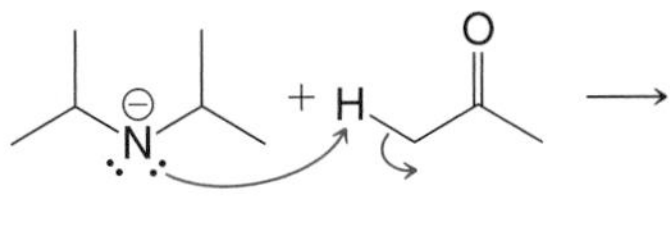

(f) 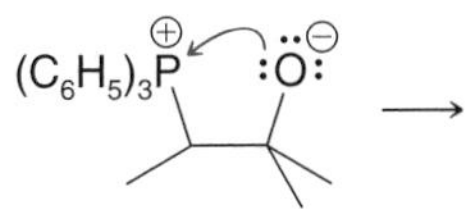

5.3 Draw the transition state for the reactions in Equations 5-7, 5-8, 5-10, 5-11, 5-13, 5-16, 5-17, and 5-19.

5.4 Draw the curved arrow notation for each elementary step, and draw the products. (*Hint:* How can you simplify each metal-containing species?)

a Proton transfer: $CH_3OK + HCN \rightarrow ?$
b Nucleophilic addition: $CH_3CO_2CH_3 + LiAlH_4 \rightarrow ?$
c E2: $CH_3CH_2MgBr + (CH_3)_2CHBr \rightarrow ?$
d S_N2: $CH_3CO_2Na + C_6H_5CH_2Cl \rightarrow ?$

5.5 For each of the following elementary steps, draw the curved arrow notation, and draw the products, paying attention to stereochemistry. Indicate whether a single stereoisomer or a mixture of stereoisomers is produced. If a mixture is produced, determine whether the stereoisomers will be produced in equal or unequal amounts.

(a) 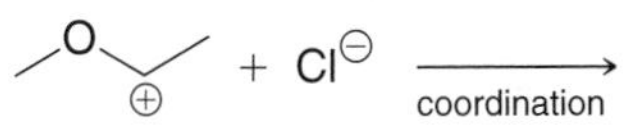

(b) 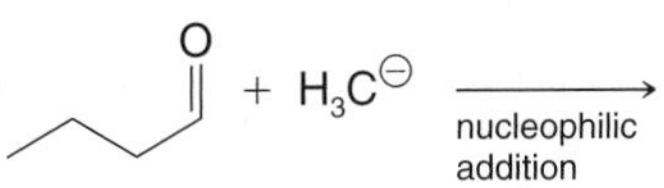

(c) 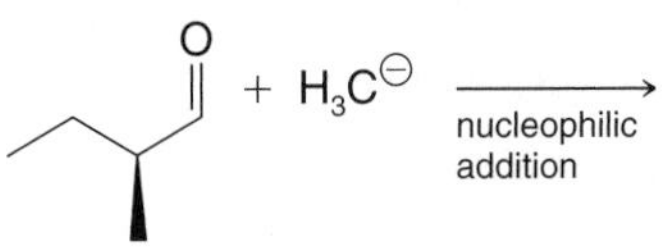

5.6 For each of the following elementary steps, draw the curved arrow notation, and draw the products, paying attention to stereochemistry. (*Hint:* How can you simplify each metal-containing species?)

(a) 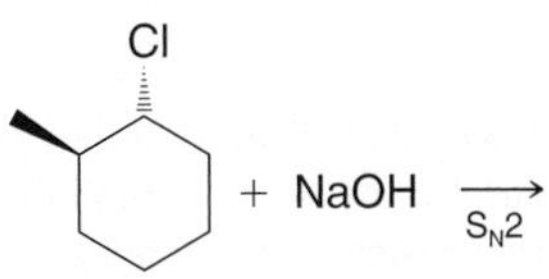

(b) 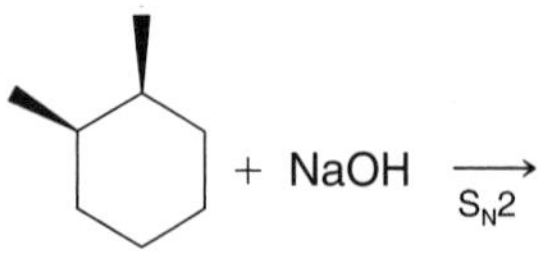

(c) 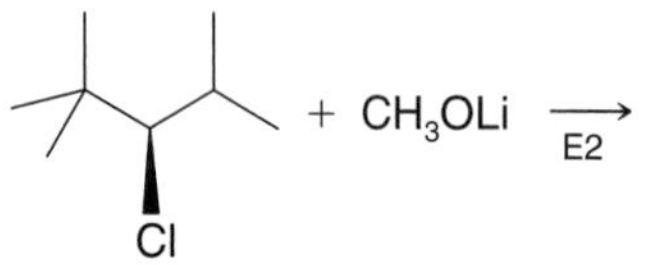

5.7 Predict the products of each of the following reactions. (*Hint:* Consider the notion of *electron-rich to electron-poor.*)

a $C_6H_5MgBr + CO_2$, then add $H_2O \rightarrow$?
b $CH_2{=}CHCH_2Br + CH_3Li \rightarrow$?
c $H_2O + LiAlH_4 \rightarrow$?
d $(CH_3)_2C{=}NCH_3 + CH_3CH_2MgBr$, then add $NH_4^+ \rightarrow$?

5.8 Draw the complete mechanism for the following sequence of reactions. Using the mechanism, predict the products, including stereochemistry if appropriate. (*Note:* D_2O is *deuterium-labeled water*. D is deuterium, which is H with an additional neutron; its chemical reactivity can be treated identically to that of H.)

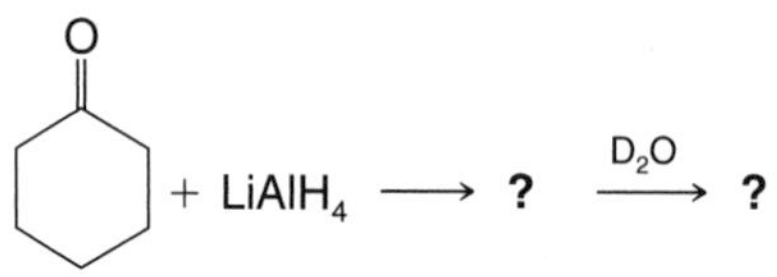

(*Note:* This reaction is a twist on one that you have seen before, which uses $LiAlH_4$ and H_2O. You should know that a problem similar to this one has appeared on the MCAT exam. It is quite likely that you will not learn this exact reaction in your traditional organic chemistry course, so this problem forces you to extend your knowledge of mechanisms. Those students who rely primarily on flash cards will not know how to answer this question because it appears to be an entirely new reaction—one they did not have a flash card for.)

5.9 Predict the products of the following sequence of reactions. (See Problem 5.8.)

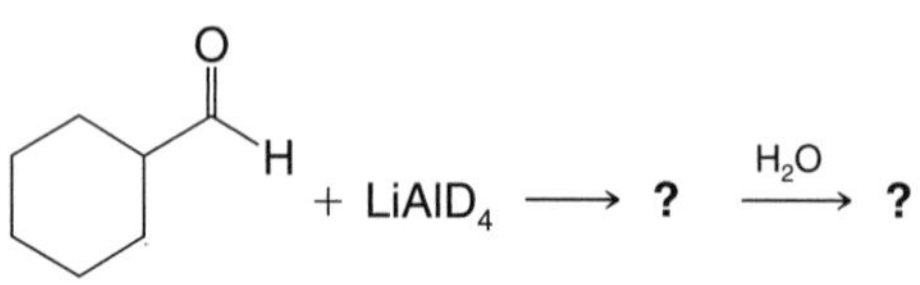

5.10 ^{18}O denotes an oxygen atom that has 8 protons and 10 neutrons (for a total of 18 amu). It therefore has 2 neutrons more than the usual oxygen atom, which has 8. If such an atom is found as the oxygen atom in water, the result is $^{18}OH_2$, called *oxygen-18-labeled water*. Suppose that oxygen-18-labeled water is used in the second reaction below. What is the product of such a reaction?

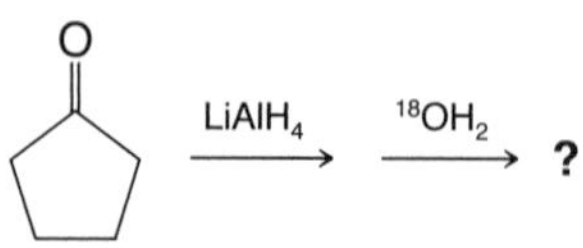

5.11 Given each reactant, provide the missing curved arrows and products for each elementary step that is described. (*Note:* The organic product of each step is to be used as a reactant in the next step.)

a $(CH_3)_3CCHBrCH_2C(CH_3)_3$
- **i** E2 step involving HO^-.
- **ii** Electrophilic addition involving HCl.
- **iii** 1,2-methyl shift.
- **iv** Coordination involving Cl^-.

b $CH_3CH_2CCl{=}O$
- **i** Nucleophilic addition of CH_3Li.
- **ii** Nucleophilic elimination.
- **iii** Nucleophilic addition of CH_3Li.
- ● **iv** Proton transfer involving water.

c C_6H_5MgBr
- **i** S_N2 step involving CH_3Br.
- **ii** Electrophilic addition of Br^+.
- **iii** Electrophilic elimination of H^+.

6 Charge Stability: Charge Is Bad!

When you complete this chapter, you should be able to:

- Explain why the presence of a formal charge in a species in general makes that species unstable.
- Explain how the stability of a charge depends on the size and electronegativity of the atom on which the charge appears.
- Predict the relative stabilities of similar ions based on the resonance delocalization of the charge.
- Describe what inductive effects are and how they can increase or decrease the concentration of charge.
- Recognize electron-withdrawing and electron-donating groups and predict how they impact the stability of a given charge.
- Rank the order of stability of multiple species based on charge stability.
- Use arguments of charge stability to determine:
  - The relative strengths of Brønsted acids and bases.
  - The relative strengths of nucleophiles.
  - The strongest resonance contributor of a charged species.
- Identify common protic and aprotic solvents and predict relative nucleophile strengths in each type of solvent.

Your Starting Point

Answer the following questions to assess your knowledge about charge stability.

1. Which do you think is in general more stable—an uncharged species or a species that carries a charge? ______
2. The farther to the right an atom appears in the periodic table, does electronegativity increase or decrease? ______
3. The farther down an atom appears in the periodic table, does electronegativity increase or decrease? Does the size of the atom increase or decrease? ______
4. What kind of atom can accommodate a negative charge better—one that is more electronegative or one that is less electronegative? A larger atom or a smaller one? ______
5. Which is more stable—a high concentration of charge or a low concentration of charge? ______
6. In which case does a reaction tend to take place more readily—when the products are more stable than the reactants or when the products are less stable than the reactants? ______
7. As the energy barrier for a reaction increases, what happens to the reaction rate? ______
8. Define a Brønsted acid and a Brønsted base. ______
9. What is a nucleophile? ______

Answers: 1. In general, an uncharged species is more stable than a charged one. 2. An atom's electronegativity increases the farther to the right it appears in the periodic table. 3. An atom's electronegativity decreases and its size increases the farther down it appears in the periodic table. 4. A negative charge is better accommodated by a more electronegative atom. A negative charge is also better accommodated by a larger atom. 5. A low concentration of charge is more stable than a high concentration. 6. A reaction takes place more readily when the products are more stable than the reactants. 7. As the energy barrier increases, the reaction rate decreases. 8. A Brønsted acid is a proton donor, and a Brønsted base is a proton acceptor. 9. A nucleophile is a species that has an atom with a lone pair and an excess of negative charge, and it tends to seek out and form a bond to an electron-poor atom other than H.

6.1 Introduction

Chapter 5 introduced us to 10 of the most common elementary steps encountered in organic reactions. There we learned how to recognize such elementary steps and how to describe them using curved arrow notation and the notion of *electron-rich to electron-poor*. However, we have yet to discuss *why* a particular set of products would want to be produced in the first place. In other words, we have yet to discuss the **driving force** for reactions.

Although a number of factors contribute to the driving force of a reaction, one of the most important ones deals with the stability (or lack thereof) associated with formal charges, which we call **charge stability.** Part of the reason that charge stability is so important is that formal charges are prevalent in reaction mechanisms. You will notice that in all of the elementary steps introduced in Chapter 5, formal charges appear on the reactant side, or the product side, or both. And in general, the atom on which the formal charge appears changes during the course of the reaction.

The impact that charge stability has on a reaction's driving force stems from the fact that *opposite charges attract*; a negatively charged species has a tendency to seek out and form a bond to a positively charged species, and vice versa. The greater the concentration of charge—either positive or negative—on a given species, the greater the tendency for that species to react. In other words, *charged species are inherently unstable* with respect to chemical reaction, and *the greater the concentration of charge is, the more unstable that species is.* To help you remember this, we say

Charge is bad, and more charge is worse.

In light of the importance of charge stability, we will examine several different types of ions that are common to organic chemistry, encompassing a variety of different structures. We will look at various means by which to evaluate how good (stable) or bad (unstable) the charge on that ion is, based on how well the species can *accommodate* that charge. Ultimately, you will be able to determine the relative reactivity of different species in a particular reaction based on just their Lewis structures. And that alone is quite a powerful tool!

Of the different applications we will work with in this chapter, we will spend the most time on predicting relative acid and base strengths, for two reasons: (1) You are probably already somewhat familiar with acid-base (or proton transfer) reactions from general chemistry, and (2) those reactions are among the simplest and most common reactions that you will face in organic chemistry. As we go through that application, think of the various ways in which you might be tempted to memorize—and then avoid them! For example, students often try to memorize relative acid and base strengths of organic molecules (by memorizing pK_a values) without understanding why they are what they are. As discussed in Chapter 1, doing this is yet another thing that can spell out disaster.

In addition to relative acid and base strengths, we will examine how charge stability plays a role in other elementary steps as well. Finally, we will learn how charge plays a role in determining the best resonance contributor of a species.

6.2 Atomic Ions

The simplest ions are atomic ions—single atoms that bear a positive or negative charge. Learning how to first determine the relative reactivity of atomic ions enables us to determine the relative reactivity of molecular ions, as discussed in the next section. To begin, let's look at the ions—both positive and negative—of some of the most common atoms in organic chemistry, including C, N, O, S, F, Cl, Br, and I.

Negative ions of the above atoms are C^-, N^-, O^-, S^-, F^-, Cl^-, Br^-, and I^-. Each of those ions bears a negative charge and is therefore inherently unstable with regard to chemical reaction—remember, *charge is bad*. However, their stabilities are not all the same because, as we will see, some can accommodate the charge better than others.

Differences in the stability of such ions arise from two major factors: (1) size of the atom and (2) electronegativity. First, with regard to the size of the atom, let's compare F^- and Cl^-. The Cl atom appears just below the F atom in the periodic table. This means that the valence electrons in Cl occupy a larger shell (specifically, the third shell) than the valence electrons in F (the second shell). As a result, the Cl^- ion is substantially larger than the F^- ion. Therefore, despite the fact that both ions bear the same total -1 charge, the negative charge in Cl^- has more room to spread out than does the negative charge in F^-. In other words, F^- has a more concentrated negative charge. Recalling that *more charge is worse*, we can say that Cl^- is happier—it can better accommodate the negative charge than F^-.

This specific example with F^- and Cl^- can be generalized as follows.

For two atoms that appear in the same column of the periodic table, the one that appears lower is significantly larger and can therefore handle a charge better.

As such, Br^- is more stable than Cl^-.

QUICK CHECK

Which ion is more stable, I^- or Br^-? Explain.

Answer: Because I and Br are in the same column of the periodic table and I is larger than Br, I^- is more stable than Br^-.

The above generalization is true for positive charges as well. So, among F^+, Cl^+, Br^+, and I^+, I^+ is the most stable—it is the largest atom and can therefore spread out the charge the most.

What about charges appearing on atoms from different columns of the periodic table, but from the same row? Consider, for example, O^- and F^-. You might recall from general chemistry that F^- is in fact smaller than O^-, given that atomic size decreases slightly in going from left to right across the periodic table. However, that difference in atomic size is only slight and therefore does not have a major impact on charge stability. A much more important factor in this case is *electronegativity*. Remember that electronegativity is a measure of how much an atom likes electrons, which are negatively charged particles. Therefore, electronegativity corresponds to how well an atom can accommodate a negative charge. Realizing, then, that F is more electronegative than O, we conclude that F^- is more stable than O^-. In general, then,

For two atoms that appear in the same row of the periodic table, the one that is more electronegative can handle a negative charge better.

While the size of an atom affects the stability of positive and negative charges in the same way, electronegativity affects the stability of positive and negative charges differently. This is because electronegativity is a measure of how much an atom likes *negative* charge. Therefore, the more electronegative an atom is, the more it *dislikes* positive charge. Turning this around, we can say that the less electronegative an atom is, the less it dislikes positive charge. In general,

For two atoms that appear in the same row of the periodic table, the one that is less electronegative can handle a positive charge better.

For example, because O is less electronegative than F, O^+ is more stable than F^+.

QUICK CHECK

Which ion is more stable, O^+ or N^+? Explain.

Answer: Because N and O are in the same row of the periodic table and N is less electronegative than O, N^+ is more stable than O^+.

So far we have looked at straightforward cases in which the comparisons are among atoms in the same row or in the same column of the periodic table. You may be curious whether there is a way to compare the stability of ions from different rows *and* different columns. The answer is "yes," but some comparisons are straightforward, and others are not. For example, if we are comparing the stability of O^- and I^-, we can consider the stability of a third ion that has a row or column in common with *both* ions—in this case, F^- is a good choice. We know that F^- is more stable than O^- because of electronegativity arguments, and we also know that I^- is more stable than F^- because it is a much larger atom. Therefore, I^- must be more stable than O^-.

On the other hand, if we compare the stability of F^- with that of S^-, we run into some difficulty. Let's throw in O^-, a third ion that has a row or column in common with both F^- and S^-. We know that F^- and S^- are *both* more stable than O^-—the former because of electronegativity and the latter because of size. So this does not automatically give us the answer. Rather, in order to know whether F^- or S^- is more stable, we need to know which factor—size or electronegativity—is more important in governing the stability of an atomic ion. Unfortunately, the answer is not a definite one, but as a rule of thumb, we can say the following.

More often than not, the size of the atom on which a formal charge appears wins out over electronegativity.

In this case, then, S^- is more stable than F^- because S^- is larger. Be aware, though, that there can be exceptions to this rule, especially if there is a large difference in electronegativity, but only a relatively small difference in size. However, rather than calling attention to the specific exceptions, it is better to simply know the general rule.

6.3 Molecular Ions

There are two types of molecular ions to consider: (1) those that have no resonance, so that the charge is isolated, or **localized,** on a single atom; and (2) those that do have resonance, which enables the charge to be distributed over at least two atoms, or **delocalized.** In this section, we will consider the scenario in which the molecular ion has no resonance.

It is rather straightforward to compare two different molecular ions whose Lewis structures are quite similar, with the only major difference being the atom that bears the formal charge. For example, we can compare CH_3O^-, CH_3NH^-, $CH_3CH_2^-$, and CH_3S^-. The only significant difference among these ions is the placement of the negative charge on the O, N, C, and S atoms, respectively. In such a case, we can simply resort back to the arguments made with atomic ions: Namely, because O, N, and C are in the same row of the periodic table and O is the most electronegative, O can accommodate the negative charge the best. Next comes N, and finally C. Note that S is in the same column of the periodic table as O, but it is one row down and therefore significantly larger. Consequently, it can accommodate the negative charge better than the O atom can. The order of stability is then $CH_3CH_2^- < CH_3NH^- < CH_3O^- < CH_3S^-$.

Let's also consider the positively charged molecular ions $CH_3OH_2^+$, CH_3FH^+, and $CH_3SH_2^+$, where the positive charges are on the O, F, and S atoms, respectively. Because O and F are in the same row of the periodic table and O is less electronegative than F, we can say that O can handle a positive charge better. Therefore, $CH_3OH_2^+$ is more stable than CH_3FH^+. Also, because S is below O in the periodic table, S is larger and can better accommodate the positive charge, which makes $CH_3SH_2^+$ more stable than $CH_3OH_2^+$. Overall, the order of stability of these three ions is $CH_3FH^+ < CH_3OH_2^+ < CH_3SH_2^+$.

6.4 Resonance Effects

Resonance can significantly stabilize a molecular ion. Let's look at a specific example: $CH_3CO_2^-$ versus $CH_3CH_2O^-$. The Lewis structures for these ions are shown in Figure 6-1. As you can see, in both Lewis structures the −1 formal charge is on an O atom. However, $CH_3CO_2^-$ has another resonance structure in which the negative charge appears on the *other* O atom. As we learned in Chapter 2, when there are multiple resonance structures of a given molecule, the one true structure of the species is a *hybrid* of those structures, which looks something like their average. Because the negative charge in $CH_3CO_2^-$ is on one O atom in one resonance structure and on the other O atom in the second resonance structure, the hybrid species has a partial negative charge on each O atom. Compare the resonance hybrid of $CH_3CO_2^-$ to the $CH_3CH_2O^-$ ion. In the latter, there is a *full* negative charge on the O atom, whereas in the former there is a partial charge on two different O atoms.

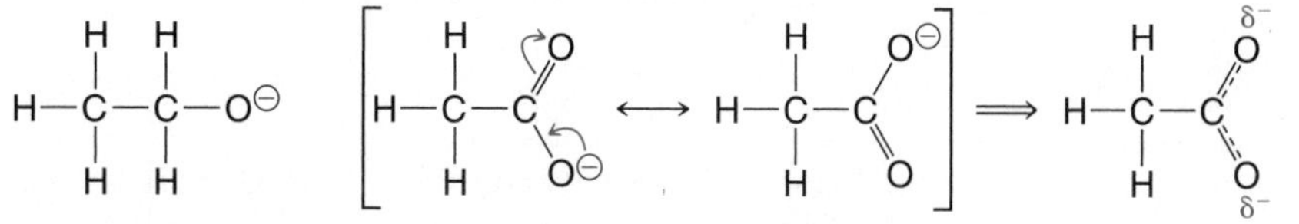

(a) (b)

FIGURE 6-1 **Resonance stabilization of ions.** (a) Lewis structure of $CH_3CH_2O^-$. (b) Resonance structures of $CH_3CO_2^-$ are shown in brackets. The resonance hybrid is shown to the right. Note that in the resonance hybrid there is a partial charge on each O atom rather than a full charge. Therefore, the charge in $CH_3CO_2^-$ is more stable than that in $CH_3CH_2O^-$.

We are now in a position to apply the concept that *more charge is worse*. Both ions have the same total charge of -1, but the $CH_3CH_2O^-$ ion has that negative charge *concentrated* on only one O atom. By contrast, in $CH_3CO_2^-$, that -1 charge is spread out over two O atoms—that is, it is *delocalized* over the two O atoms. Therefore, we can say that $CH_3CO_2^-$ is much happier, or more stable. Such an effect on the properties of a chemical species is called a **resonance effect.**

The specific example of $CH_3CO_2^-$ versus $CH_3CH_2O^-$ can be generalized as follows.

If resonance serves to delocalize (i.e., share) a charge over multiple atoms, the charge is stabilized.

Notice that this generalization does not specify a positive charge or a negative charge. As a result, it can be used to determine the relative stabilities of $CH_3CH_2OH_2^+$ and $CH_3C(OH)_2^+$, the focus of the next Quick Check exercise.

QUICK CHECK

Which ion is more stable, $CH_3CH_2OH_2^+$ or $CH_3C(OH)_2^+$? Explain.

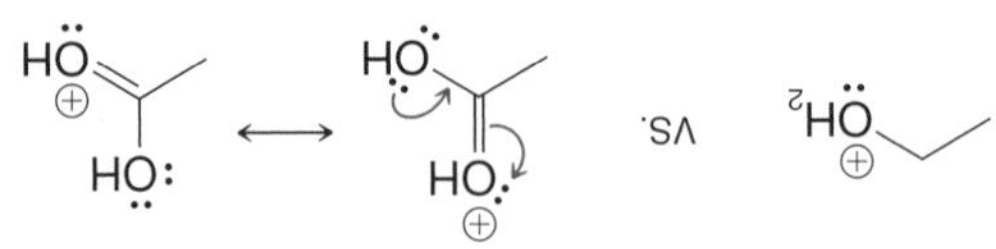

Answer: The second ion is more stable because there is a resonance structure that helps share (delocalize) the positive charge.

We can also examine how the stability of an ion is affected by the *number of resonance structures* it has. As an example, let's compare the $CH_3CO_2^-$ ion to the HSO_4^- ion. Each individual resonance structure of either ion places the -1 formal charge on an O atom. However, there are three resonance structures that can be drawn for HSO_4^- (Figure 6-2), each of which has the negative charge on a different O atom. As a consequence, in the hybrid of these resonance structures the negative charge is shared over three O atoms. Only two resonance structures can be drawn for $CH_3CO_2^-$ (Figure 6-1b); the resonance hybrid shows the negative charge shared over the two O atoms. Therefore, the negative charge on *each* O atom in HSO_4^- is smaller than that on each O atom in $CH_3CO_2^-$—more specifically, about one-third of a negative charge compared to about one-half of a negative charge. Consequently, we can say that the HSO_4^- ion has a smaller *concentration* of negative charge and thus is much happier than the $CH_3CO_2^-$ ion.

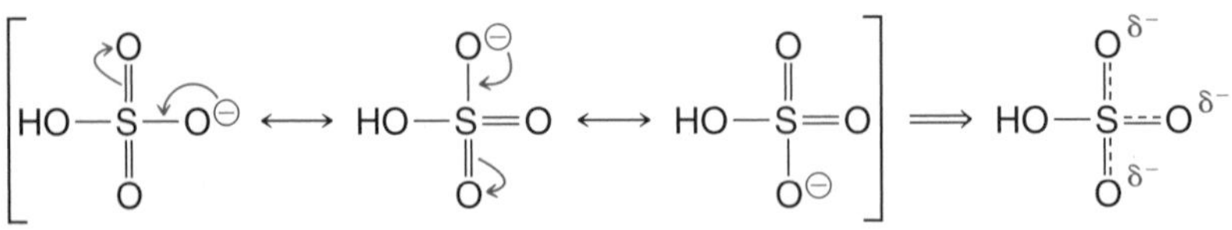

FIGURE 6-2 **Resonance delocalization in HSO_4^-.** In the resonance structures of HSO_4^- (*left*), we can see that the negative charge is shared over three different O atoms. Therefore, in the resonance hybrid (*right*), the concentration of negative charge is smaller than in $CH_3CO_2^-$, which has the negative charge delocalized over just two O atoms.

6.5 Inductive Effects

Inductive effects, like resonance effects, can significantly alter an ion's stability. Let's use an example to introduce the concept. Consider CH_3O^- and FCH_2O^-, which are structurally very similar. The only major difference is an F atom bonded to C in place of an H atom bonded to C. Which ion is more stable?

To answer this question, realize first that in both ions the O atom has the −1 formal charge. Therefore, in both cases the negative charge is accommodated equally well by the atom on which the charge appears. Furthermore, notice that neither ion has a resonance structure, meaning that the negative charge is localized in both cases. Therefore, neither charge is stabilized by resonance.

To understand what causes a difference in stability between CH_3O^- and FCH_2O^-, imagine what will happen if we begin with CH_3O^- (Figure 6-3a) and replace an H atom by an F atom, producing FCH_2O^- (Figure 6-3b). The F atom is more electronegative than the H atom, meaning that in a covalent bond F pulls electrons toward itself more than H does. Therefore, if such a swap occurs, the electrons that form the bond to C are shifted slightly toward F. Moreover, as those bonding electrons are shifted away from that C atom, it becomes somewhat electron-deficient—and thus hungry for electrons. In turn, the C atom draws the electrons of the C—O bond toward itself (Figure 6-3c). With electrons shifted away from O, the negative charge on O is diminished (Figure 6-3d). Ultimately, then, with a smaller concentration of charge on O, we can say that FCH_2O^- is the more stable ion.

We are now in a position to give a definition of *inductive effects.*

Inductive effects are the effects that come about as a result of the shifting of electrons through covalent bonds when one atom is exchanged for another atom that has different electronegativity.

In FCH_2O^-, for example, the inductive effects are described as the shifting of the various covalently bound electrons toward F, leading to a diminished negative charge on O. Alternatively, we would say that, compared to H, F is **electron-withdrawing.**

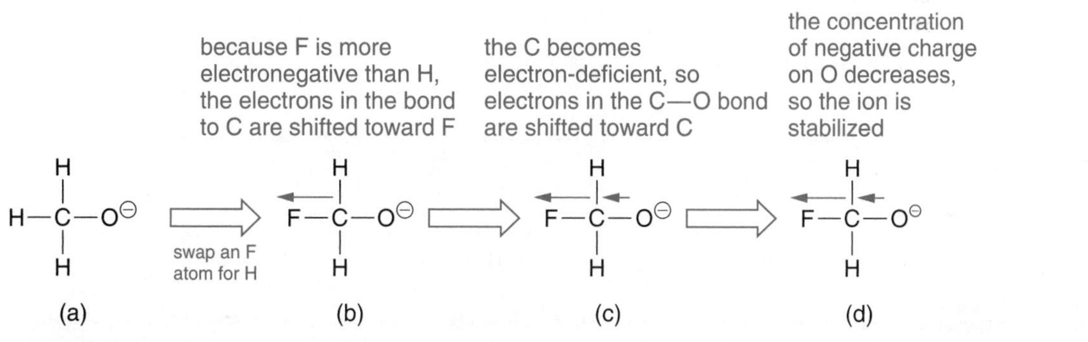
because F is more electronegative than H, the electrons in the bond to C are shifted toward F
the C becomes electron-deficient, so electrons in the C—O bond are shifted toward C
the concentration of negative charge on O decreases, so the ion is stabilized
swap an F atom for H
(a)
(b)
(c)
(d)

FIGURE 6-3 **Inductive effects in CH_3O^- versus FCH_2O^-.** In going from (a) to (b), the leftmost H atom is replaced by an F atom. The greater electronegativity of F causes electrons in the bond to C to be shifted toward F. This causes C to become somewhat electron-deficient. Therefore, in going from (b) to (c), electrons in the C—O bond are shifted toward C. As a result, (d) shows that the negative charge on O is diminished, resulting in stabilization of the ion.

Furthermore, as we just saw in the case of FCH_2O^-, F is inductively *stabilizing*. In general,

Electron-withdrawing atoms inductively stabilize nearby negative charges.

✔ **QUICK CHECK**

Which ion is more stable: $ClCH_2CH_2NH^-$ or $CH_3CH_2NH^-$? Explain.

Answer: The first ion is more stable. Compared to H, Cl is electron-withdrawing because it is more electronegative. Being electron-withdrawing, it will better stabilize nearby negative charges.

It is important to realize that electron-withdrawing groups are not always inductively stabilizing—it depends on what kind of charge is nearby. This can be seen by comparing $CH_3CH_2^+$ (Figure 6-4a) and $FCH_2CH_2^+$ (Figure 6-4b). Imagine what will happen if we begin with $CH_3CH_2^+$ and replace one of the H atoms of the CH_3 group with an F atom, producing $FCH_2CH_2^+$. As indicated in Figure 6-4b, the electrons in the bond to the uncharged C atom are shifted toward the F atom. The uncharged C atom then becomes somewhat electron-deficient and pulls electrons toward itself from the charged C atom (Figure 6-4c). In other words, some negative charge is removed from the C^+ atom, leaving behind some additional positive charge (Figure 6-4d). That increased concentration of charge makes $FCH_2CH_2^+$ even more unhappy, or more unstable, than the $CH_3CH_2^+$ ion. In such a case, the withdrawing effect of F (compared to H) serves to destabilize the ion, so we say that F is **inductively destabilizing.**

The above example comparing $CH_3CH_2^+$ to $FCH_2CH_2^+$ is a specific case that can be generalized as follows.

Electron-withdrawing atoms inductively destabilize nearby positive charges.

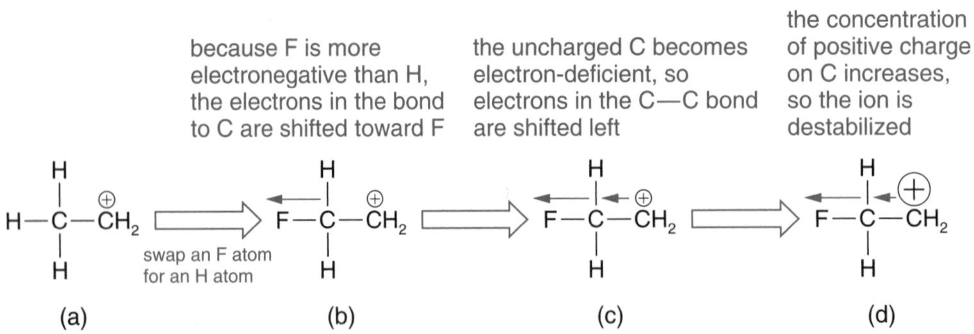
because F is more electronegative than H, the electrons in the bond to C are shifted toward F
the uncharged C becomes electron-deficient, so electrons in the C—C bond are shifted left
the concentration of positive charge on C increases, so the ion is destabilized
swap an F atom for an H atom
(a)
(b)
(c)
(d)

FIGURE 6-4 **Inductive effects in $CH_3CH_2^+$ versus $FCH_2CH_2^+$.** In going from (a) to (b), the leftmost H atom is replaced by an F atom. The greater electronegativity of F causes electrons in the bond to C to be shifted toward F. This causes the uncharged C to become somewhat electron-deficient. Therefore, in going from (b) to (c), electrons in the C—C bond are shifted left toward the uncharged C. As a result, (d) shows that the positive charge on C^+ is increased, resulting in destabilization of the ion.

QUICK CHECK

Which ion is more stable: $ClCH_2CH_2NH_3^+$ or $CH_3CH_2NH_3^+$? Explain.

Answer: The second ion is more stable. The Cl atom is electron-withdrawing compared to the H atom, so it inductively destabilizes nearby positive charges.

Like the F atom, most substituents you encounter will be inductively electron-withdrawing groups compared to the H atom. This is because most atoms we encounter have a greater electronegativity than H. However, there are a handful of substituents that are **inductively electron-donating.** By far the most common of these are **alkyl groups,** which are groups that consist entirely of C and H atoms and contain only single bonds. If we consider the simplest alkyl group, the CH_3 group, it becomes clear why this is so. The C and H atoms are very similar in electronegativity, but the C atom is slightly higher (2.5 compared to 2.2). Therefore, in each of the C—H bonds, the C atom gets a slightly greater share of the bonding electrons, creating an excess of electron density, and a slight δ^-, on the C atom. This buildup of negative charge, or electron density, on the C atom enables that C atom to donate some of its negative charge to a group to which it is connected (Figure 6-5).

With this in mind, let's compare the stabilities of CH_3^+ (Figure 6-6a) and $CH_3—CH_2^+$ (Figure 6-6b). As shown, to go from the first ion to the second, we can simply replace an H atom with a CH_3 group. Because a CH_3 group is electron-donating compared to H, such a swap causes a shift in the bonding electrons toward C^+ (Figure 6-6b). Thus, some of the positive charge is canceled by the additional negative charge (Figure 6-6c). The smaller positive charge, as a result, is more stable.

Although this example presents the specific case of a CH_3 group, the same basic idea is at play for other alkyl groups (groups consisting of only C, H, and single bonds). In general,

Alkyl groups are electron-donating and therefore stabilize nearby positive charges.

QUICK CHECK

Which ion is more stable: $(CH_3)_2NH_2^+$ or $(CH_3)_3NH^+$? Explain.

Answer: The second ion is more stable. To go from the first ion to the second, an H atom is swapped for a CH_3 group. That CH_3 group is inductively electron-donating and therefore stabilizes the nearby positive charge on N.

FIGURE 6-5 The electron-donating ability of alkyl groups. (a) The bond dipoles in an alkyl group point toward C. (b) The buildup of negative charge on C can be donated to another group to which the alkyl group is attached. Therefore, alkyl groups are electron-donating groups.

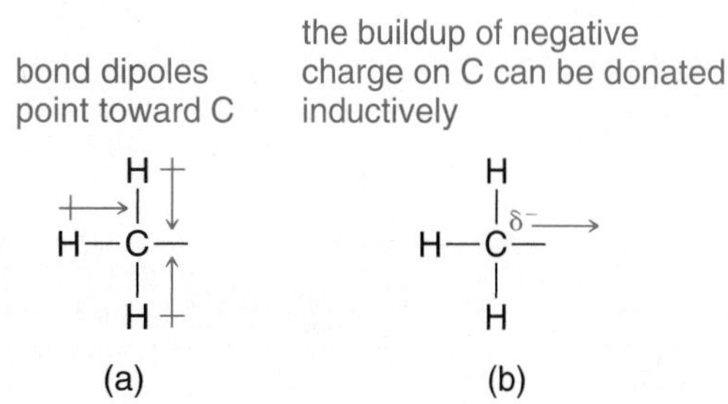
bond dipoles point toward C
the buildup of negative charge on C can be donated inductively
H
H—C—
H
δ−
(a)
(b)

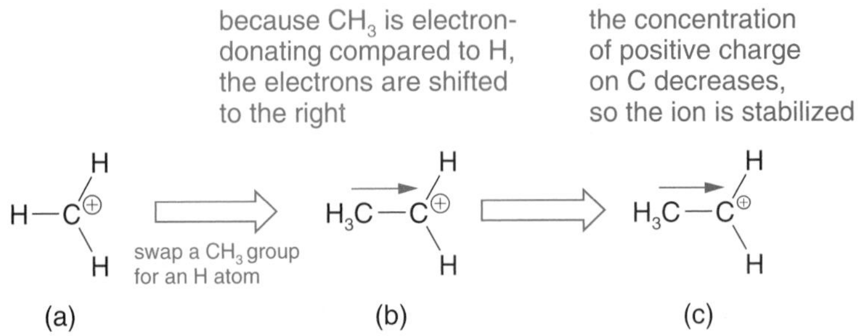
because CH3 is electron-
donating compared to H,
the electrons are shifted
to the right
the concentration
of positive charge
on C decreases,
so the ion is stabilized
swap a CH3 group
for an H atom
(a)
(b)
(c)

FIGURE 6-6 **Inductive effects in CH_3^+ versus $CH_3CH_2^+$.** In going from (a) to (b), the leftmost H atom is replaced by a CH_3 group. Because a CH_3 group is electron-donating compared to H, this swap causes a shift in the bonding electrons toward C^+. As a result, (c) shows that the positive charge on C^+ is decreased, resulting in stabilization of the ion.

By contrast, alkyl groups have exactly the opposite effect on the stability of anions. A good example is the comparison of the stabilities of HO^- (Figure 6-7a) and CH_3O^- (Figure 6-7b). In the CH_3O^- ion, the CH_3 group serves to donate negative charge to the O atom (relative to what the H atom does), which increases the concentration of negative charge on O (Figure 6-7c). Because *more charge is worse*, the CH_3O^- ion is less stable than the HO^- ion.

Once again, this example presents the specific case of a CH_3 group, but it works for other alkyl groups as well. That is,

Alkyl groups destabilize nearby negative charges.

QUICK CHECK

Which ion is more stable: $(CH_3)_2N^-$ or H_2N^-? Explain.

Answer: The second ion is more stable. In going from the second ion to the first, two H atoms are swapped for CH_3 groups, and each CH_3 group is inductively electron-donating, which destabilizes the nearby negative charge on N.

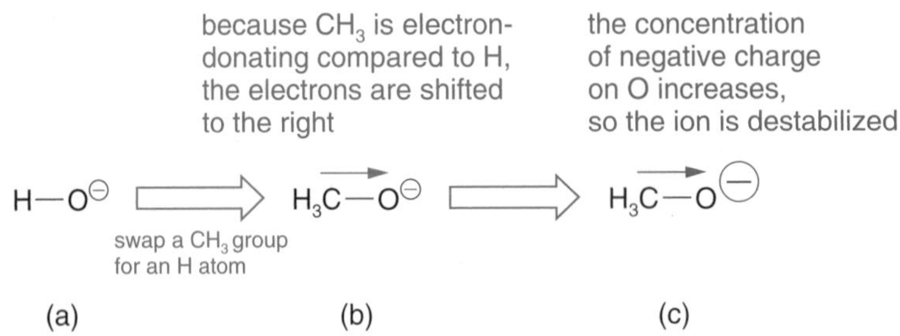
because CH3 is electron-
donating compared to H,
the electrons are shifted
to the right
the concentration
of negative charge
on O increases,
so the ion is destabilized
swap a CH3 group
for an H atom
(a)
(b)
(c)

FIGURE 6-7 **Inductive effects in HO^- versus CH_3O^-.** In going from (a) to (b), the H atom is replaced by a CH_3 group. Because a CH_3 group is electron-donating compared to the H atom, this swap causes a shift in the bonding electrons toward O. As a result, (c) shows that the negative charge on O is increased, resulting in destabilization of the ion.

Thus far, inductive effects on ion stability have been discussed only in a qualitative sense; electron-withdrawing groups stabilize nearby negative charges and destabilize nearby positive charges, whereas electron-donating groups destabilize nearby negative charges and stabilize nearby positive charges. However, we can be somewhat quantitative as well by simply applying a few straightforward principles.

The first rule has to do with the *number* of groups responsible for the inductive effects.

The greater the number of electron-withdrawing or electron-donating groups, the more pronounced the effect.

Consider, for example, $CF_3CH_2^+$ versus $CH_2FCH_2^+$. In both cases, the F atoms, which are electron-withdrawing, serve to destabilize the nearby positive charge. That effect is more pronounced in the first ion than in the second due to the greater number of F atoms. Thus, $CH_2FCH_2^+$ is the more stable ion.

The next rule has to do with distance.

Inductive effects become significantly weaker with each additional bond between the formal charge and the electron-withdrawing or electron-donating group.

For example, let's compare $CH_3CH_2CHFCH_2NH^-$ and $CH_3CHFCH_2CH_2NH^-$. In both cases, the electronegative F atom stabilizes the negative charge. In the first ion, however, the F atom is one bond closer to the negative charge, so that ion is the more stable one.

Finally, let's consider the identity of the group responsible for the inductive effects.

The stronger an electron-withdrawing or electron-donating group, the more pronounced the effect.

The $CH_3CHFCH_2O^-$ and $CH_3CHClCH_2O^-$ ions, for example, are nearly identical, with the only difference being an F atom in place of a Cl atom. Both the F and the Cl atoms are electron-withdrawing and therefore stabilize nearby negative charges. However, because F is more electronegative than Cl, the effect is stronger in $CH_3CHFCH_2O^-$. Therefore, $CH_3CHFCH_2O^-$ is more stable than $CH_3CHClCH_2O^-$.

When applying this rule to a situation where alkyl groups are involved, it is important to know that the size of the alkyl group has very little impact on the group's electron-donating ability. In other words,

All alkyl groups have about the same electron-donating ability.

For example, $CH_3CH_2O^-$ and $CH_3CH_2CH_2CH_2O^-$ have about the same stability because in each case an O^- is bonded to a single alkyl group. (In reality, there will be a slight difference, but nothing you need to worry about for this course.)

A related concern is how to determine relative charge stability when an ion has both an electron-withdrawing group and an electron-donating group. To help in such situations, it is useful to know that electron-donating effects of alkyl groups are relatively weak. Therefore, we can apply the following rule.

Electron-withdrawing effects are typically stronger than electron-donating effects.

To illustrate this point, let's compare $CH_3CH_2CH_2OH_2^+$ to $(CH_3)_2CFCH_2OH_2^+$, shown in Figure 6-8. In going from the first ion to the second, two H atoms on the same C atom are replaced by a CH_3 group and an F atom. The CH_3 group is electron-donating, whereas the F atom is electron-withdrawing. Therefore, whereas the nearby positive charge is stabilized by the CH_3 group, it is destabilized by the F atom. According to the above rule, the F wins out, so the second ion is less stable than the first.

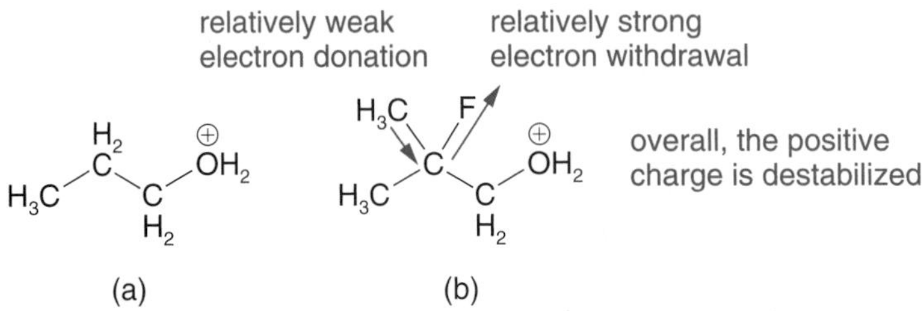

FIGURE 6-8 Relative strengths of electron-donating and electron-withdrawing effects. Compared to the ion in (a), the one in (b) has a CH_3 group and an F atom in place of two H atoms. The CH_3 group is electron-donating, whereas the F atom is electron-withdrawing. Because electron-withdrawing effects are generally stronger, the F atom wins out. Therefore, the ion in (b) is less stable than the one in (a).

QUICK CHECK

Which ion is more stable: $H_2NCH_2CH_2CH_2O^-$ or $CH_3CH(OH)CH_2O^-$? Explain.

Answer: The second ion is more stable for two reasons. In both ions, an electron-withdrawing group stabilizes the negative charge—an NH_2 group in the first ion and an OH group in the second. However, the OH group is closer to the negative charge, which contributes to a larger effect. Furthermore, OH is a stronger electron-withdrawing group than NH_2, owing to the fact that O is more electronegative than N.

6.6 Putting It All Together

So far, we have examined a variety of factors that dictate the stability of a species. We have learned that it makes a difference whether or not a species carries a charge. We have also learned that both the size and the electronegativity of the atom on which we find the charge are important. Furthermore, ion stability is affected by resonance—the greater the sharing of the charge over different atoms, the greater the stability. Finally, nearby electron-withdrawing and electron-donating substituents can affect ion stability by inductive effects. Given all of these factors, it can become quite overwhelming to evaluate the stability of a particular species, especially when more than one factor contributes at the same time.

For situations like this, it is advantageous to know the relative importance of factors that contribute to charge stability. As it turns out, *the order in which this chapter introduced the factors is essentially the order of priority*. Therefore, whether or not a species is charged is generally the most important factor to consider. If one species is charged and another is not, then the charged species is typically more unstable with respect to chemical reaction. If we are comparing two ions that have the same charge, we next examine the atoms on which we find those charges. If there is a tie up to this point, we further consider how much resonance serves to share the charges on different atoms. And finally, we examine differences in inductive effects.

In light of this, whenever you are trying to determine the relative stabilities of two species, it helps to go through the following list of tie-breaking questions, in order.

TIE-BREAKING QUESTIONS FOR DETERMINING CHARGE STABILITY

1. Do the species have the same charge?
2. Are the charges on the same type of atom?
3. Are the charges equally delocalized by resonance?
4. Do the charges experience the same inductive effects?

When you get to a question for which the answer is "no," you know which factor will determine the difference in charge stability.

As an example of how to put together what we have learned thus far, let's determine the order of stability of the following four species:

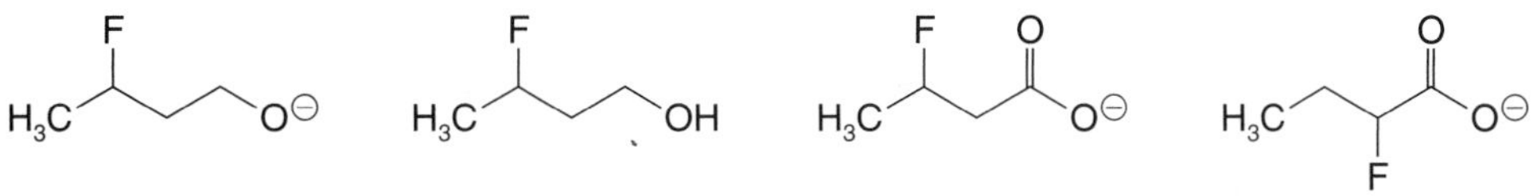

I II III IV

As you will see, such "ranking" problems are somewhat common in organic chemistry.

We begin with the first tie-breaking question from above: Do the species have the same charge? The answer is "no"—structure II is uncharged, while the others are all negatively charged. Therefore, right away, we know that structure II is the most stable.

Of the remaining structures—structures I, III, and IV—we ask the second tie-breaking question: Are the charges on the same type of atom? The answer is "yes," as all negative charges are on O atoms. So we proceed to the third tie-breaking question: Are the charges equally delocalized by resonance? The answer is "no." Structures III and IV have resonance structures that allow the negative charge to be shared over two O atoms, whereas in structure I the negative charge is localized on just the one O atom. Therefore, structure I is less stable than either structure III or structure IV. Presently, our order of stability is I < (III & IV) < II.

TIME TO TRY

Draw the resonance structures mentioned for structures III and IV, showing that the negative charge is shared.

Finally, we move to the fourth tie-breaking question: Do the charges experience the same inductive effects? The answer is "no." Both structures have an F atom near the negative charge, but F is closer to the charge in structure IV, making the inductive effects more pronounced. Because F is inductively electron-withdrawing, it will stabilize the nearby negative charge, making structure IV more stable than structure III. Our final order of stability is therefore I < III < IV < II.

6.7 Application: Strengths of Acids and Bases

Similar to the problem given in the last section, you can expect questions in organic chemistry that ask you to rank the order of acid or base strengths, given only their Lewis structures. To do this, you could memorize the pK_a values (i.e., the numerical values that correspond to acid strength) of a dozen or more types of acids and memorize the effects that electron-withdrawing and electron-donating groups have on acid and base strengths (they are opposite each other). You could then say, to first approximation, that the strength of the acid of interest is approximately the same as one that is structurally similar, whose pK_a you memorized. For a better approximation, you could then apply the inductive effects you memorized—does the substituent make the acid stronger or weaker?

At this point, you should realize that this is the *wrong* way to do it. The better way is to derive relative acid and base strengths in a systematic way, doing something you already know how to do—determining relative stabilities of ions.

To begin, recall that a Brønsted acid is a proton (H^+) donor, whereas a Brønsted base is a proton acceptor. Therefore,

The more that a species wants to donate a proton, the stronger it is as an acid.

The more that a species wants to accept a proton, the stronger it is as a base.

These statements may seem simple, but realize that the idea of a species wanting to undergo a reaction actually has to do with how much a reactant would prefer to be a product of the reaction—and therefore depends on the relative stabilities of the reactants and products. Specifically, the more stable (lower in energy) the products are relative to the reactants, the more the reactants prefer to be products, and vice versa.

It helps to think of this in terms of an analogy involving a path on a hill, where the reactants represent the beginning of the path and the products represent the end of it. If the products are lower in energy (more stable) than the reactants, the reaction is said to be **downhill,** whereas if the reverse is true, it is **uphill.** Therefore,

The more **downhill** the reaction is, the more the reactants prefer to be products.

The more **uphill** the reaction is, the less the reactants prefer to be products.

These ideas are summarized graphically in Figure 6-9. In curve (a), the products are much lower in energy, and therefore much more stable than the reactants, so the reaction is very downhill. This means that the reactants really want to undergo the reaction to become products. In curve (b), the products are slightly more stable than the reactants, so the corresponding reaction is slightly downhill, meaning that the reactants only somewhat want to

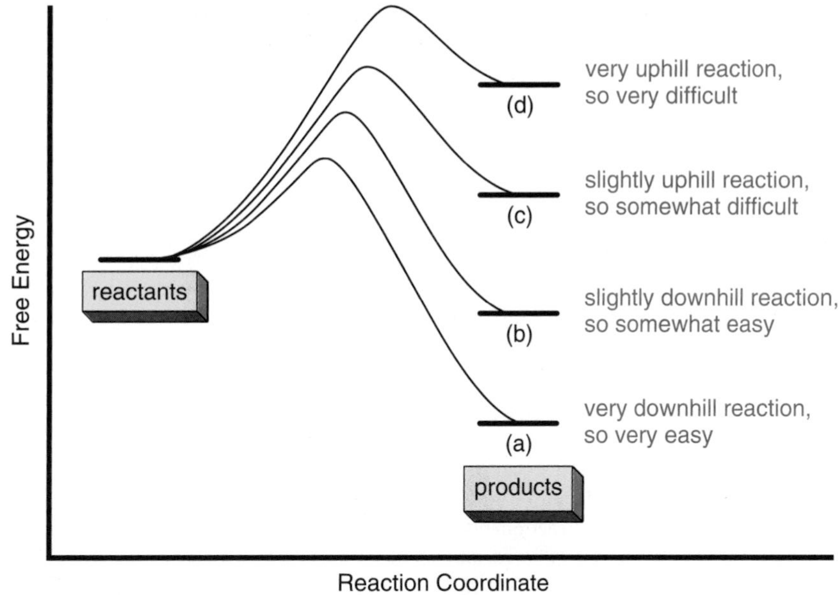

FIGURE 6-9 **Uphill and downhill nature of reactions.** The stability of the products, relative to the reactants, corresponds to how much reactants want to undergo a reaction to become products. (a) The products are much more stable than the reactants (i.e., the reaction is very downhill), so the reactants really want to undergo the reaction. (b) The products are slightly more stable than the reactants (i.e., the reaction is slightly downhill), so the reactants somewhat want to undergo the reaction. (c) The products are slightly less stable than the reactants (i.e., the reaction is slightly uphill), so the reactants somewhat do not want to undergo the reaction. (d) The products are much less stable than the reactants (i.e., the reaction is very uphill), so the reactants really do not want to undergo the reaction.

undergo the reaction. In curves (c) and (d), the products are less stable than the reactants, making the corresponding reactions increasingly uphill. Therefore, the reactants have less of a desire to undergo the reaction.

Returning to the discussion of proton transfer reactions, we can now say the following.

The more downhill (less uphill) a proton transfer reaction is, the stronger the acid and/or the stronger the base is.

The more uphill (less downhill) a proton transfer reaction is, the weaker the acid and/or the weaker the base is.

Now we are in a position to make predictions about relative acid and base strengths. Let's begin by asking this question: Which is the stronger acid, H_2O or H_2S? With what we have just learned, we can reword the question to ask: Which species reacts as an acid in a more downhill (or less uphill) reaction? So let's begin by writing out complete reactions in which each acid behaves as an acid—that is, donates a proton. There must also be a base present, to accept the proton, but because we want to consider only the differences brought about by the different acids, we make sure that the base is the same in both cases—we choose water. The reactions are shown in Equations 6-1a and 6-1b.

$$\begin{aligned} &HO{-}H + H_2O \rightarrow HO^- + H_3O^+ \quad \text{(a)} \\ &HS{-}H + H_2O \rightarrow HS^- + H_3O^+ \quad \text{(b)} \end{aligned} \qquad (6\text{-}1)$$

TIME TO TRY

In Equations 6-1a and 6-1b, draw in the pertinent lone pairs of electrons, and add the curved arrow notation.

The next step is to construct the free-energy diagrams for these reactions, but in order to do so, we must know the relative stabilities of each set of reactants and products. Right away, you should notice that in each reaction the products have two charges, whereas the reactants have none. Therefore, both reactions are expected to be uphill, as shown in Figure 6-10. Next, realize that each reaction has a common reactant, H_2O, so we can ignore its contribution to the overall stability of the reactants. Similarly, because H_3O^+ appears as a product in both reactions, we can ignore its contribution to the overall stability of the products. These simplifications are indicated in Figure 6-10 by crossing out H_2O and H_3O^+.

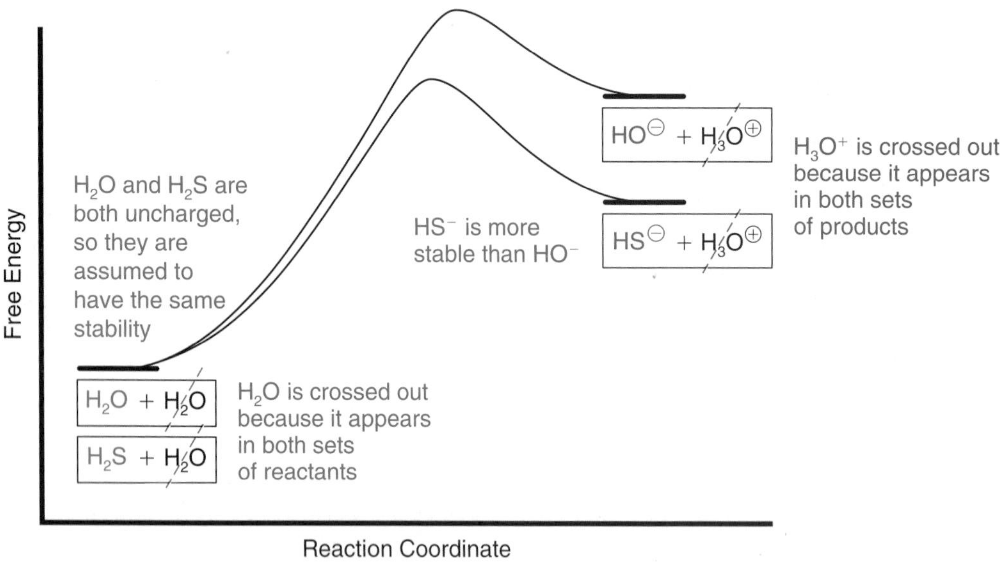

FIGURE 6-10 Free-energy diagrams for the reactions in equations 6-1a and 6-1b. The products are higher in energy than the reactants because each set of products has a greater number of charges. We ignore the contributions by H_2O toward the stability of the reactants, as well as the contributions by H_3O^+ toward the stability of the products, so these species are crossed out. The remaining HS^- is lower in energy than HO^- because the S atom can better handle the negative charge. The remaining H_2O and H_2S molecules on the reactant side appear at the same energy because they are both uncharged. Therefore, H_2O reacts as an acid in the more uphill reaction, making H_2O the weaker acid.

The difference on the product side boils down to HO^- versus HS^-. Because these are both ions, we can determine which is more stable by asking the tie-breaking questions from Section 6.6. Question 1 results in a tie because the two species have the same -1 charge. The tie is broken by question 2 because the charges are on different atoms—an O atom in HO^- and an S atom in HS^-. Given that S is below O in the periodic table, S is significantly larger, allowing it to handle the negative charge better. Thus, HS^- is more stable than HO^-. In Figure 6-10, this is indicated by placing HS^- lower in energy.

Finally, on the reactant side, the difference boils down to H_2O versus H_2S. These are both uncharged species, so by the rules we have learned in this chapter, we do not know how to determine which one is more stable. However, because both molecules are uncharged, we know that they are relatively stable compared to ions. Furthermore, there should be relatively little difference in stability between H_2O and H_2S compared to the differences we encounter for ions. We can take this a step further to make the following very useful assumption.

When comparing reactions that involve both charged and uncharged species, we can assume that all the uncharged species have about the same stability.

For this reason, the reactants of Equations 6-1a and 6-1b start at the same place in Figure 6-10.

Having constructed the free-energy diagrams for the proton transfer reactions, we can finally answer the question about the relative acid strengths of H_2O and H_2S. Notice that both proton transfer reactions are uphill, but the one in which H_2S acts as the acid is less uphill than the one in which H_2O acts as the acid. Therefore, H_2S is the stronger acid.

The previous problem had us compare the strengths of two uncharged acids. We can also make predictions about the strengths of two charged acids, such as NH_4^+ and $CH_3NH_3^+$. Again, we must start by writing out the complete proton transfer reactions in which the species at hand behave as acids, with water as the base. These are shown in Equations 6-2a and 6-2b.

$$H_3N^+{-}H + H_2O \rightarrow H_3N + H_3O^+ \quad \text{(a)}$$
$$H_3CNH_2^+{-}H + H_2O \rightarrow H_3CNH_2 + H_3O^+ \quad \text{(b)} \qquad \text{(6-2)}$$

TIME TO TRY

In Equations 6-2a and 6-2b, draw in the pertinent lone pairs of electrons, and add the curved arrow notation.

Next, we construct the free-energy diagrams for these reactions in a single plot. Notice that in each reaction a positive charge appears in the reactants as well as the products. Therefore, the reactants and products will have similar energies. However, because O is more electronegative than N, it is better to have the positive charge on N. This is indicated in Figure 6-11 by having both sets of products slightly higher in energy than the reactants.

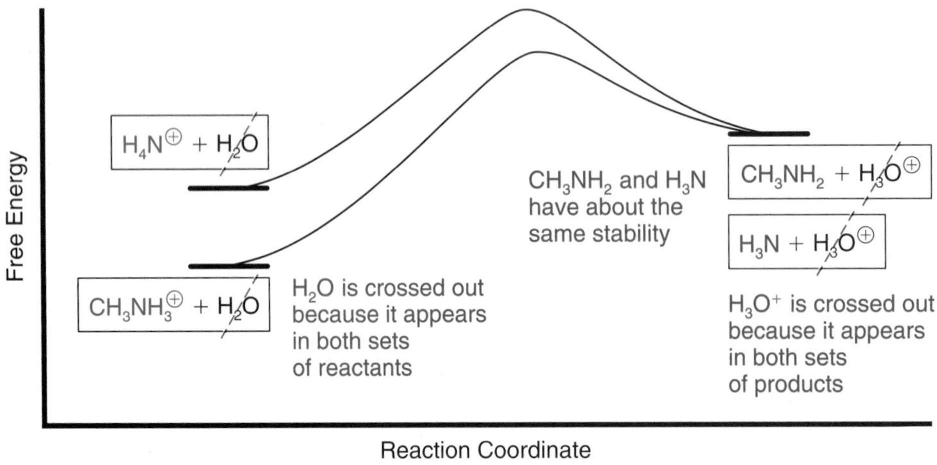

FIGURE 6-11 Free-energy diagrams for the reactions in equations 6-2a and 6-2b. Each reaction is slightly uphill because the O atom in the products does not handle the positive charge as well as the N atom in the reactants does. The H_2O and H_3O^+ species are crossed out because they appear in both reactions. The remaining CH_3NH_2 and H_3N molecules appear at the same place on the product side because they are both neutrals. On the reactant side, $CH_3NH_3^+$ is lower in energy than H_4N^+ because the CH_3 group stabilizes the positive charge by inductive effects. $CH_3NH_3^+$ is therefore the weaker acid because it acts as an acid in the more uphill proton transfer reaction.

Just as we did previously, we can then ignore the contributions by H_3O^+ on the product side and H_2O on the reactant side because these species appear in both reactions. This is indicated in Figure 6-11 by crossing out the respective species.

The remaining CH_3NH_2 and H_3N molecules appear at the same place on the product side of the diagram because they are both uncharged and are therefore assumed to have about the same stability. We can determine the relative stabilities of the remaining H_4N^+ and $CH_3NH_3^+$ ions on the reactant side using the tie-breaking questions from Section 6.6. Question 1 results in a tie because the two species have the same +1 charge. Question 2 also results in a tie because the positive charge appears on N in both cases. Also, question 3 results in a tie because neither $CH_3NH_3^+$ nor H_4N^+ has an additional resonance structure that would serve to share the positive charge over multiple atoms. The tie is broken with question 4, which pertains to inductive effects. The CH_3 group is electron-donating, which helps to stabilize nearby positive charges, so $CH_3NH_3^+$ is more stable than H_4N^+.

Having constructed the free-energy diagrams, we can now determine the stronger acid. Notice that both reactions are uphill, but the reaction in which H_4N^+ acts as the acid is the one that is less uphill, making H_4N^+ the stronger acid.

Let's now turn our attention to determining the stronger of two bases. The approach to doing so is exactly the same as the approach to determining relative acid strengths except for one difference—what we write out for the complete reactions. Suppose, for example, we are asked to determine the stronger base, $CH_3CH_2NH^-$ or CH_3CONH^-. The first step is to write out the complete proton transfer reactions in which the two species act as bases—that is, pick up a proton. For these reactions, as with any proton transfer reactions, an acid must be present to donate a proton. But since our focus is on the different bases, we choose the acids to be the same—water. The reactions are shown in Equations 6-3a and 6-3b.

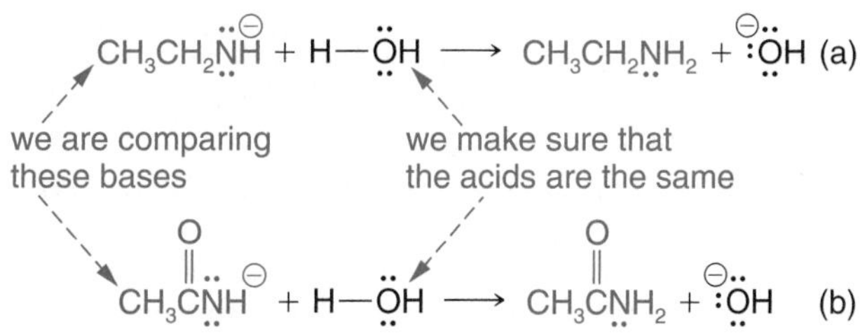

(6-3)

TIME TO TRY

Add the curved arrow notation for the reactions in Equation 6-3.

To construct the free-energy diagrams for these reactions, notice first that in each reaction a negative charge appears in the reactants as well as the products. Therefore, the reactants and products should appear at roughly the same energy, as shown in Figure 6-12. However, because O can handle a negative charge better than N (greater electronegativity), the products are more stable than the reactants.

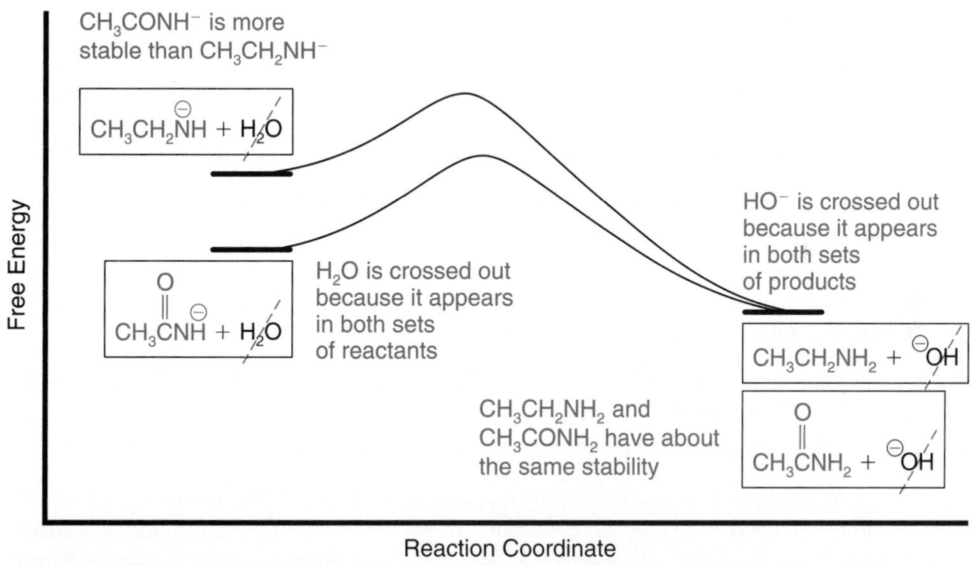

FIGURE 6-12 Free-energy diagrams for the reactions in equations 6-3a and 6-3b. Each reaction is slightly downhill because O can handle a negative charge better than N can. The H_2O and HO^- species are crossed out because they appear in both reactions. The remaining $CH_3CH_2NH_2$ and CH_3CONH_2 molecules appear at the same place on the product side because they are both uncharged. On the reactant side, the CH_3CONH^- ion is lower in energy than the $CH_3CH_2NH^-$ ion because the charge in the former is resonance-delocalized. $CH_3CH_2NH^-$ is therefore the stronger base because it acts as a base in the more downhill proton transfer reaction.

Next, we can cross out H_2O and HO^- in each reaction because they are common to both reactions. This leaves $CH_3CH_2NH^-$ and CH_3CONH^- on the reactant side and $CH_3CH_2NH_2$ and CH_3CONH_2 on the product side. The species on the product side are uncharged, so they appear at roughly the same energy in the diagram. The species on the reactant side, however, are ions, and their stabilities can be distinguished by the tie-breaking questions from Section 6.6. Both species have the same -1 charge, so question 1 results in a tie. Both charges appear on N, so question 2 also results in a tie. The tie is broken with question 3 because, whereas $CH_3CH_2NH^-$ has no resonance structures, CH_3CONH^- has a resonance structure in which the negative charge is shared on O. This makes CH_3CONH^- more stable. Therefore, we can see in the figure that the reaction in which $CH_3CH_2NH^-$ acts as a base is the more downhill reaction, making $CH_3CH_2NH^-$ the stronger base.

TIME TO TRY

Draw the other resonance structure for CH_3CONH^-, mentioned above.

A related problem in organic chemistry is one that asks you to determine the most acidic or the most basic site of a molecule. Suppose, for example, that we are asked to determine the most basic site of $HOCH_2CH_2NH_2$. This is very similar to the type of problem in which we are asked to determine the stronger of two bases—except that in this case the bases are both part of the same molecule.

As we have done previously, we begin by writing the complete reactions, showing each site acting as a base and using water as an acid that is common to both reactions. These are shown in Equations 6-4a and 6-4b.

$$HOCH_2CH_2NH_2 + H{-}OH \rightarrow HOCH_2CH_2NH_3^+ + {}^-OH \quad \text{(a)}$$
$$HOCH_2CH_2NH_2 + H{-}OH \rightarrow H_2^+OCH_2CH_2NH_2 + {}^-OH \quad \text{(b)} \qquad (6\text{-}4)$$

TIME TO TRY

In Equations 6-4a and 6-4b, draw in the pertinent lone pairs of electrons, and add the curved arrow notation.

To construct the free-energy diagrams, notice that each reaction has two charged species in the products, but none in the reactants. Therefore, we expect both reactions to be significantly uphill, as shown in Figure 6-13. Next, realize that the reactants are exactly the same in each reaction, so they appear at precisely the same energy in the diagram. On the product side, the HO^- species are crossed out because they appear in both reactions. The difference thus comes down to the relative stabilities of the different product ions, $HOCH_2CH_2NH_3^+$ and $H_2^+OCH_2CH_2NH_2$. Because N can handle a positive charge better than O, the first ion is the more stable one. Finally, with the free-energy diagrams constructed, we can see that the N atom is the more basic site because it accepts a proton in the less uphill reaction.

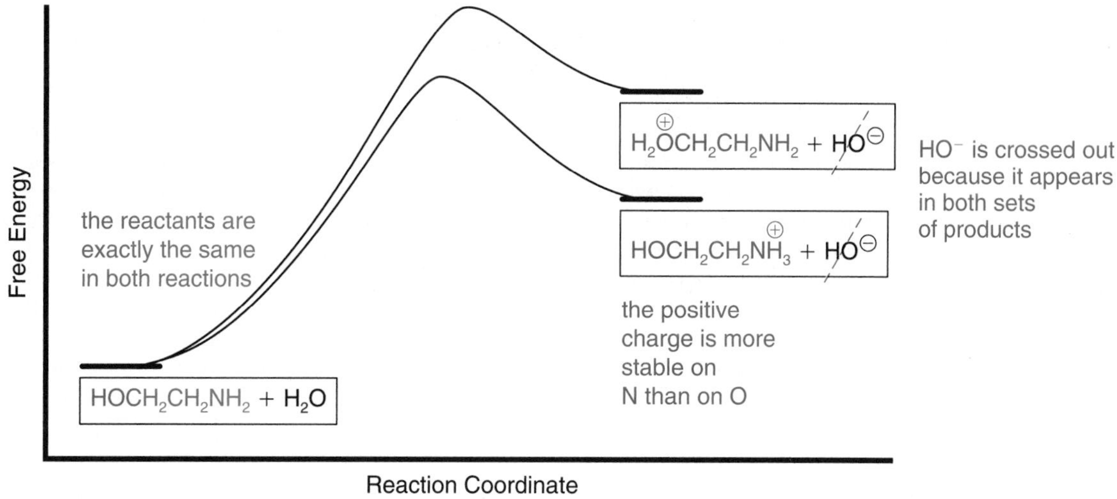

FIGURE 6-13 Free-energy diagrams for the reactions in equations 6-4a and 6-4b. Each reaction is substantially uphill because each set of products has two more charges than the reactants. The reactants are exactly the same in both cases, so they appear at precisely the same energy. The HO^- species on the product side are crossed out because they appear in both reactions. The $HOCH_2CH_2NH_3^+$ product ion is lower in energy than the $H_2^+OCH_2CH_2NH_2$ product ion because N can handle a positive charge better than O can. The N atom in the reactant is therefore the more basic site because it can accept a proton in the less uphill proton transfer reaction.

6.8 Application: Strengths of Nucleophiles and the Hammond Postulate

Section 5.4 introduced us to *nucleophiles*, which are identified as species that have an atom with both a partial or full negative charge and a lone pair of electrons. Thus, as we saw, a nucleophile tends to seek out and form a bond to an atom that has a partial or full positive charge. This is exemplified by what we see in an S_N2 reaction, shown generically in Equation 6-5.

$$Nu:^- + R—L \rightarrow Nu—R + :L^- \qquad (6\text{-}5)$$

Here Nu^- represents the nucleophile, R—L represents the substrate, and L^- is the leaving group.

TIME TO TRY

In Equation 6-5, write δ^+ next to the electron-poor atom, and add the curved arrow notation.

Just like acids and bases, nucleophiles have different strengths. However, the strength of a nucleophile is defined differently than acid or base strength. Whereas the strength of an acid or base corresponds to how uphill or downhill a proton transfer reaction is, the strength of a nucleophile, called its **nucleophilicity**, corresponds to the *rate* at which the nucleophile undergoes an S_N2 reaction.

For two nucleophiles that participate in an S_N2 reaction with the same substrate, R—L, the nucleophile that reacts *faster* is the stronger nucleophile.

There are lists of relative nucleophilicities that you can memorize, but it is much more powerful (and much less frustrating) to use your understanding of charge stability to learn relative nucleophile strengths. To do so, recall from Section 5.2A that the rate of an elementary step depends on the size of the energy barrier between reactants and products—the smaller the energy barrier, the faster the reaction. Therefore, determining the stronger of two nucleophiles is simply a matter of determining which reaction has the smaller energy barrier.

From the previous section, we know how to use charge stability to predict which reaction is the more downhill (or less uphill) one, but we have yet to see how to predict which has the smaller energy barrier. Fortunately, a version of the *Hammond postulate* provides a very useful correlation.

THE HAMMOND POSTULATE

For two reactions that proceed by the same elementary step, the one that is more downhill (or less uphill) tends to have the smaller energy barrier.

To give you a feel for why this should be so, you can use a strip of paper to represent a reaction free-energy diagram, as illustrated in Figure 6-14. If you bend the paper strip as shown in Figure 6-14a, the representation is of an uphill reaction. The energy barrier is represented by the height difference between the top of the paper and the left side, as indicated by the vertical double-headed arrow. Holding the left side in place, as you lower the right side, the representation is of an increasingly more downhill reaction. As that happens, notice that the representation of the energy barrier becomes smaller.

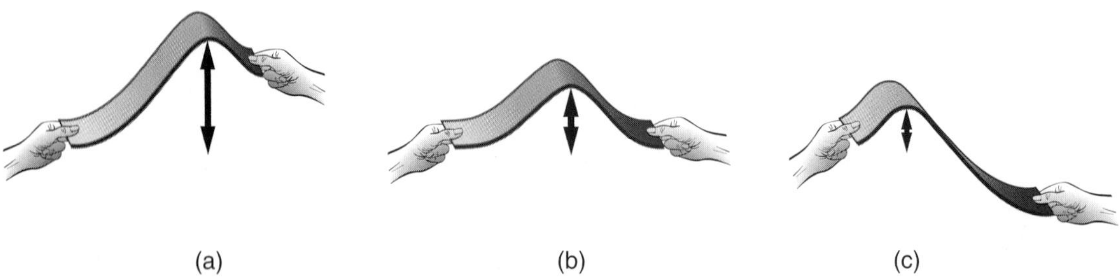

FIGURE 6-14 **The Hammond postulate.** A strip of paper represents the free-energy diagram for an elementary step. In going from (a) to (b) to (c), the right side of the paper is lowered, representing an increasingly more downhill reaction. As that happens, the height difference between the top of the paper and the left side becomes smaller, representing a decrease in the size of the energy barrier.

TIME TO TRY

Carry out the "experiment" shown in Figure 6-14 to see for yourself!

LOOK OUT

You must be careful when applying the Hammond postulate to chemical reactions. As noted, it should be used only to compare reactions that proceed by the same elementary step. When comparing elementary steps that are different, there are many cases where the more downhill reaction is slower. Moreover, some exceptions to our general rule do exist, which typically involve *steric hindrance* (recall that we introduced steric hindrance in Section 5.11). We will examine a few of these exceptions later in this book. By and large, though, the Hammond postulate is remarkably accurate.

With the Hammond postulate in mind, we can now make the following generalization about relative nucleophile strengths.

For two nucleophiles that participate in an S_N2 reaction with the same substrate, R—L, the more downhill (or less uphill) reaction tends to be the one that involves the stronger nucleophile.

Therefore, to determine the more downhill of two reactions, we simply carry out the same exercise we used to determine relative acid and base strengths.

As an example, let's determine the relative nucleophile strengths of $CH_3CH_2O^-$ and $CH_3CO_2^-$. Just as we did in predicting relative acid and base strengths, *we must write out the whole reactions.* The two reactions to compare are shown in Equations 6-6a and 6-6b.

$$CH_3CH_2O^- + R{-}L \rightarrow CH_3CH_2O{-}R + L^- \quad (a)$$
$$CH_3CO_2^- + R{-}L \rightarrow CH_3CO_2{-}R + L^- \quad (b) \qquad (6\text{-}6)$$

TIME TO TRY

In Equations 6-6a and 6-6b, draw in the pertinent lone pairs of electrons, and add the curved arrow notation.

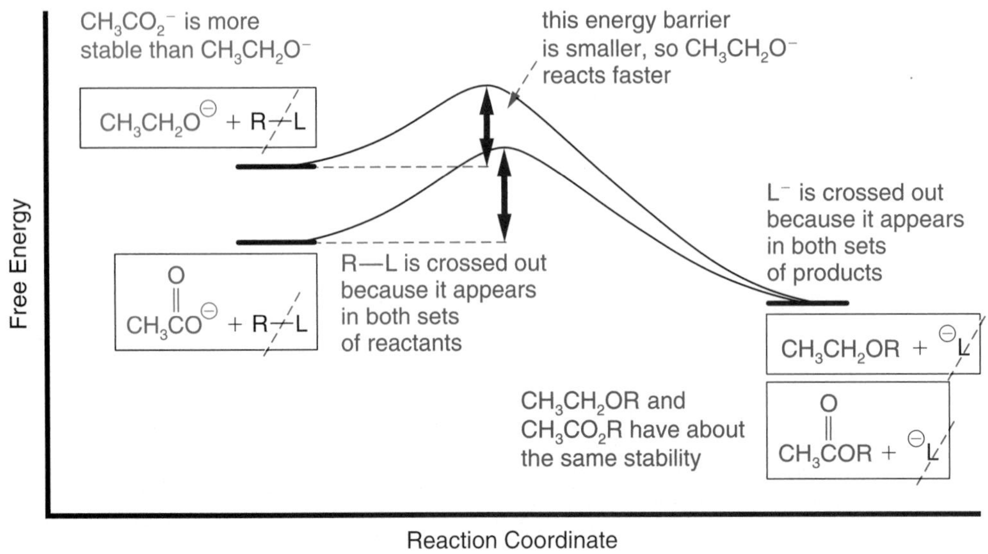

FIGURE 6-15 **Relative strengths of the nucleophiles in equations 6-6a and 6-6b.** The substrates, R—L, and the leaving groups, L^-, are crossed out because they are the same in both reactions. Of the remaining anions on the reactant side, $CH_3CO_2^-$ is more stable due to resonance delocalization of the charge. The remaining products are assumed to have roughly the same stability because they are both uncharged. Because the reaction involving $CH_3CH_2O^-$ as the nucleophile is more downhill, the energy barrier is smaller (according to the Hammond postulate), so the reaction proceeds faster, making $CH_3CH_2O^-$ the stronger nucleophile.

To construct the free-energy diagrams, begin by noticing that in each reaction there is one negatively charged species on the reactant side and one on the product side. Therefore, the reactants and products are expected to have similar energies, as shown in Figure 6-15. (Depending on the exact identity of L^-, the reactions could be somewhat downhill or somewhat uphill.) Next, notice that we can ignore R—L on the reactant side, as well as L^- on the product side, because these species appear in both reactions. As such, these species are crossed out in Figure 6-15. The two uncharged products that remain are taken to have about the same stability, which is why they appear at the same energy in the figure. The negatively charged reactants that remain can be distinguished by the tie-breaking questions from Section 6.6. Both have the same −1 charge, so question 1 results in a tie. Both charges appear on an O atom, so question 2 also results in a tie. The tie is broken with question 3 because $CH_3CO_2^-$ has a resonance structure that allows the negative charge to be shared over two O atoms, whereas $CH_3CH_2O^-$ does not. (This was previously shown in Figure 6-1a.) Consequently, $CH_3CO_2^-$ is the more stable ion, such that $CH_3CH_2O^-$ is the nucleophile that is involved in the more downhill S_N2 reaction. According to the Hammond postulate, then, $CH_3CH_2O^-$ will have the smaller energy barrier and will therefore react faster, making it the stronger nucleophile.

What about two uncharged nucleophiles, such as H_2O and H_2S? We begin by writing the reactions that show each species acting as a nucleophile (Equations 6-7a and 6-7b).

$$H_2O + R{-}L \rightarrow H_2O^{+}{-}R + L^{-} \quad (a)$$
$$H_2S + R{-}L \rightarrow H_2S^{+}{-}R + L^{-} \quad (b) \qquad (6\text{-}7)$$

TIME TO TRY

In Equations 6-7a and 6-7b, draw in the pertinent lone pairs of electrons, and add the curved arrow notation.

In drawing the free-energy diagrams (Figure 6-16), notice that each set of products is significantly less stable than each set of reactants due to the two additional charges in the products. Next, we cross out R—L and L^- because they appear in both reactions. The remaining nucleophiles on the reactant side appear at the same energy because they are both uncharged. On the product side, we can distinguish the stabilities of the remaining positively charged ions,

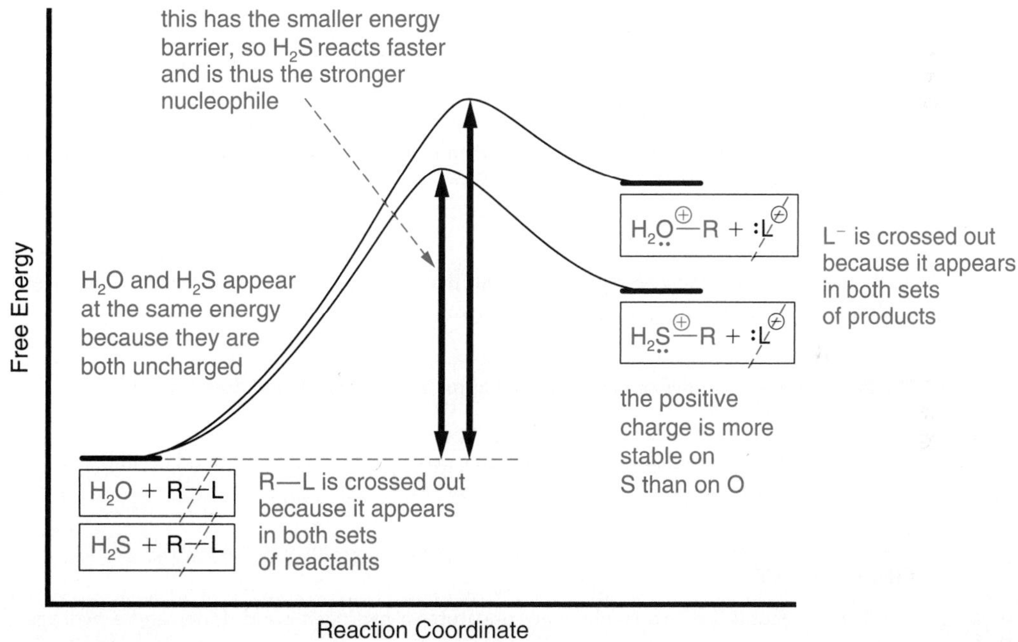

FIGURE 6-16 **Relative strengths of the nucleophiles in equations 6-7a and 6-7b.** The products are substantially higher in energy than the reactants because of the additional charges. The substrates, R—L, and the leaving groups, L^-, are crossed out because they are the same in both reactions. The remaining nucleophiles are assumed to have roughly the same stability because they are both uncharged. Of the remaining positively charged ions on the product side, H_2S^+—R is more stable because the S atom is larger and can thus handle a charge better. Because the reaction involving H_2S as the nucleophile is less uphill, the energy barrier is smaller (according to the Hammond postulate), so H_2S reacts faster, making it the stronger nucleophile.

using the tie-breaking questions from Section 6.6. Both species have the same +1 charge, so question 1 results in a tie. The tie is broken by question 2 because the charge appears on different atoms. The S atom is larger, so it can better handle the charge, making $H_2S^+{-}R$ more stable than $H_2O^+{-}R$. With this, we can see that the reaction involving H_2S as the nucleophile is less uphill than the one involving H_2O as the nucleophile. Therefore, H_2S will react faster, making it the stronger nucleophile.

6.9 Application: Solvent Effects on Nucleophile Strength

The previous section showed you how to determine the relative strength of a nucleophile based on only its Lewis structure. There we applied what we knew about relative stabilities of charges to determine relative strengths of nucleophiles. However, in determining the relative nucleophile strengths in this manner, we did not consider the roles solvent molecules might play. And these roles of solvents—**solvent effects**—are important because most organic reactions take place in some variety of solvent.

WHY SHOULD I CARE?

The type of solvent in which a reaction is carried out can have a dramatic effect on the outcome of that reaction. This is perhaps most evident with the competition that takes place among the reactions we will examine in Chapter 8. When two reagents are mixed, one solvent could lead to the formation of one set of products, and a different solvent could lead to the formation of another set of products. It is important to understand the origin of such solvent effects in order to know the roles they play in chemical reactions.

There are essentially two types of solvents that we must consider when determining nucleophile strength: (1) protic solvents and (2) aprotic solvents.

Protic solvents are defined as solvents that have a hydrogen-bond donor.

Aprotic solvents are defined as solvents that *do not* have a hydrogen-bond donor.

TIME TO TRY

Review Section 3.6D to recall what constitutes a hydrogen-bond donor and a hydrogen-bond acceptor.

Examples of protic solvents are H_2O, CH_3OH, CH_3CH_2OH, and $CH_3CH_2NH_2$. In each of those molecules, the H atom that is covalently bonded to the O or N atom bears a large δ^+ as a result of the high electronegativity of O and N atoms. Furthermore, that H atom is very tiny, so it can approach other species closely without steric hindrance being a problem. Therefore, if

FIGURE 6-17 Solvation of an anion by a protic solvent. The HO^- ion is solvated by H_2O as a result of the many ion-dipole interactions between HO^- and the individual molecules of H_2O. Because each H has a large partial positive charge and is very small, such ion-dipole interactions stabilize HO^- very strongly.

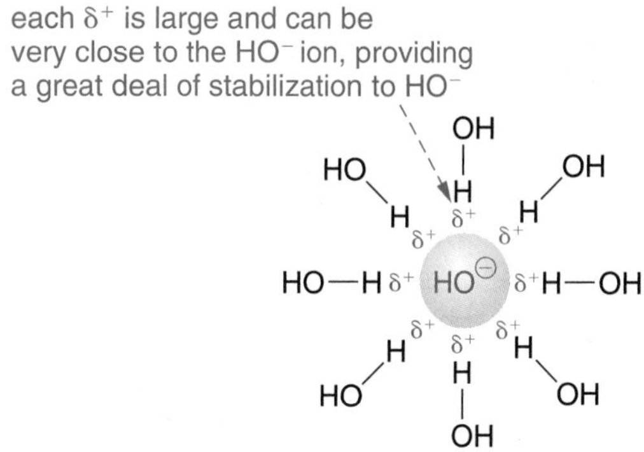

a negatively charged nucleophile, such as HO^-, is dissolved in a protic solvent, such as H_2O, then the H_2O molecules will stick to the HO^- ion *very* strongly. This is a result of the ion–dipole interaction between the full negative charge on HO^- and the large δ^+ on the H atom in H_2O. The end result is that each HO^- ion will have several H_2O molecules strongly attached to it, as illustrated in Figure 6-17. In other words, the HO^- ion is *strongly solvated* by H_2O and is therefore highly stabilized.

TIME TO TRY

We first discussed solvation in Section 3.9. Review that discussion to recall the basics of solvation.

A similar phenomenon happens with other protic solvents and other anions.

Protic solvents tend to solvate, and therefore stabilize, anions *very* strongly.

With this in mind, recall from Section 6.8 that a nucleophile's strength corresponds to the rate at which it can undergo an S_N2 reaction—the stronger the nucleophile, the faster the S_N2 reaction. Further, by applying the Hammond postulate, we can identify the stronger of two nucleophiles as the one that promotes the more downhill (or less uphill) S_N2 reaction. Suppose, then, that we want to compare the relative nucleophile strengths of HO^- in a protic solvent and in a situation in which there is no solvent. As we did in Section 6.8, we write the specific reaction to examine (Equation 6-8).

$$HO^- + R{-}L \rightarrow HO{-}R + L^- \qquad (6\text{-}8)$$

With no solvent, the reaction will proceed with some energy barrier, as shown by the dashed curve of Figure 6-18. In the presence of a protic solvent, however, HO^- is stabilized strongly by solvation. Thus, the reactant energy is lowered substantially, making the reaction less downhill, as indicated by the solid curve. (You might notice that L^- would also be stabilized, but

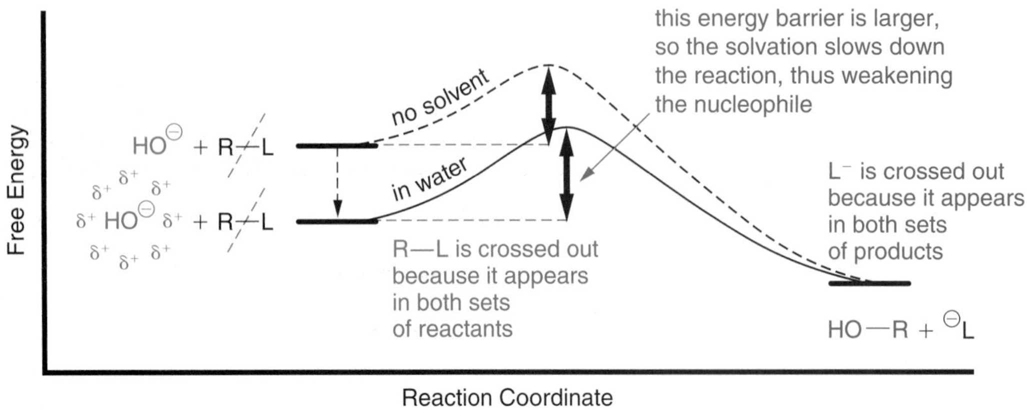

FIGURE 6-18 Nucleophile strengths in protic solvents. The dashed curve represents the free-energy diagram of the S_N2 reaction in Equation 6-8 with no solvent. The solid curve represents the free-energy diagram of the same reaction taking place in water, a protic solvent. In water, HO^- is strongly solvated (represented by the partial positive charges) and therefore strongly stabilized. The result is a lowering of the reactant energy (indicated by the dashed, thin black arrow), and a corresponding increase in the size of the energy barrier. Thus, water serves to slow the S_N2 reaction, meaning that the nucleophile has been weakened.

typical leaving groups are not solvated as well as nucleophiles are, so the product energies have been left alone.) According to the Hammond postulate, the energy barrier should increase as a result, thus slowing down the reaction. This corresponds to a *weakening* of the nucleophile.

The above result is typical of negatively charged nucleophiles in protic solvents.

Protic solvents tend to substantially weaken negatively charged nucleophiles.

However, this weakening of the nucleophile strength is not the same for all negatively charged nucleophiles.

The weakening of a negatively charged nucleophile by a protic solvent is much more pronounced when the negative charge appears on a smaller atom than when it appears on a larger atom.

Why should this be so? The reason is that having a negative charge on a smaller atom represents a significantly greater *concentration* of negative charge. This dramatically increases the strength of each ion–dipole interaction and thus greatly enhances the effects of solvation.

It is important to know that, with the possibility of having such large differences in solvation, the relative strengths of some negatively charged nucleophiles in protic solvents are actually different from what they are with no solvent. Consider, for example, HO^- versus HS^-. The S_N2 reaction to consider for HO^- was shown previously in Equation 6-8. The one for HS^- is shown in Equation 6-9.

$$HS^- + R{-}L \rightarrow HS{-}R + L^- \tag{6-9}$$

With no solvent, we can determine the relative strengths of the two nucleophiles as we did in Section 6.8. This is shown in Figure 6-19 (the corresponding curves are labeled "no solvent"). On the reactant side, we can cross out R—L, and on the product side, we can cross out L^-. Further, we can assume that R—SH and R—OH have about the same stability because they are both uncharged. Therefore, the main difference between the two reactions comes down to HS^- versus HO^-. Because S is larger than O, HS^- is more stable than HO^-. Consequently, as we can see with the dashed curves in Figure 6-19, the reaction with HO^- is more downhill, making HO^- the stronger nucleophile in the absence of solvent.

In a protic solvent, however, the story is a little different. From what we concluded above, we know that HO^- will be solvated much more strongly than HS^- (because O is a much smaller atom than S). This corresponds to a much greater stabilization of HO^- than of HS^-. As a result, HO^- actually becomes more stable than HS^-, as shown by the solid curves in Figure 6-19, so HS^- is the nucleophile that participates in the more downhill S_N2 reaction. Thus, in a protic solvent, HO^- is a weaker nucleophile than HS^-.

Such a reversal of relative nucleophile strengths is not limited to HO^- and HS^-.

In general, a protic solvent will cause the reversal of the relative nucleophile strengths of two negatively charged nucleophiles if the negative charges appear on different atoms from the same column of the periodic table.

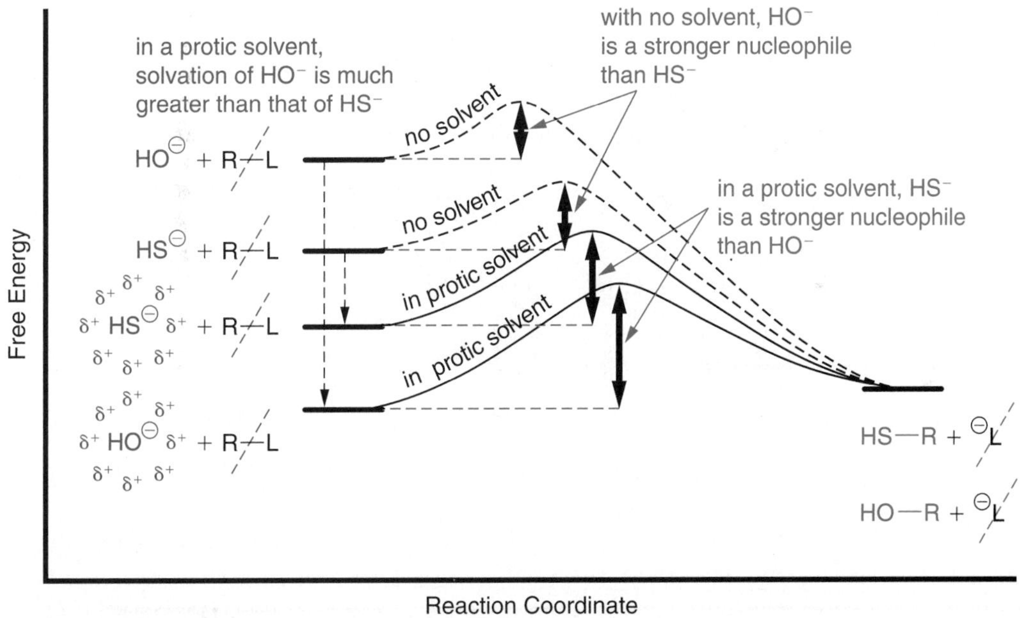

FIGURE 6-19 Relative nucleophile strengths of HO^- and HS^- with and without a protic solvent. The dashed curves represent the S_N2 reactions in Equations 6-8 and 6-9 taking place with no solvent. The solid curves represent the same reactions taking place in a *protic* solvent. With no solvent, HS^- is more stable than HO^-, making HO^- the stronger nucleophile. In a protic solvent, HO^- is more stable than HS^- as a result of the much stronger solvation of HO^-, making HS^- the stronger nucleophile.

✔ **QUICK CHECK**

With no solvent, F^- is a stronger nucleophile than Cl^-, and H_3P is a stronger nucleophile than H_3N. In water, which nucleophile in each pair is stronger?

Answer: Because water is a protic solvent, we expect a reversal of the relative nucleophile strengths of F^- and Cl^-, given that a negative charge is on atoms in different rows of the same column of the periodic table. Therefore, Cl^- is stronger. We do not expect any reversals for H_3P and H_3N because both are uncharged.

What about the situation in which two nucleophiles have a negative charge on different atoms from the same row of the periodic table? For example, consider HO^- and F^-. Absent any solvent, HO^- is a stronger nucleophile than F^-, as shown by the dashed curves in Figure 6-20—being in the same row, the O and F atoms are similar in size, but F^- is the more stable anion due to the greater electronegativity of F.

In the presence of a protic solvent, the anions are both strongly stabilized by solvation, yielding the solid curves in Figure 6-20. In this case, however, that stabilization is similar in magnitude for the two nucleophiles (indicated by similar lengths of the downward-pointing thin black arrows), because the O and F atoms are similar in size—that is, the *concentration* of charge is about the same on O and F. Therefore, in the protic solvent, F^- remains more stable than HO^-, and as such, HO^- remains the stronger nucleophile.

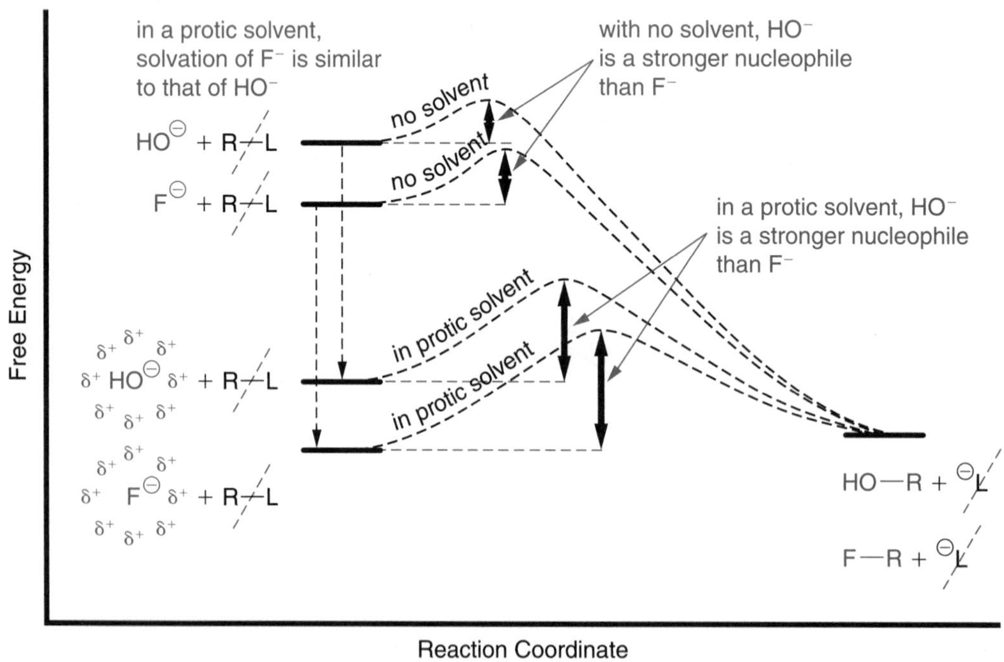

FIGURE 6-20 **Relative nucleophile strengths of HO^- and F^- with and without a protic solvent.** The dashed curves represent the S_N2 reactions involving HO^- and F^- as the nucleophiles, taking place with no solvent. The solid curves represent the same reactions taking place in a protic solvent. With no solvent, F^- is more stable than HO^-, making HO^- the stronger nucleophile. In a protic solvent, HO^- and F^- are both solvated strongly, but roughly to the same extent, because the O and F atoms are similar in size. Therefore, F^- remains more stable than HO^-, and HO^- remains the stronger nucleophile.

The specific example of HO^- and F^- can be generalized as follows.

In general, a protic solvent will *not* cause the reversal of the relative nucleophile strengths of two negatively charged nucleophiles if the negative charges appear on different atoms in the same row of the periodic table.

QUICK CHECK

Absent any solvent, HS^- is a stronger nucleophile than Cl^-. Which one is a stronger nucleophile with CH_3CH_2OH as the solvent?

Answer: The solvent is ethanol, which is protic. Therefore, both HS^- and Cl^- will be solvated strongly, but roughly to the same extent. Thus, no reversals are expected, so HS^- should remain the stronger nucleophile.

As another example, let's look at similar nucleophiles that differ in their charges—H_2N^- and H_3N. Absent any solvent, we know that H_2N^- is the much stronger nucleophile, stemming from the fact that H_2N^- is much less stable than H_3N. In a protic solvent, H_2N^- is indeed stabilized quite a lot from solvation, just as we have seen with other negatively charged nucleophiles. By comparison, H_3N is not solvated much because it is uncharged. However, that difference in solvation is not enough to make H_2N^- more stable than H_3N, so no reversal of nucleophile strength takes place.

In general, negatively charged nucleophiles will be stronger than similar uncharged nucleophiles, both absent any solvent and in a protic solvent.

This is a testament to how important the presence of a charge is in the stability of a species.

Let's now turn our attention to *aprotic solvents*—solvents that do not have hydrogen-bond donors. Some of the most common aprotic solvents you will encounter are **acetone** [$(CH_3)_2C{=}O$], **dimethyl sulfoxide** [DMSO, $(CH_3)_2S{=}O$], and ***N,N*-dimethylformamide** [DMF, $(CH_3)_2NCH{=}O$]. These are shown below. Notice that in each of these compounds no H atoms are covalently bonded to any of the three highly electronegative atoms (N, O, and F).

common aprotic solvents

none of these compounds has a hydrogen-bond donor (NH, OH, or FH)

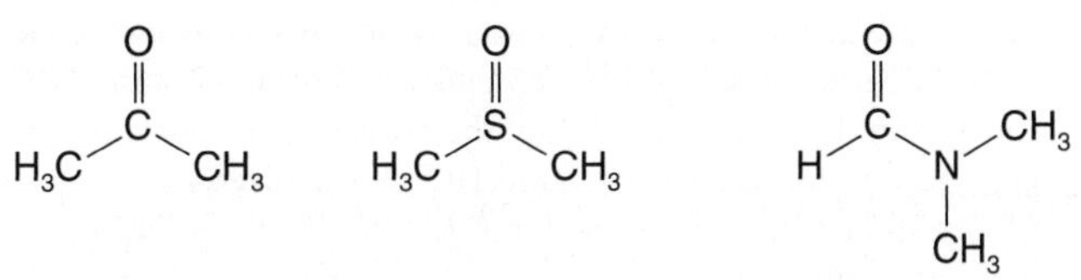

acetone | dimethyl sulfoxide (DMSO) | *N,N*-dimethylformamide (DMF)

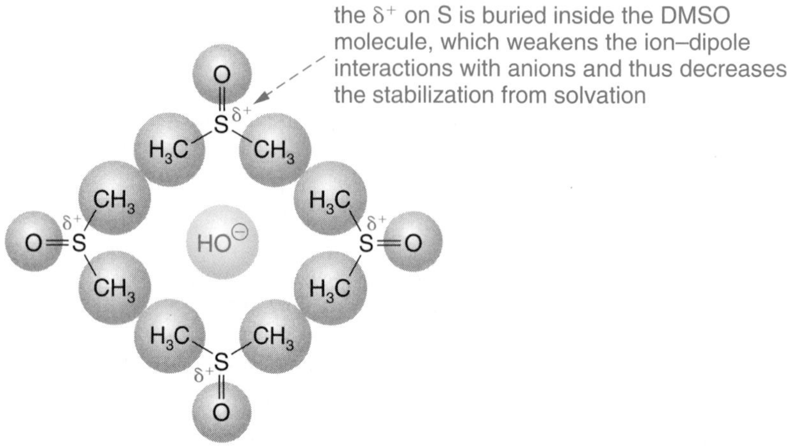
the δ+ on S is buried inside the DMSO molecule, which weakens the ion–dipole interactions with anions and thus decreases the stabilization from solvation
HO⊖
O=S
S=O
H3C
CH3
δ+

FIGURE 6-21 **The diminished solvation by aprotic solvents.** The DMSO molecule has its δ^+ located on S, which is not very accessible to the HO^- anion. Therefore, DMSO does not solvate HO^- very well.

These molecules do, however, all have net molecular dipoles. Therefore, as a result of ion–dipole interactions, each of these solvent molecules will stick fairly well to an anion like HO^-. Consequently, ionic compounds are capable of dissolving in these solvents. But these solvent molecules will not stick nearly as well as the molecules of protic solvents will. The reason has to do with how accessible the relevant δ^+ is. Recall that in protic solvents there is a large δ^+ located on the H atom of a hydrogen-bond donor. And because that H atom is very small, the δ^+ is very easily accessible—it can get very close to a negatively charged atom. This is not true for any of the aprotic solvents—in each case, the important δ^+ is located on a non-H atom, so it is essentially buried inside the molecule. For example, in DMSO, the δ^+ is located on S, which is surrounded by two CH_3 groups and the O atom. Therefore, as shown in Figure 6-21, this makes it difficult for negatively charged species to approach the δ^+, which, in turn, makes the ion–dipole interactions relatively weak.

The specific example with DMSO and HO^- can be generalized as follows.

Aprotic solvents typically do not solvate anions very strongly and therefore do not tremendously weaken nucleophiles.

A consequence of this can be seen by examining the relative nucleophile strengths of HO^- and HS^- with no solvent and in the presence of an aprotic solvent. As shown by the dashed curves in Figure 6-22 (and as shown previously by the dashed curves in Figure 6-19), HO^- is the stronger nucleophile in the absence of a solvent because the O atom is much smaller than the S atom. In an aprotic solvent, the moderate solvation that takes place serves to stabilize each nucleophile somewhat—and HO^- is stabilized more than HS^- due to the smaller size of O. Unlike what we saw with protic solvents, however, the extent of solvation in DMSO is not enough to make HO^- more stable than HS^-. Therefore, HO^- remains the stronger nucleophile. In general, then,

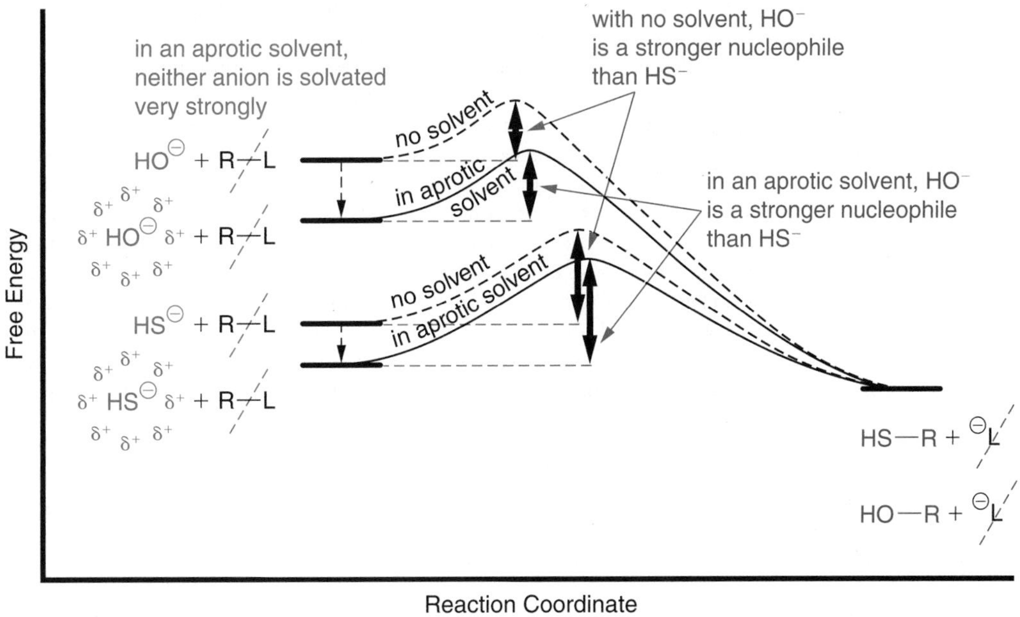

FIGURE 6-22 **Relative nucleophile strengths of HO^- and HS^- with and without an aprotic solvent.** The dashed curves represent the S_N2 reactions involving HO^- and HS^- as the nucleophiles, taking place with no solvent. The solid curves represent the same reactions taking place in an aprotic solvent. With no solvent, HS^- is more stable than HO^-, making HO^- the stronger nucleophile. In an aprotic solvent, HO^- and HS^- are solvated, but not very strongly because the δ^+ in the solvent molecule is not very accessible. Therefore, HS^- remains more stable than HO^-, so HO^- remains the stronger nucleophile.

Relative strengths of nucleophiles in aprotic solvents are the same as those observed in the absence of any solvent. Thus, aprotic solvents do not cause reversals of relative nucleophile strengths.

6.10 Application: The Best Resonance Contributor

Resonance was introduced in Chapter 2. There we mentioned that individual resonance contributors are imaginary species and that the real, true species is the resonance hybrid—an average of all of the resonance contributors of the species. However, in taking the average, not all of the resonance contributors are equal. It turns out, and should make sense, that

The resonance hybrid of a species looks most like the best (i.e., most stable) resonance contributor.

But, given a set of possible resonance contributors for a species, the question becomes, How do you determine which is the most stable one, so that you can figure out which one the hybrid behaves most like? In general, this involves evaluating differences in charge stability, so we can apply the tie-breaking questions from Section 6.6.

To demonstrate this, let's examine the three pairs of resonance contributors in Figure 6-23 and, in each case, determine which is the better contributor.

For the species in Figure 6-23a, we are given two resonance structures to compare. In the first structure, none of the atoms bears a formal charge. In the second structure, there are two formal charges—a $+1$ charge on N and a -1 charge on O. Thus, the tie is broken by question 1 of the tie-breaking questions from Section 6.6. Indeed, because *charge is bad*, we can say that the first structure is the better resonance contributor and that CH_3CONH_2 looks much more like the first structure than the second.

Both species in Figure 6-23b have a -1 formal charge, but the charge is on different atoms—C in the first structure and O in the second. Therefore, the tie is broken with question 2 from Section 6.6. Because O is more electronegative than C, O can handle the negative charge better, making the second structure the better resonance contributor. That is, the resonance hybrid looks more like the second resonance structure than the first.

Finally, let's examine the two resonance structures in Figure 6-23c. Questions 1 and 2 from Section 6.6 result in a tie because both resonance structures bear the same $+1$ charge on a C atom. Normally, we would proceed to question 3, which asks us to consider the delocalization

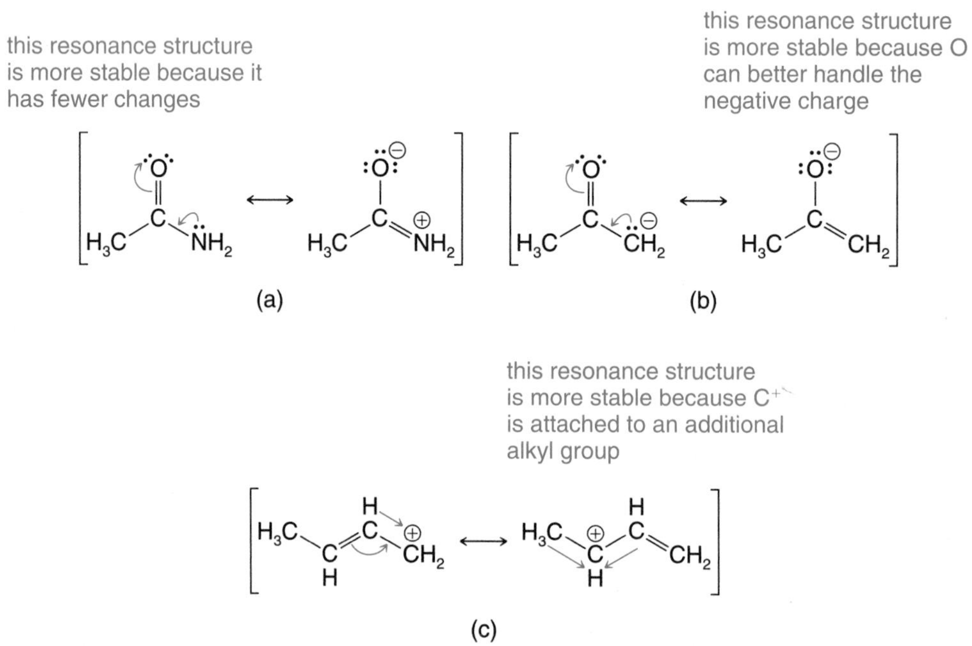
this resonance structure
is more stable because it
has fewer changes
this resonance structure
is more stable because O
can better handle the
negative charge
(a)
(b)
this resonance structure
is more stable because C+
is attached to an additional
alkyl group
(c)

FIGURE 6-23 **Relative stabilities of resonance structures.** (a) The first resonance structure is more stable because it has two fewer charges than the second structure. (b) The second resonance structure is more stable because O is more electronegative and can therefore handle a negative charge better than C. (c) The second resonance structure is more stable because there is an additional electron-donating alkyl group attached to C^+.

of the charge via resonance. However, in the problem at hand, we are examining *individual* resonance structures—in a particular resonance structure, we can think of resonance as being "turned off." Therefore, we skip question 3 and move to question 4, which has us consider inductive effects. As indicated in the figure, the C^+ in the second structure is attached to an additional alkyl group, which is electron-donating and therefore serves to stabilize the positive charge. Therefore, the second structure is the better resonance contributor.

WHAT DID YOU LEARN?

6.1 Two resonance contributors of CH_2OH^+ are provided. It turns out that the resonance contributor whose positive charge is on the O atom is a better resonance contributor than the one whose positive charge is on the C atom, despite the fact that C is less electronegative than O. Can you explain why? (*Hint:* Recall from Chapter 2 what makes a viable resonance contributor.)

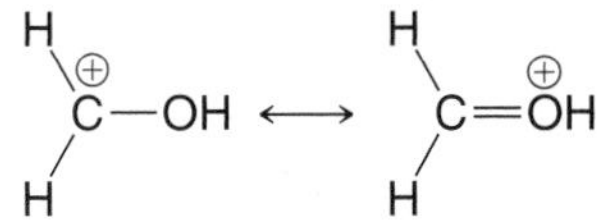

6.2 Rank the following in order of increasing base strength:

a F^-
b HO^-
c Br^-
d CH_3NH^-
e NH_2^-
f NH_3
g H_2O
h $CH_3CH_2^-$

6.3 Rank the following in order of increasing acid strength:

a $CH_3CH_2CH_2CO_2H$
b $CH_3CH_2CHClCO_2H$
c $CH_3CH_2CH(CH_3)CO_2H$
d $ClCH_2CH_2CH_2CO_2H$
e $CH_3CH_2CCl_2CO_2H$
f $CH_3CH_2CH(CH_3)CH_2OH$
g $CH_3CHClCH_2CO_2H$
h $CH_3CH_2CF_2CO_2H$

6.4 Based on arguments from this chapter, is HNO_3 a stronger or a weaker acid than CH_3CO_2H? (In HNO_3, N is bonded to each O, and H is bonded to one O.) (*Hint:* When the H^+ atom is removed from each acid, a -1 formal charge is generated on an O atom. For which species is that -1 charge better stabilized?)

6.5 Draw all resonance contributors of the following ion, and rank them in order of increasing contribution to the overall resonance hybrid. (*Hint:* There are four resonance contributors in all.)

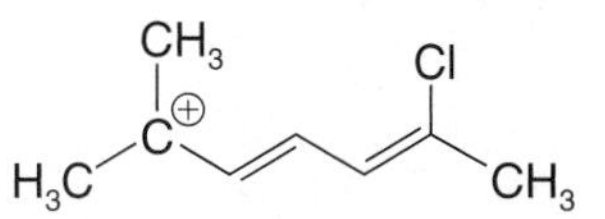

6.6 Which of the following molecules would you expect to be a stronger acid? (The potentially acidic hydrogens are explicitly shown.) Explain.

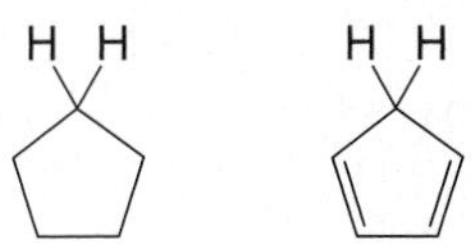

6.7 The first step in the Fischer esterification reaction (the topic of Section 7.7) is believed to be a proton transfer step in which a molecule such as the one below is protonated (accepts an H^+ ion). Which of the two O atoms in this molecule is more likely to be protonated? Why?

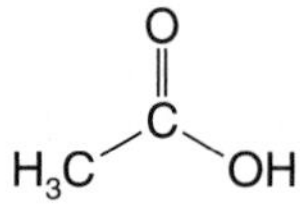

6.8 Following are two examples of *acid derivatives.*

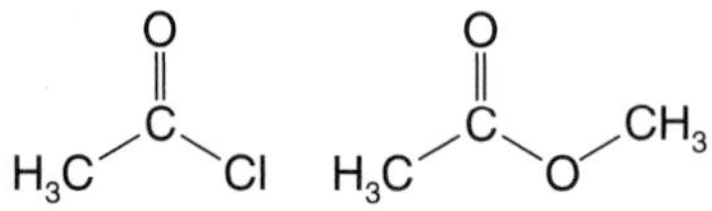

Both of these reagents can undergo nucleophilic substitution with a nucleophile, X^-, as shown below. If the nucleophile is the same in both reactions, which reaction do you expect to be faster? Why?

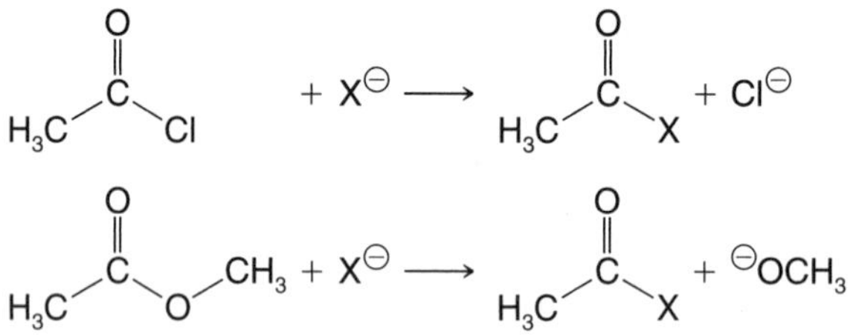

6.9 Rank the following species in order of increasing nucleophile strength, absent any solvent:

a Cl^- **b** H_2O **c** CH_3^- **d** $CH_3CO_2^-$ **e** H_2S **f** HS^- **g** I^- **h** $CH_3CH_2O^-$

6.10 Rank the species in the previous problem in order of increasing nucleophile strength with DMSO as the solvent.

6.11 One of the species in Problem 6.9 will be protonated by CH_3CH_2OH. Which one is it?

6.12 Except for the species identified in Problem 6.11, rank the species in Problem 6.9 in order of increasing nucleophile strength with CH_3CH_2OH as the solvent.

6.13 Rank the following heterolysis reactions in order of increasing reaction rate:

(a) 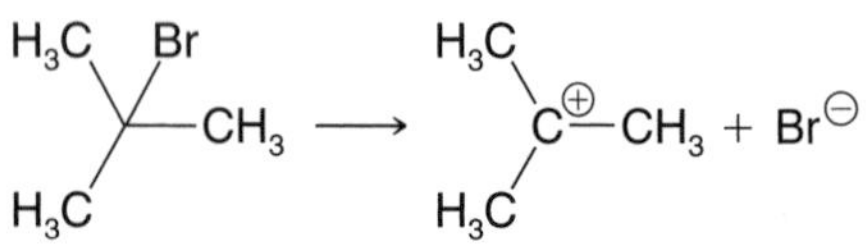

(b) 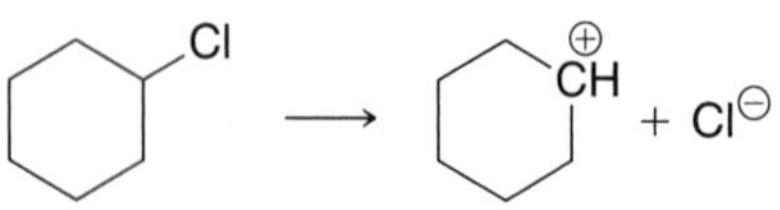

(c) 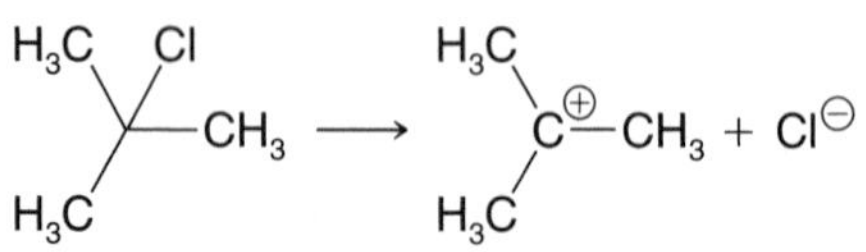

7 Reaction Mechanisms 2:

S_N1 and E1 Reactions and Rules of Thumb for Multistep Mechanisms

When you complete this chapter, you should be able to:

- Draw the mechanism of an S_N1 reaction and an E1 reaction.
- Recognize the difference between a reaction mechanism and an overall reaction and sum the steps of a multistep mechanism to determine the overall reaction.
- Identify overall reactants, overall products, intermediates, and transitions states of a reaction mechanism.
- Draw a reaction energy diagram that is consistent with the number of steps in a mechanism.
- Distinguish the products of an S_N1 reaction from those of an S_N2 reaction based on stereochemistry.
- Predict when a carbocation rearrangement will take place in a reaction and incorporate such a step into the reaction mechanism.
- Incorporate proton transfer steps into a reaction mechanism in order to ensure that the mechanism is reasonable.
- Explain why termolecular steps are unreasonable and avoid incorporating them into reaction mechanisms.
- Draw the mechanism for a keto-enol tautomerization reaction under neutral, basic, or acidic conditions and explain the relative reaction rates under the various conditions.
- Analyze a multistep mechanism as to why it may be reasonable, but other variations may not be.

Your Starting Point

Answer the following questions to assess your knowledge about multistep mechanisms.

1. What types of species appear in an overall reaction? What types of species do not? ______

2. How many transition states are in a mechanism consisting of four elementary steps? ______

3. When a planar C atom becomes a stereocenter, how many configurations of the C atom are produced? ______

4. Which is a more stable carbocation—one in which the positive charge is localized or one in which the positive charge is delocalized by resonance? One with two alkyl groups adjacent to the C^+ atom or one with three alkyl groups? ______

5. Classify each of the following as a strong base, weak base, strong acid, or weak acid: H_2O, HO^-, H_3O^+, HCl, H_2SO_4, NH_3, H_2N^-, NH_4^+, CH_3CO_2H. ______

6. What is a unimolecular step? A bimolecular step? A termolecular step? ______

7. When a species gains a formal charge, does it become more stable or less stable? ______

Answers: 1. Only overall reactants and overall products appear in an overall reaction. Intermediates and transition states should not appear. 2. Four transition states—one for each elementary step. 3. Two configurations are produced (*R* and *S*) because the new bond can be formed on either side of the plane. 4. A carbocation is more stable if the positive charge is resonance-delocalized and/or if it has more alkyl groups attached to C^+. 5. Strong bases: HO^-, H_2N^-. Weak bases: H_2O, NH_3. Strong acids: H_3O^+, HCl, H_2SO_4. Weak acids: H_2O, NH_4^+, CH_3CO_2H. 6. Unimolecular, bimolecular, and termolecular steps are defined as elementary steps that have, respectively, one, two, and three separate reactant species. 7. Charged species are inherently unstable, so if a species gains a formal charge, it becomes less stable.

7.1 Introduction

In Chapter 5, we were introduced to 10 of the most common elementary steps found in organic reaction mechanisms. We learned how to use curved arrow notation to keep track of what happens to valence electrons in such elementary steps, and with this notation, we further saw the tendency of electrons to flow from electron-rich sites to electron-poor sites. In Chapter 6, we examined charge stability, a major factor that governs the extent to which reactants want to become products.

In this chapter, we will continue with the discussion of reaction mechanisms, focusing specifically on how individual elementary steps can be combined in various ways to construct *multistep* mechanisms. In doing so, we will introduce two particular reactions as prototypes—S_N1 and E1 reactions. We will see that describing such reactions as multiple steps rather than a single step will have significant impacts on the free-energy diagrams as well as on the stereochemistry of those reactions.

Using S_N1 and E1 reactions as models, we will also introduce rules of thumb for multistep mechanisms. Perhaps the most important of these deals with ways in which proton transfer steps can be incorporated into multistep mechanisms. Another important rule of thumb deals with carbocation rearrangements. Ultimately, as we will see in the applications at the end of this chapter, these rules of thumb apply to a variety of other multistep reactions as well.

7.2 Elementary Steps as Part of Multistep Mechanisms: S_N1 and E1 Reactions

Two important elementary steps that were introduced in Chapter 5 are the bimolecular nucleophilic substitution (S_N2) and the bimolecular elimination (E2) steps. Recall that in an S_N2 step a nucleophile (Nu^-) replaces the leaving group (L) on a substrate (R—L). As shown in Equation 7-1, the Nu—R bond is formed at the same time that the R—L bond breaks. However, the same overall reaction can take place in a *stepwise* fashion, as shown in Equation 7-2. In such a case, the leaving group can depart first, in a heterolysis step (Equation 7-2a), resulting in the formation of a carbocation, R^+. In a second step, the nucleophile can form a bond to R^+, in a coordination step (Equation 7-2b), yielding Nu—R as a product.

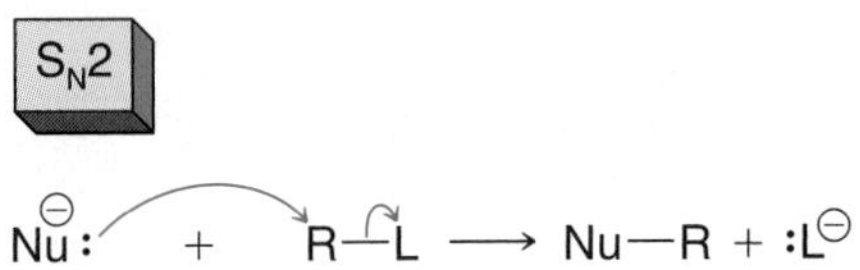

(7-1)

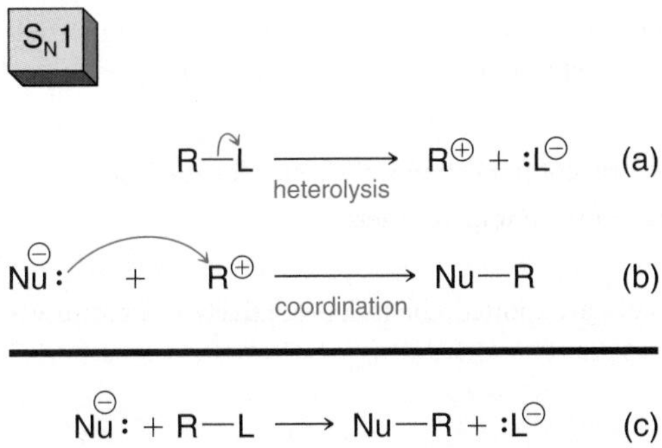

(7-2)

Similarly, recall that in an E2 step (Equation 7-3) a base removes a proton from the substrate at the same time the leaving group departs, resulting in the formation of a new C=C double bond. Such a reaction, too, can take place stepwise, as shown in Equation 7-4. In the first step (Equation 7-4a), the leaving group departs, in a heterolysis step, just as in Equation 7-1. The second step (Equation 7-4b) is an electrophilic elimination, whereby H^+ departs from the carbocation to form the new C=C double bond, and simultaneously a base forms a bond to that H^+.

E2

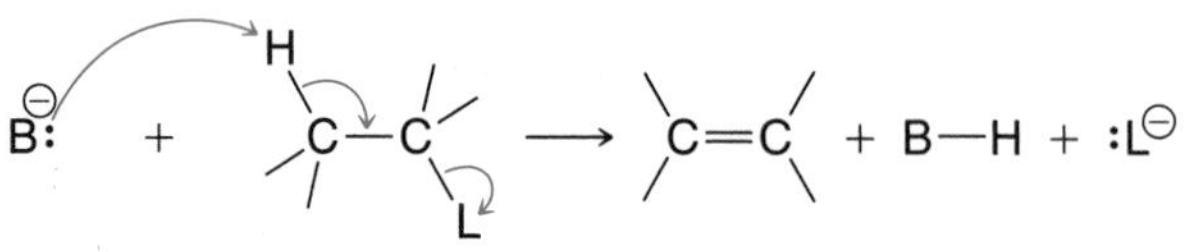

(7-3)

E1

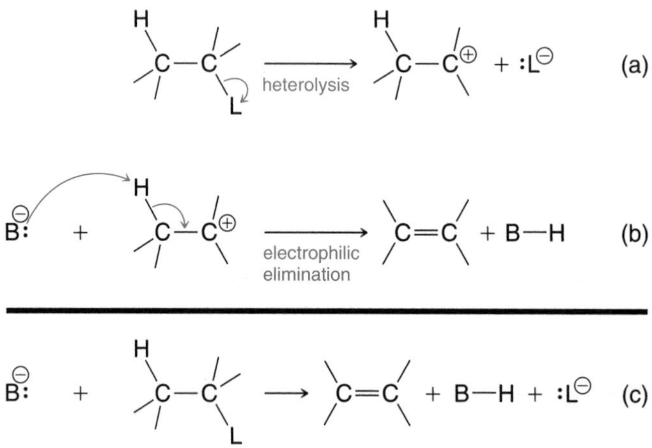

(7-4)

Given a multistep mechanism such as the ones above, it is important to be able to determine the *overall* reaction. It first requires identifying all *intermediates*.

An intermediate of a reaction mechanism is a species that appears as a product in one elementary step and a reactant in another.

Once the intermediates are spotted, the task is relatively straightforward.

To write an overall reaction from a sequence of balanced elementary steps,

- Cross out each intermediate that appears.
- Write the remaining reactants from the elementary steps as overall reactants.
- Write the remaining products from the elementary steps as overall products.

Looking back at the S_N1 reaction in Equation 7-2, notice that the carbocation that is produced in the first step of the mechanism (Equation 7-2a) is also a reactant in the second step (Equation 7-2b), so it is an intermediate and should be crossed out. The remaining species are written in the overall reaction shown in Equation 7-2c—Nu^- and R—L are overall reactants, and Nu—R and L^- are overall products. Similarly, in Equation 7-4, the carbocation is an intermediate and is crossed out. The remaining reactants and products are written in the overall reaction in Equation 7-4c.

TIME TO TRY

Cross out and label the intermediates that appear in the mechanisms in Equations 7-2 and 7-4.

LOOK OUT

To save time, your professor and textbook will rarely write out multistep mechanisms as they appear in Equations 7-2 and 7-4—that is, with one completely balanced elementary step on each line. Usually, an entire multistep mechanism will be written on a single line, and some species will be omitted (depending on their importance). For example, the E1 mechanism in Equation 7-4 might be written as follows.

E1

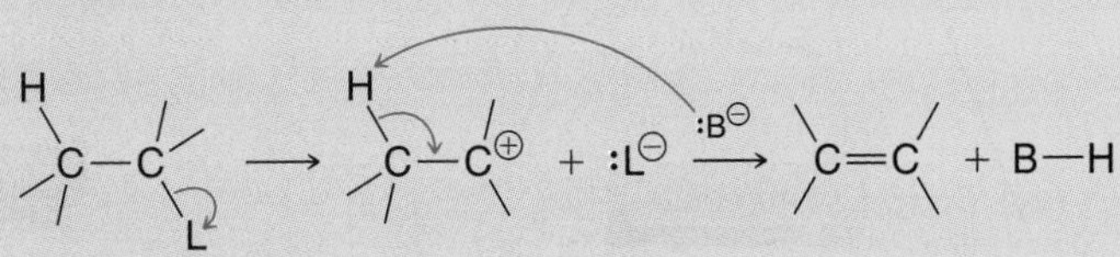

Notice that the base is not shown as a reactant in the first step because it is not important at that time. Rather, it is shown as a reactant in the second step (above the reaction arrow) because that's the step in which it becomes important. Similarly, notice that L^- is not shown as a product in the second step because it departed in the first step.

✔ QUICK CHECK

Write out the S_N1 mechanism from Equation 7-2 on a single line.

Answer:

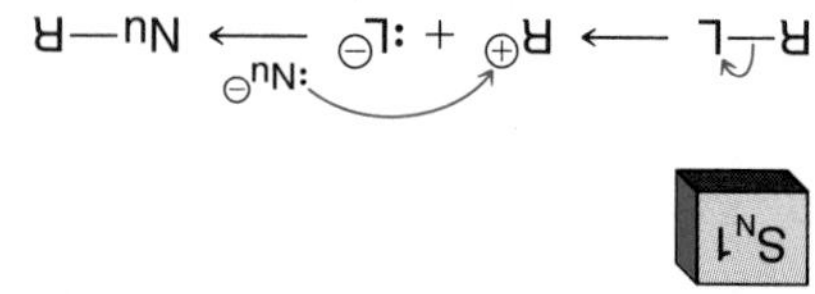

In addition to writing an overall reaction, it is important to be able to draw a reaction free-energy diagram for a multistep mechanism. Realizing that a multistep mechanism is just a set of back-to-back elementary steps, we can think of fusing together the free-energy diagrams of each of the elementary steps, resulting in a single, smooth curve. Recall from Section 5.2A that each elementary step will proceed through its own transition state, which appears at a maximum in energy. Therefore, an overall free-energy diagram will typically have several bumps and wiggles. Importantly,

The free-energy diagram of a multistep mechanism will exhibit one energy maximum for each elementary step, corresponding to each transition state.

For example, an S_N1 mechanism consists of two elementary steps, so the free-energy diagram will exhibit two energy maxima—one for the transition state of the first step and one for the transition state of the second step. This is shown in Figure 7-1. A very similar picture arises with the free-energy diagram of an E1 reaction, which is shown in Figure 7-2.

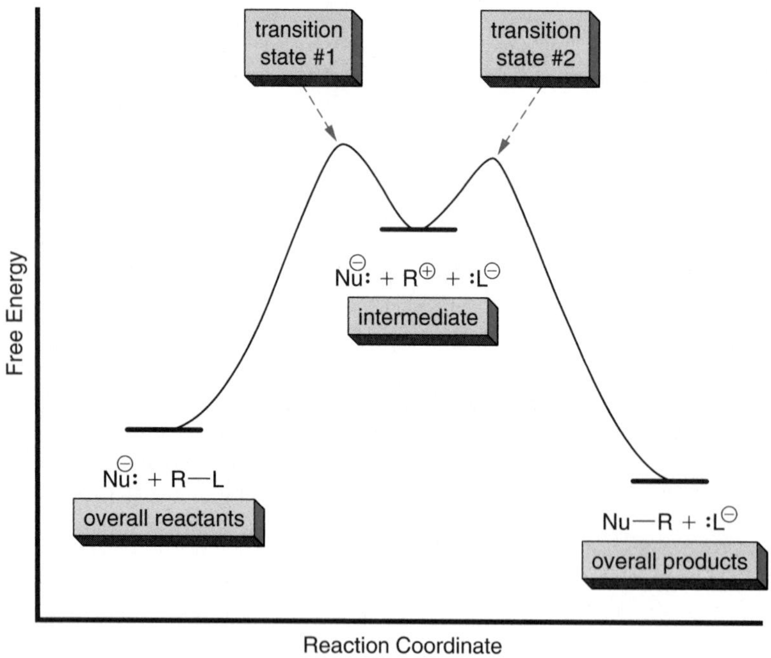

FIGURE 7-1 Free-energy diagram for an S_N1 reaction. Going from left to right along the reaction coordinate represents the smooth conversion of the overall reactants into overall products. A transition state appears at the energy maximum for each step of the two-step mechanism. The intermediate appears at an energy minimum between the two transition states.

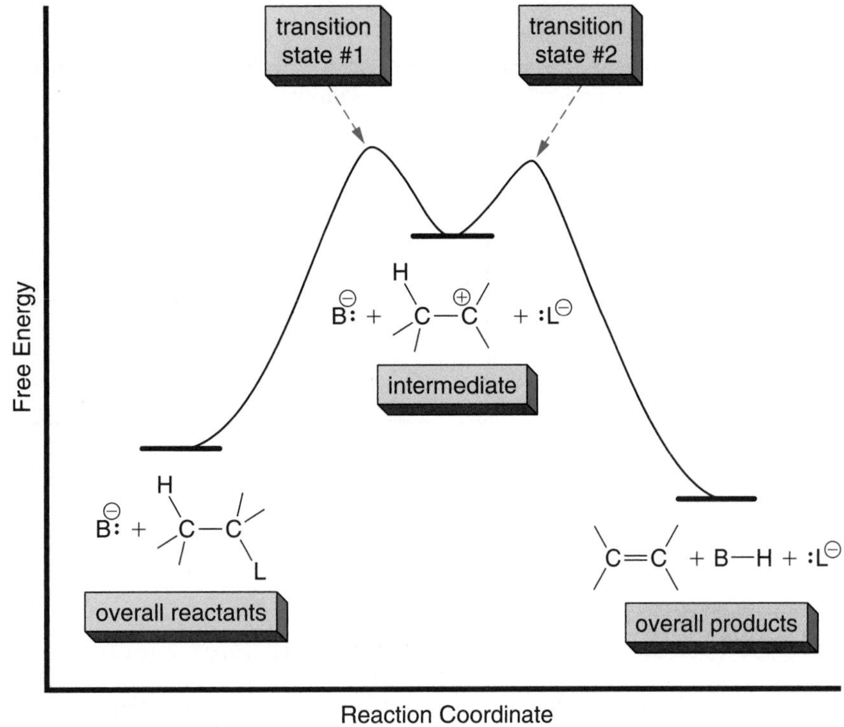

FIGURE 7-2 **Free-energy diagram for an E1 reaction.** Going from left to right along the reaction coordinate represents the smooth conversion of the overall reactants into overall products. A transition state appears at the energy maximum for each step of the two-step mechanism. The intermediate appears at an energy minimum between the two transition states.

Notice that for S_N1 and E1 reactions, the intermediate appears between the two transition states in the free-energy diagram and furthermore is lower in energy than the transition states on either side. This is generally true, regardless of the number of steps in the mechanism.

In a free-energy diagram of a multistep mechanism, each intermediate appears as an energy minimum between two transition states.

For the specific case of S_N1 and E1 reactions, the intermediate stage of the reaction is also higher in energy than either the overall reactants or the overall products. In other words, heterolysis steps are very uphill and highly unfavorable. This is because the step involves the loss of an octet from C and also the formation of two new formal charges.

WHY SHOULD I CARE?

The fact that the intermediate of an S_N1 or E1 reaction is higher in energy than the reactants or the products has important implications for the *rate-determining step* of the reaction. As we will see in Chapter 8, the rate-determining step controls the outcome of a competition that exists between these reactions as well as between S_N2 and E2 reactions.

✔ **QUICK CHECK**

Given the free-energy diagram below, how many transition states are there? How many intermediates? How many elementary steps?

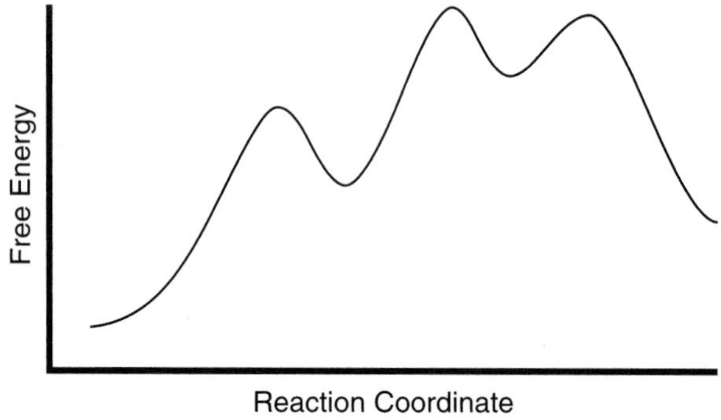
Free Energy
Reaction Coordinate

Answer: Three transition states—one for each energy maximum. Two intermediates—one for each energy minimum between transition states. Three elementary steps—one for each transition state.

7.3 Consequences of Single-Step Versus Multistep Mechanisms

As you examine the *overall* products of an S_N1 reaction (Equation 7.2c), they might appear to be no different from those of an S_N2 reaction (Equation 7.1). Similarly, the overall products of an E1 reaction (Equation 7.4c) might appear to be no different from those of an E2 reaction (Equation 7.3). In some cases, this will be true—but not always. As we will see in this section, the stereochemistry of the two reactions could differ. Alternatively, an intermediate might undergo various types of reactions, making certain outcomes available to multistep mechanisms that are not available to single-step ones.

7.3A STEREOCHEMISTRY OF S_N1 VERSUS S_N2 AND E1 VERSUS E2

Recall from Section 5.11 that S_N2 reactions are *stereospecific*. In situations where the new Nu—C bond is part of a stereocenter in the products, one configuration is produced exclusively—either *R* or *S*. The reason has to do with the restriction of the nucleophile's approach, which must be from the side opposite the leaving group. This is shown again in Equation 7-5.

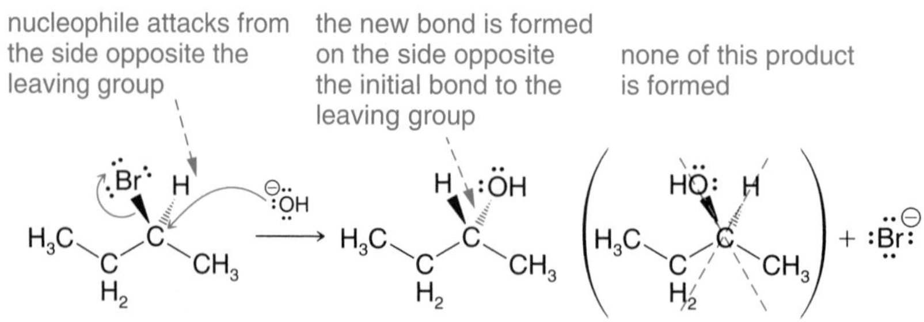
nucleophile attacks from the side opposite the leaving group
the new bond is formed on the side opposite the initial bond to the leaving group
none of this product is formed
Br
H
H₃C
C
CH₃
H₂
OH
+ :Br:

(7-5)

This stereospecificity is not required of S_N1 reactions. As shown in Equation 7-6, the stereocenter is produced in the second step of the mechanism, where the C atom that gains the bond is initially trigonal planar. Therefore, as we learned in Section 5.10, the new bond can form on either side of the C's plane, producing a mixture of configurations at that stereocenter. As a result,

In an S_N1 reaction, if the new Nu—C bond that is formed is part of a stereocenter, a mixture of stereoisomers will be produced.

In this case, the products are enantiomers of each other, so there must be an equal mixture of the two produced—that is, a racemic mixture.

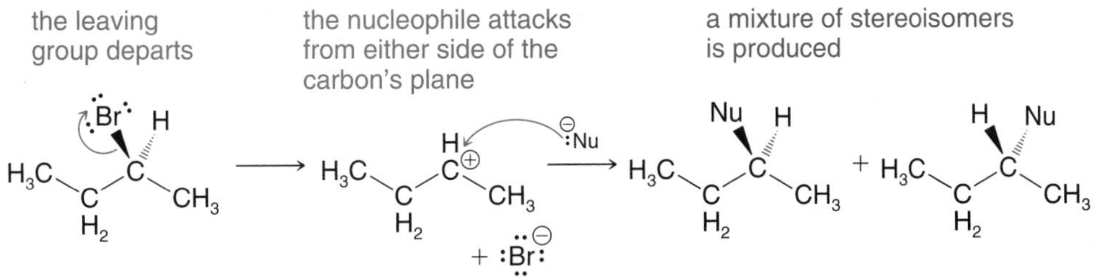

the leaving group departs
the nucleophile attacks from either side of the carbon's plane
a mixture of stereoisomers is produced

(7-6)

Although the complete discussion will be left for your full-year organic chemistry course, we call attention to the fact that a similar story unfolds when we compare E1 and E2 reactions. Namely,

An E2 reaction is stereospecific because it occurs in a single step.

An E1 reaction will produce a mixture of stereoisomers because it takes place in two steps.

When this topic comes up in your organic class, make sure you pay particular attention to the specific reasons why!

7.3B CARBOCATION REARRANGEMENTS IN S_N1 AND E1 REACTIONS

As alluded to above, an intermediate in a multistep mechanism might be able to participate in a variety of different reactions, which could impact what is produced as an overall product. In the case of S_N1 and E1 reactions, the presence of a carbocation intermediate brings about the possibility of carbocation rearrangements. Recall from Section 5.7 that there are two main types of carbocation rearrangements to be aware of—1,2-hydride shifts and 1,2-methyl shifts. Examples of each of these steps are shown in Equations 7-7 and 7-8, respectively.

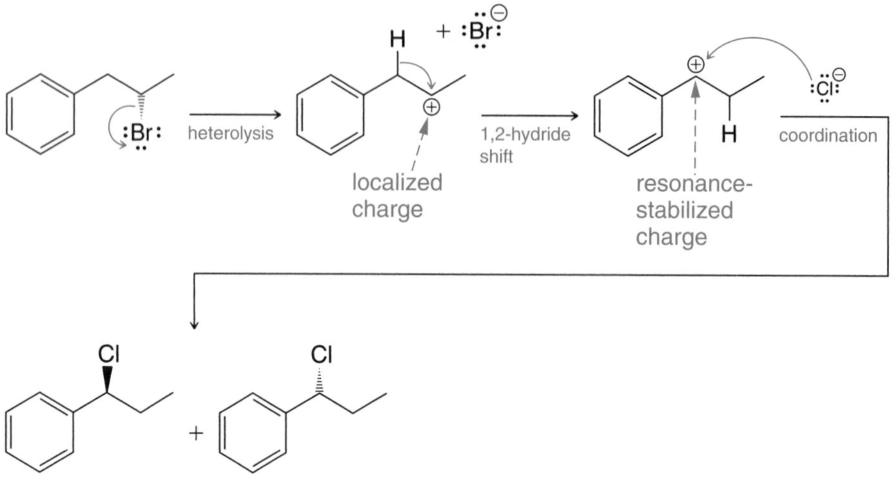

(7-7)

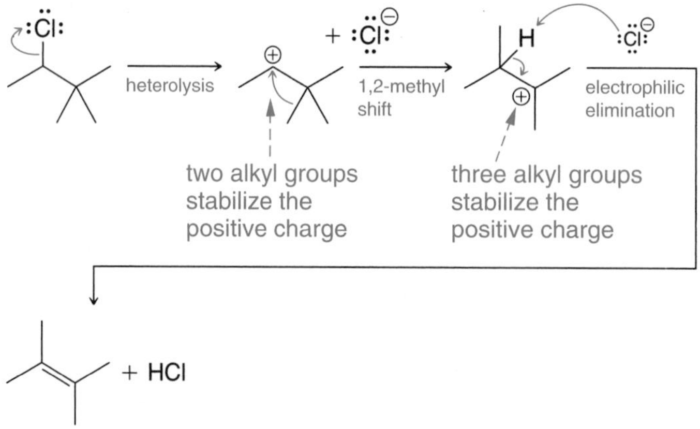

(7-8)

In Equation 7-7, the first and third steps comprise the normal S_N1 mechanism—heterolysis followed by coordination. However, a 1,2-hydride shift occurs before the coordination step has a chance to. The reason is that carbocation rearrangements are quite fast. In general,

Carbocation rearrangements are much faster than other elementary steps with which they might compete.

Similarly, in Equation 7-8, the first and third steps comprise the normal E1 mechanism—heterolysis followed by electrophilic elimination of H^+. Once again, before the normal second step of an E1 mechanism has a chance to take place, a 1,2-methyl shift occurs.

Despite their fast rate, carbocation rearrangements do not always take place.

A carbocation rearrangement takes place when a 1,2-hydride shift or 1,2-methyl shift results in a carbocation that has significantly greater stability than the initial carbocation.

For example, in Equation 7-7, notice that the 1,2-hydride shift that occurs results in greater stability of the carbocation—the initial carbocation has a localized positive charge, whereas the rearranged carbocation has resonance structures available that allow the positive charge to be shared on atoms of the ring. Also, notice that the carbocation in Equation 7-8 gains significant stability by undergoing a 1,2-methyl shift—the rearranged carbocation has an additional alkyl group that is electron-donating to the positively charged C.

By contrast, no carbocation rearrangement takes place in the S_N1 reaction in Equation 7-9.

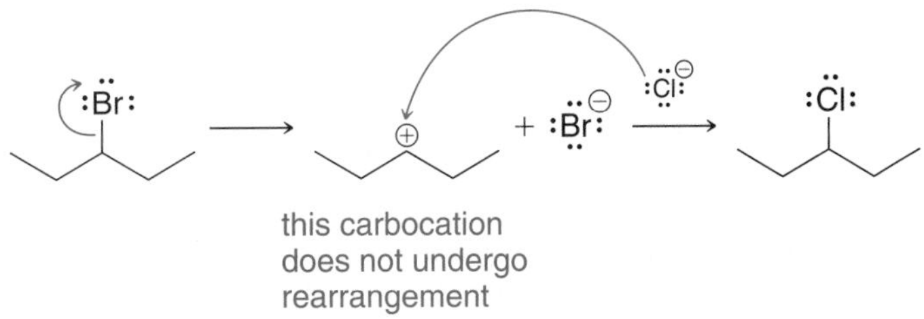
:Br:
+ :Br:
:Cl:
:Cl:
this carbocation
does not undergo
rearrangement

(7-9)

We can see why simply by writing out the products of a 1,2-hydride shift and a 1,2-methyl shift for the particular carbocation that is produced, as shown in Equations 7-10 and 7-11.

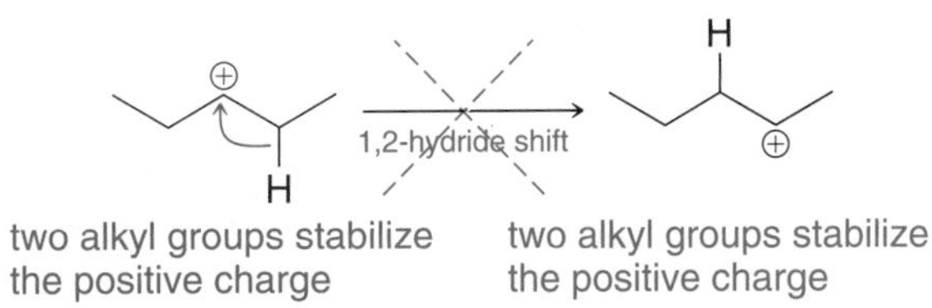
H
1,2-hydride shift
H
two alkyl groups stabilize
the positive charge
two alkyl groups stabilize
the positive charge

(7-10)

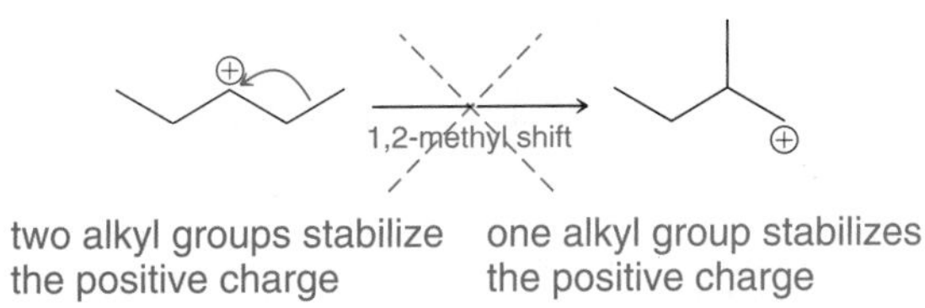
1,2-methyl shift
two alkyl groups stabilize
the positive charge
one alkyl group stabilizes
the positive charge

(7-11)

TIME TO TRY

In Equations 7-10 and 7-11, circle each alkyl group stabilizing the positively charged C in both the reactant and the product.

LOOK OUT

Although carbocation rearrangements are dealt with here specifically in the context of S_N1 and E1 reactions, they can take place any time a carbocation intermediate appears in a mechanism. Be on the lookout for such reactions!

✔ QUICK CHECK

Assuming the following substrates undergo an S_N1 or E1 reaction, determine which mechanisms, if any, will include a carbocation rearrangement.

Answers: For either an S_N1 or an E1 mechanism, a carbocation is produced upon heterolysis of the bond between C and the leaving group. Once the carbocation is produced, we need to check for possible 1,2-hydride shifts and 1,2-methyl shifts. These are shown below.

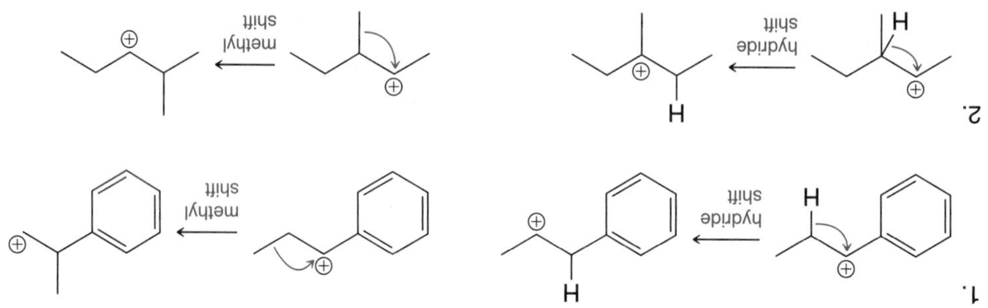

1. A hydride shift converts a resonance-stabilized positive charge to a localized one, which represents a decrease in stability. The same is true of a methyl shift. Therefore, no carbocation rearrangements will take place.
2. A hydride shift converts a carbocation stabilized by two alkyl groups to one that is stabilized by three such groups, which represents an increase in stability. By contrast, the initial and final carbocations of a methyl shift both have positive charges stabilized by two alkyl groups, so there is no increase in stability. Therefore, only a 1,2-hydride shift will take place.

7.4 Proton Transfers as Part of Multistep Mechanisms

Proton transfers are the most common elementary step found in multistep mechanisms. This is not only because of the abundance of protons (H^+) in organic molecules, but also because of the rate of such elementary steps.

Proton transfer steps are typically very fast.

The fast nature of such steps has to do with the fact that protons are very small and very light, which tends to make it quite easy for them to hop from one atom to another.

Reactions that proceed by multistep mechanisms often take advantage of such fast steps by shuttling protons around within a molecule so as to provide the best route for the reactants to get to products. Realize, however, that such proton transfer steps are not incorporated randomly into a mechanism—some ways of doing so are reasonable, and others are not. The remainder of this section introduces rules of thumb that allow you to judge whether or not a proton transfer step is reasonable.

The first rule of thumb involves the compatibility of certain species with the reaction conditions that are given.

Under acidic conditions, no strong bases should appear in a mechanism.

Under basic conditions, no strong acids should appear in a mechanism.

This raises the question, How do we recognize a species as a *strong base or a strong acid*? The answer is relatively straightforward if you remember that HO^- and H_3O^+ are the benchmarks.

A species is a strong base if it is similar in strength to or stronger than HO^-.

A species is a strong acid if it is similar in strength to or stronger than H_3O^+.

According to this definition, a variety of bases and acids are classified in Figure 7-3 as either strong or weak. You should commit to learning these classifications, but not by rote memorization. Rather, you should do so by applying the systematic method involving charge stability presented in Section 6.7 in order to determine the strength of each base or acid relative to the benchmark.

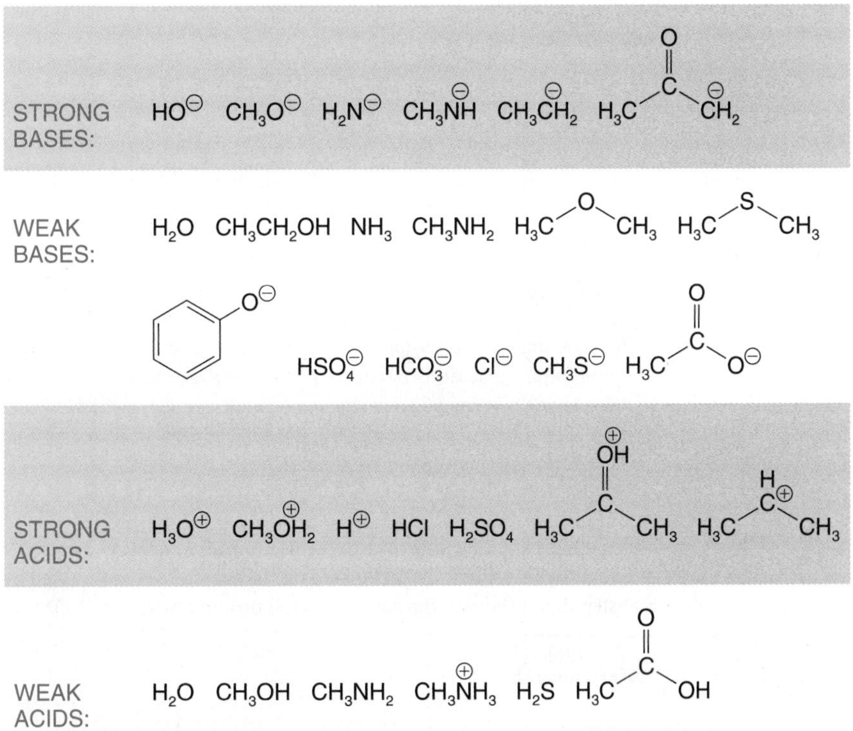

FIGURE 7-3 Strengths of various bases and acids. Bases are classified as being strong if their strength is roughly the same as or stronger than HO^-. Otherwise, they are weak bases. Acids are classified as being strong if their strength is roughly the same as or stronger than H_3O^+. Otherwise, they are weak acids.

By examining the species in Figure 7-3, you can see that there is a trend pertaining to the charge of the species. Specifically, negatively charged species tend to be strong bases, whereas positively charged species tend to be strong acids. Uncharged species are usually weak bases or weak acids. However, there are important exceptions to take note of. For example, Cl^- and $CH_3CO_2^-$ are weak bases, despite having a negative charge. That Cl atom is larger than O, which leads to stabilization of the negative charge and makes Cl^- a much weaker base than HO^-. Resonance, on the other hand, is what stabilizes the negative charge in $CH_3CO_2^-$. As another example, $CH_3NH_3^+$ is a weak base, despite the positive charge. This is because N is less electronegative than O, so the positive charge is better stabilized in $CH_3NH_3^+$ than in H_3O^+. Finally, there are a number of uncharged species that are strong acids, including HCl and H_2SO_4.

With this knowledge of which bases and acids are classified as strong, consider the reaction in Equation 7-12, in which an OH group is replaced by Cl.

$$O{=}CHCH_2CH_2OH + NaCl \xrightarrow[H^+]{} O{=}CHCH_2CH_2Cl \qquad (7\text{-}12)$$

Recall that NaCl can be simplified to Cl^-, so we could imagine this reaction taking place by an S_N2 step, as shown in Equation 7-13. However, this is unreasonable according to the above rules of thumb. The reason is that the reaction takes place under acidic conditions, as indicated by the H^+ that appears in the overall reaction in Equation 7-12. But the mechanism in Equation 7-13 shows that HO^- is produced, which is a strong base.

unreasonable

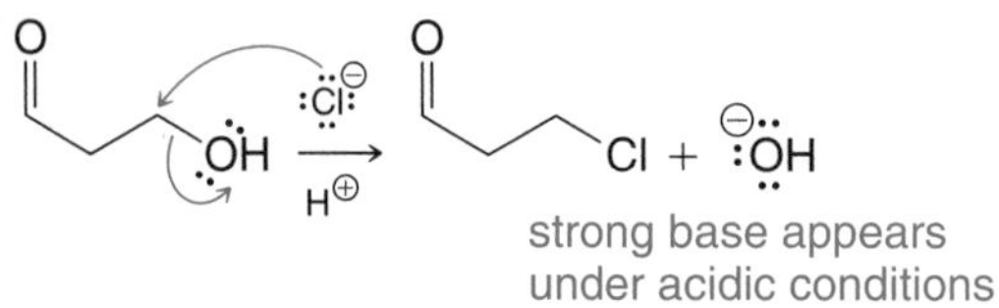

(7-13)

LOOK OUT

Many students are tempted to say that the reaction in Equation 7-12 takes place under *basic* conditions because of the OH that appears on the reactant side. However, that OH group is not the same thing as ^-OH, which is a strong base. Instead, that alcohol, which can be abbreviated R—OH, is about as acidic as water, H—OH, so it is in fact a weak acid.

To obtain a similar substitution without producing HO^-, two steps are required, as shown in Equation 7-14. First, the OH group on the substrate is protonated by the acid. A subsequent S_N2 step then produces the correct overall product, without any strong base appearing.

reasonable

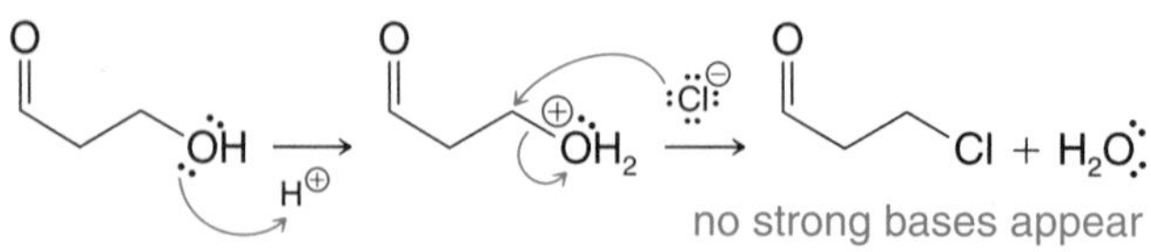

(7-14)

As another example, consider the overall reaction in Equation 7-15, in which Cl is replaced by OH.

$$C_6H_5CH_2Cl + NaOH \xrightarrow[H_2O]{} C_6H_5CH_2OH \qquad (7\text{-}15)$$

As indicated, the reaction takes place in water, so one mechanism that can be drawn is shown in Equation 7-16, where water displaces the Cl^- leaving group. A proton transfer in the second step would then lead to the overall product.

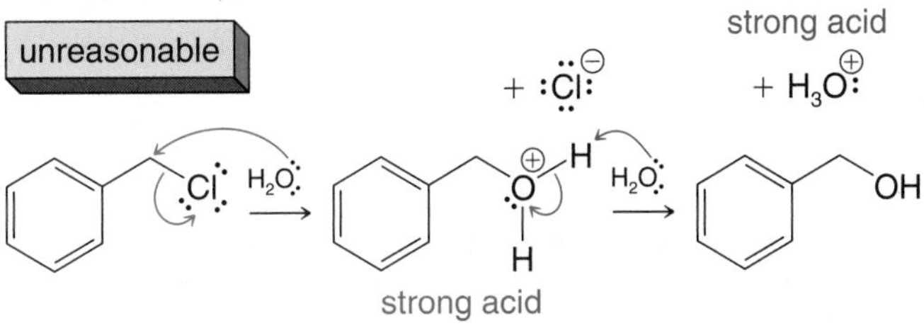

(7-16)

However, such a mechanism is unreasonable because of the appearance of strongly acidic species under basic conditions (indicated by the presence of NaOH) in Equation 7-15. You should recognize H_3O^+ as a strong acid, appearing as a product in the second step. In addition, the product of the first step is a strong acid—its structure is simplified to ROH_2^+, which is about as acidic as H_3O^+.

By contrast, the mechanism in Equation 7-17 is reasonable. We can simplify NaOH to HO^-, which then behaves as a nucleophile in a simple S_N2 step.

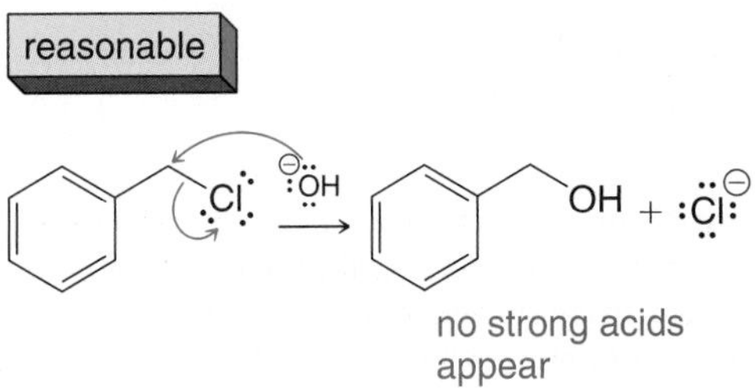

(7-17)

✔ **QUICK CHECK**

Consider the following overall reaction.

$$CH_3CH_2OCH_2CH_3 + H_2O \xrightarrow[H^+]{} 2\,CH_3CH_2OH$$

The following proposed mechanism is unreasonable. Explain why.

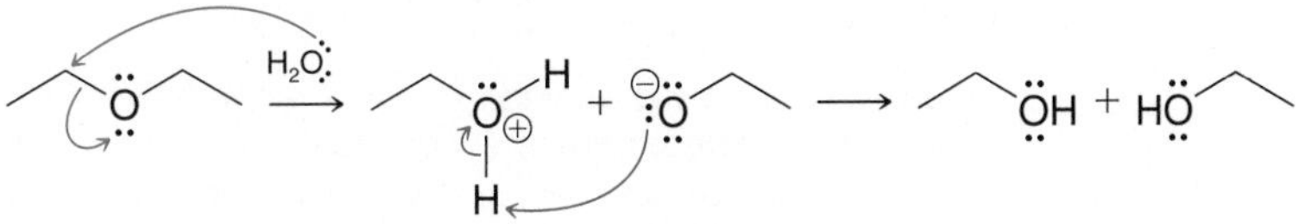

Propose an alternate mechanism that is reasonable.

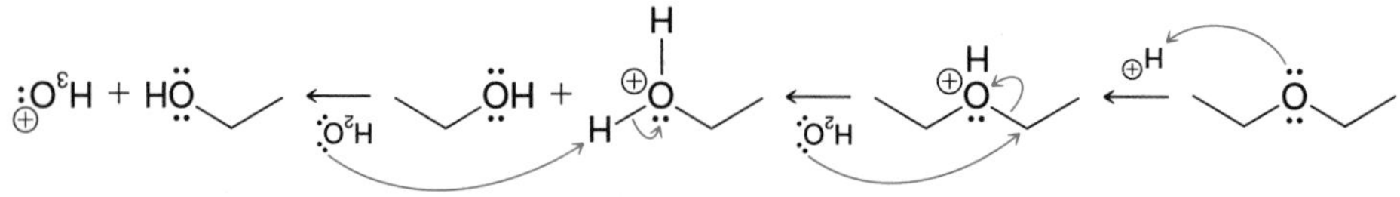

Answer: The above-proposed mechanism is not reasonable because one of the products of the first step, $CH_3CH_2O^-$, is a strong base—such species are not allowed under acidic conditions. However, if the O atom of the starting compound is protonated first, an S_N2 reaction can proceed without producing a strong base.

Even if the species that appear in a mechanism are compatible with the reaction conditions, there is another rule of thumb to consider when incorporating proton transfer steps into a mechanism.

Internal proton transfer reactions are generally unreasonable.

One of the major reasons for this is that the distance the proton has to travel is far too great.

As an example, consider the overall reaction in Equation 7-18.

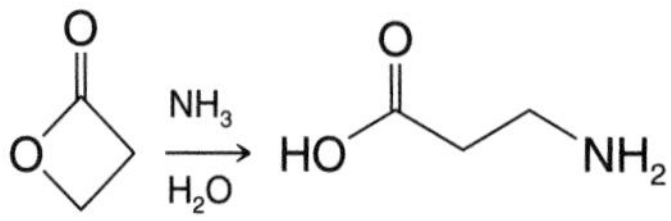

(7-18)

A possible mechanism for this is shown in Equation 7-19. First, NH_3 acts as a nucleophile in an S_N2 step, forming a bond to C, while the C—O bond breaks. The intermediate that is produced has a positively charged N and a negatively charge O, so to get from there to the overall product, a proton must be given up by N and a proton must be gained by O. Equation 7-19 shows this happening in a single step. However, this is a violation of the above rule.

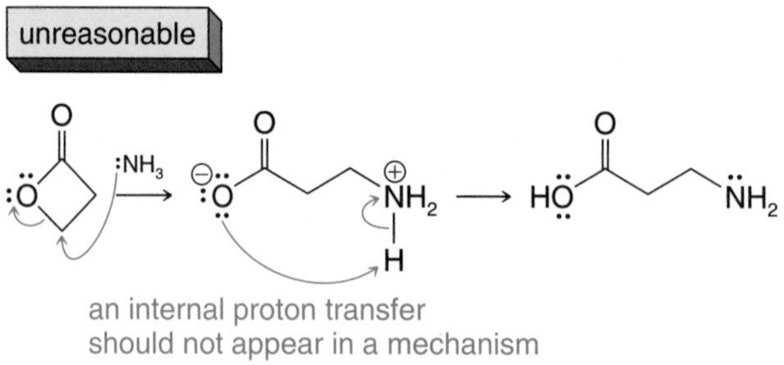
unreasonable
an internal proton transfer
should not appear in a mechanism

(7-19)

A more reasonable way to get a proton from N to O is to involve the solvent—in this case, water—as shown in Equation 7-20. Water is slightly basic, so in the second step of the mechanism, it can remove a proton from N. Water is also slightly acidic, so in the third step, a second molecule of water can donate a proton to O, thus producing the overall product. The assistance of the solvent in this way comprises what is called a **solvent-mediated proton transfer.**

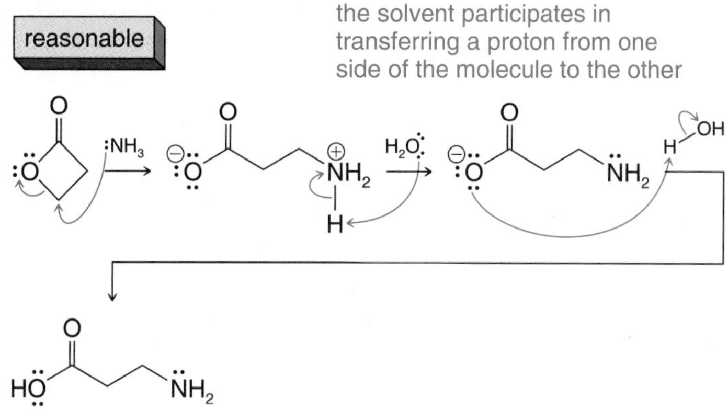
reasonable
the solvent participates in transferring a proton from one side of the molecule to the other
:NH3
H2O:
H
OH
NH2
H
HO
NH2

(7-20)

In general, the transfer of a proton from one side of a molecule to another requires a solvent-mediated proton transfer.

In your full year of organic chemistry, you will encounter many reactions where this happens.

A final rule of thumb involving proton transfers in multistep mechanisms is really a matter of convenience.

Only proton transfer steps that lead to the overall products are shown in a mechanism.

With this in mind, examine, once again, the mechanism in Equation 7-14, which we decided was reasonable. The first step is a proton transfer, in which the O atom that is part of the OH group gains a proton. However, notice that there is also an O atom that is part of the C=O bond. Realizing this, a common question students ask is "Why can't the O atom that is part of the C=O group gain the proton?" As it turns out, that O atom in fact *can* and *does* gain a proton during the course of the reaction, at least temporarily. But when it does, the result is effectively a dead end. (Why this is a dead end is beyond the scope of this book, but you should expect to learn why in your traditional course.) Therefore, the only thing that the species can do is give up the proton it just acquired, thus regenerating the initial reactant. The complete mechanism that includes these two additional proton transfer steps is shown in Equation 7-21. However, because the first two steps do not represent progress toward products, they are simply omitted from the mechanism.

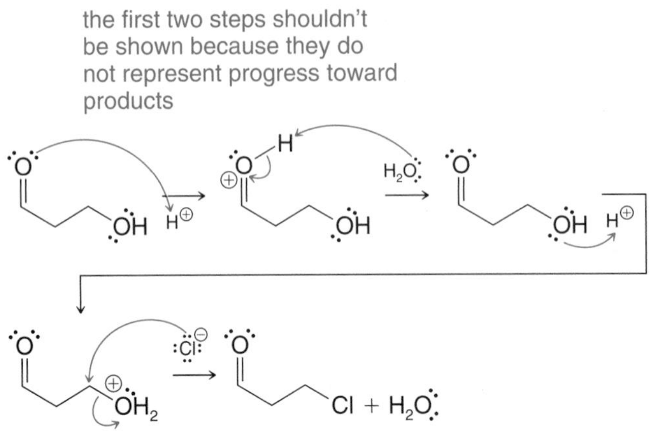

the first two steps shouldn't be shown because they do not represent progress toward products
H
H₂O
OH
Cl
Cl + H₂O
OH₂

(7-21)

7.5 Molecularity of Elementary Steps

In addition to the rules of thumb in the previous section, which are specific to proton transfer steps, there is a more general rule of thumb to know. To introduce it, let's once again consider the reasonable mechanism we saw in Equation 7-20. After the ring has been opened in the first step, the second and third steps are responsible for transferring an H^+ from N to O. In Equation 7-20, this was shown to take place in two steps, each involving a solvent molecule.

Instead, we can envision such a solvent-mediated proton transfer that takes place in a single step, as shown in Equation 7-22.

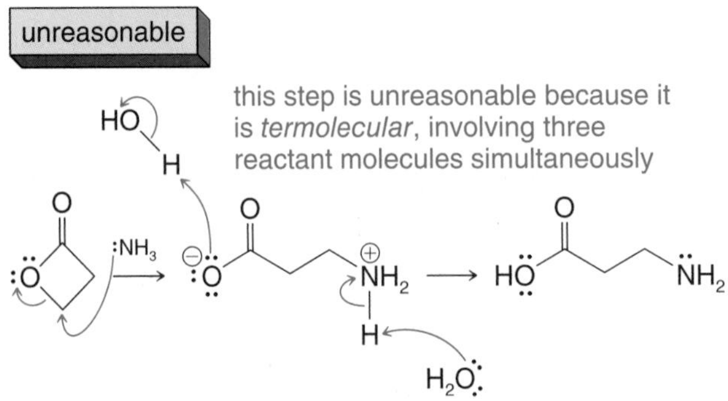

unreasonable
this step is unreasonable because it is *termolecular*, involving three reactant molecules simultaneously
HO
H
NH₃
NH₂
H₂O

(7-22)

This proposed mechanism, however, is unreasonable. The reason has to do with the fact that the second step is an elementary step that involves three reactant species—it is referred to as a **termolecular step**. The second step therefore indicates that one water molecule is picking a proton up at precisely the same time that the second water molecule is giving one up. For all of this to take place simultaneously, the three reactant species (the two water molecules and the species with a positive and a negative charge) must collide at precisely the same time. The likelihood of this happening is vanishingly small, so steps like this should not appear in a mechanism.

The above situation can be generalized as follows.

A *termolecular* step is generally unreasonable.

What remains reasonable, then, are **bimolecular steps** and **unimolecular steps**—elementary steps that involve two reactant species and one reactant species, respectively.

TIME TO TRY

Identify each of the three steps in the mechanism in Equation 7-20 as unimolecular, bimolecular, or termolecular.

PICTURE THIS

As mentioned above, the reason that termolecular steps are unreasonable is that the likelihood of three reactant species colliding at precisely the same time is vanishingly small. As an analogy, imagine that you and two friends are standing at three corners of a billiards table, each with a billiard ball (representing a molecule). If you each roll your ball at the same time toward the center of the table, it is extremely unlikely that all three of them will collide at precisely the same time. What is much more likely to happen instead is that two of the balls will collide first, and then some time later (perhaps very shortly afterward), the third ball will collide with one of the other two, representing two successive bimolecular steps.

7.6 Application: Tautomerization Reactions—Neutral, Acidic, and Basic Conditions

One of the simplest multistep mechanisms you will encounter in organic chemistry belongs to a reaction called a **keto-enol tautomerization,** an example of which is shown in Equation 7-23. Specifically, this reaction is an equilibrium between two isomers—a **keto form** and an **enol form**—which are called **tautomers** of each other. As indicated, in going from the keto form to the enol form, the C atom adjacent to the C=O group loses an H atom, and the O atom gains an H atom.

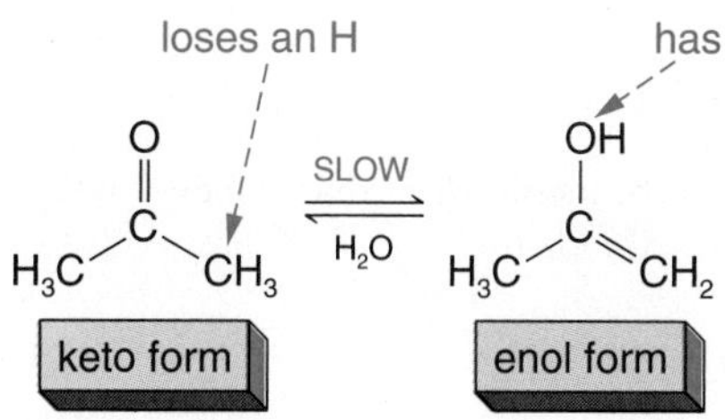

(7-23)

In Section 7.4, we saw that the transfer of an H atom from one part of a molecule to another can be accomplished by a *solvent-mediated* proton transfer. In this case, water is the solvent. Therefore, a molecule of water can remove an H^+ ion from the C atom in the first step, and in a subsequent step, another water molecule can donate an H^+ ion to the O atom. Such a mechanism is shown in Equation 7-24.

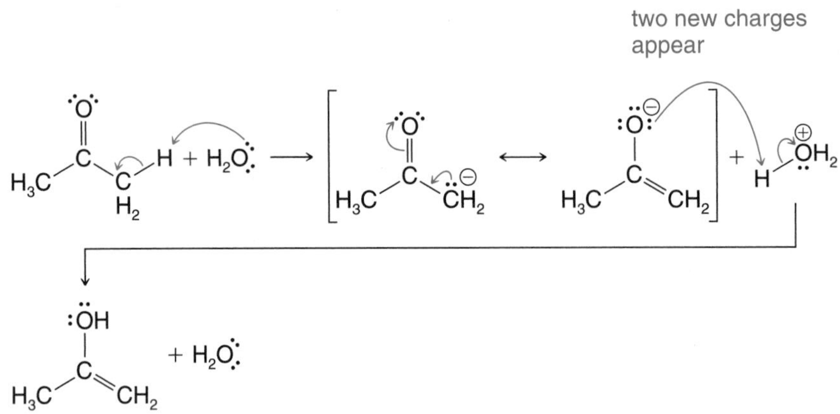

(7-24)

A key aspect of this mechanism is the resonance that exists in the intermediate that is produced from the first step. In the first resonance structure, the negative charge appears on the C atom, and in the second, it appears on the O atom. The negative charge on the O atom makes it relatively electron-rich, which is what facilitates the gain of an electron-poor proton in the second step.

LOOK OUT

Many students make the mistake of envisioning the resonance arrow in Equation 7-24 as a reaction arrow and therefore incorrectly think that the mechanism consists of three elementary steps. However, you must remember that, when a species has resonance, none of its resonance structures actually exists. Rather, the one, true structure is the resonance hybrid. Nevertheless, you will find that reaction mechanisms are almost always shown with resonance structures instead of a resonance hybrid because individual resonance structures allow you to account for valence electrons much more easily.

TIME TO TRY

Draw the resonance hybrid of the intermediate that is shown in Equation 7-24.

The reaction in Equation 7-23 is shown to take place under neutral conditions, but it can take place under basic or acidic conditions as well. The conditions for Equation 7-25, for example, are basic, due to the presence of HO^-.

basic conditions

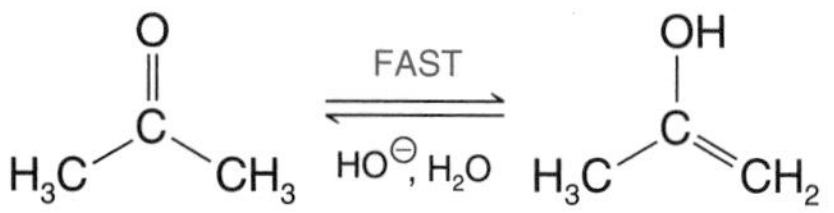

(7-25)

Interestingly, by changing the reaction conditions in these ways, the mechanism cannot be the same as the one we saw in Equation 7-24. The reason is that the first step of the mechanism in Equation 7-24 shows that a strong acid, H_3O^+, is produced. As we discussed previously in Section 7.4, a strong acid should not appear in a mechanism when the conditions are basic.

Despite the fact that the reactions in Equations 7-23 and 7-25 must proceed by different mechanisms, the same general things must happen in each reaction—that is, an H^+ ion must be lost from the C atom and an H^+ ion must be gained by the O atom. However, in order to avoid having H_3O^+ produced under these basic conditions, H_2O cannot be the species that acts as a base to remove the proton in the first step. Instead, HO^- must act as the base. As shown in Equation 7-26, when HO^- removes the proton in the first step, the product is H_2O, a weak acid, which is allowed to appear under basic conditions.

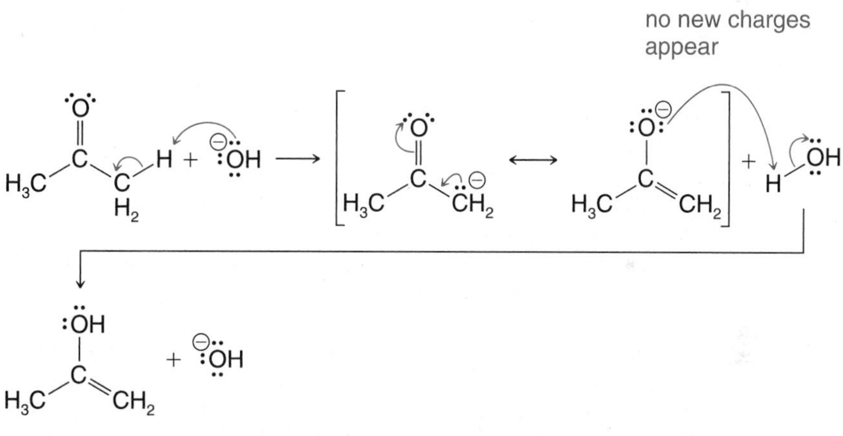

(7-26)

In addition to neutral and basic conditions, tautomerization can take place under acidic conditions, as shown in Equation 7-27. Once again, an H^+ ion must be removed from the C atom and an H^+ ion must be added to the O atom but neither of the previous mechanisms is reasonable. This is because in each mechanism a strong base appears. In Equation 7-26, that strong base is HO^-. In Equation 7-24, it is the negatively charged intermediate (see Figure 7-3).

acidic conditions

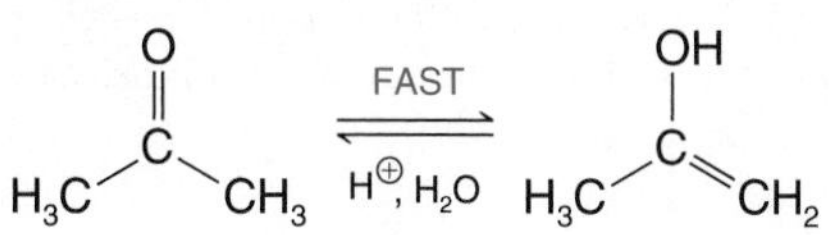

(7-27)

Because the negatively charged intermediate in Equation 7-24 is a strong base, you should quickly realize that under acidic conditions the first step cannot be the removal of a proton from the reactant. However, seeing that in the overall reaction the O atom must gain an H^+ ion the first step could be the addition of an H^+ ion to the O atom as shown in Equation 7-28. The positively charged intermediate is a strong acid (see Figure 7-3), but that's okay under acidic conditions. In the second step, water removes an H^+ ion from the C atom to produce the enol product. (Notice that the curved arrow notation directly leads to a weakly contributing resonance structure of the enol, but that the resonance structures are one and the same species.)

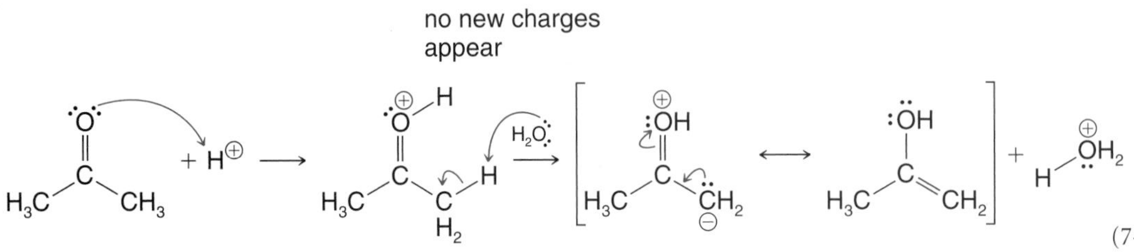

(7-28)

With the different reaction conditions leading to different mechanisms, we find that the reaction rates are not all the same. As indicated previously in Equations 7-23, 7-25, and 7-27, the tautomerization reaction is slow under neutral conditions, but fast under either basic or acidic conditions. We say that the reaction is **base-catalyzed** and is also **acid-catalyzed**. Why should this be so?

The answer can be understood through the various mechanisms. Notice that in the mechanism for neutral conditions (Equation 7-24), the first step produces two new charges—both reactant species are uncharged, but one product of the first step has a positive charge, and the other has a negative charge. As we learned in Chapter 6, those charges are "bad." Producing them requires energy, making the process substantially uphill and slow.

By contrast, in the mechanism that takes place under basic conditions (Equation 7-26), there is no step that produces an *additional* charge. In the overall reactants, there is a single formal charge of -1, and the same is true of the intermediates and the overall products. Without the generation of a new charge, none of the steps is dramatically uphill and slow. Similarly, in the mechanism for acidic conditions (Equation 7-28), there is no step in which a new charge is produced—at each stage, there is a single formal charge of $+1$—so none of the steps proceeds dramatically slowly.

WHY SHOULD I CARE?

Acid- and base-catalyzed reactions are relatively common in organic chemistry. Knowing how mechanisms can change under these different conditions contributes greatly to our understanding of the general reasons why certain reactions are catalyzed by the addition of an acid or base, while others are not.

7.7 Application: Dealing with Relatively Lengthy Mechanisms—Fischer Esterification and Imine Formation

The mechanisms that we have examined thus far have been relatively short, typically consisting of three or fewer total steps. However, many reactions you will encounter in organic chemistry proceed by mechanisms that are significantly longer. When you encounter such reactions, it is important that you pay particularly close attention to the mechanism. You must commit to understanding why the mechanism that is presented to you is reasonable, while other variations of the mechanism (typically not given to you) are unreasonable. Therefore, this section is devoted to dissecting a couple of reactions that have relatively lengthy mechanisms—*Fischer esterification* and *imine formation*. The line of thinking that we use here is one that you should do your best to apply to any new mechanism that you encounter.

An example of a Fischer esterification reaction is shown in Equation 7-29.

Fischer esterification

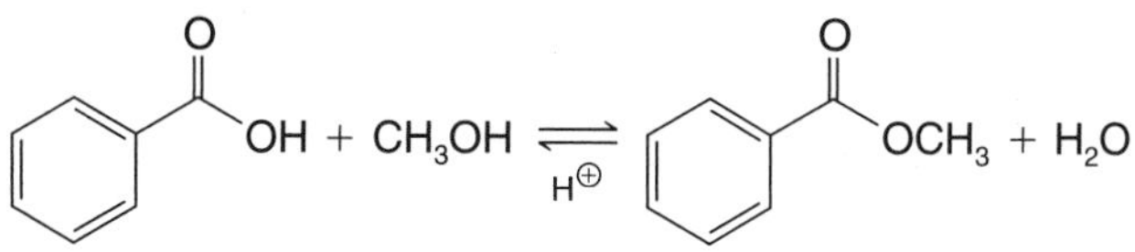

(7-29)

The specific example shown is not the only one. In general,

In a **Fischer esterification,** a carboxylic acid (a molecule with a C—CO_2H) reacts with an alcohol (a molecule with a C—OH group) under acidic conditions to produce an ester (a molecule with a CO_2—C group).

TIME TO TRY

In Equation 7-29, circle the carboxylic acid, alcohol, and ester functional groups.

The mechanism for this reaction is shown in Equation 7-30. As you can see, it consists of six steps, which are numbered below the reaction arrows.

The first things you should do when encountering such a mechanism are to identify precisely what is taking place in each elementary step and to rationalize how the combination of steps accomplishes the overall reaction. In this case, step 1 is a proton transfer, in which the O atom of the C═O group gains a proton. Step 2 is a nucleophilic addition, in which CH_3OH acts as a nucleophile and adds to the C atom of the C═O group, forcing a pair of electrons from the double bond to be kicked up to the O atom. Steps 3 and 4 are proton transfers—in step 3, the H^+ ion is lost from the O atom of the OCH_3 group, and in step 4, an H^+ ion is added to the OH group. Step 5 is nucleophilic elimination, in which a molecule of H_2O leaves from the C atom, and simultaneously the C═O bond is regenerated. Finally, step 6 is a proton transfer, in which the H^+ ion is lost from the O atom of the C═O group, resulting in the overall product.

TIME TO TRY

In Equation 7-29, next to each of the numbers below the reaction arrows, specify the name of that elementary step. (Try to do so without looking at the previous paragraph.)

Having identified the steps in this way, you can now see that four of the six steps are proton transfers. The other two steps—steps 2 and 5—are the addition of CH_3OH and the elimination of H_2O, respectively, which result in an overall substitution of one group for the other.

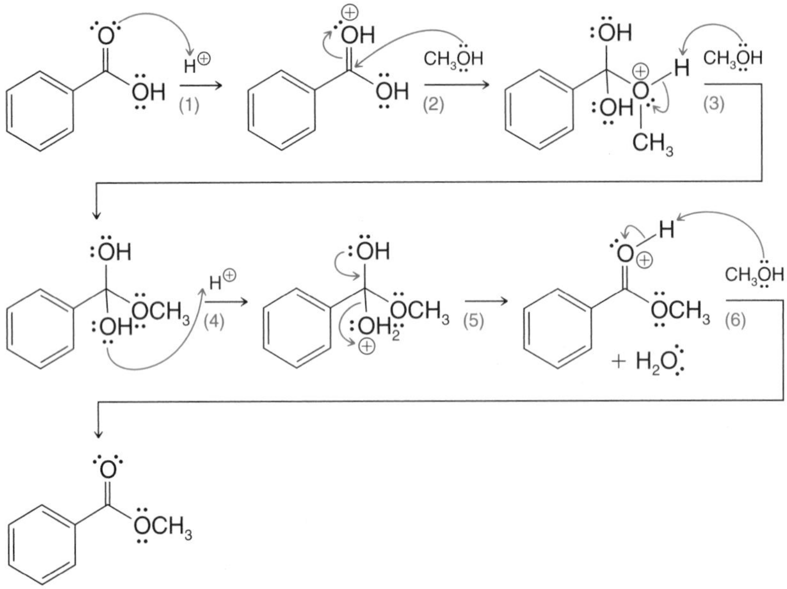

(7-29)

LOOK OUT

Notice in the mechanism in Equation 7-29 that CH_3OH is used as a reactant in three steps, but in the balanced overall equation, Equation 7-28, only one molecule of CH_3OH appears. This apparent discrepancy often does not sit well with students initially, but it is not a problem. The reason is that, when a compound participates in a reaction, there is not just a single molecule of it present. Rather, you can expect something on the order of a mole of the compound, which has 6×10^{23} molecules. Therefore, if one molecule is used in one particular step, there are still plenty more to be used in a subsequent step.

Now it's time to start asking questions pertaining to why this sequence of steps is reasonable, while other imaginable steps are not. The first question might be, Why can't the mechanism consist of just a single step? The answer is not because there is no way to draw the curved arrow notation for a single-step mechanism. On the contrary, such a mechanism is shown in Equation 7-30. However, you should quickly realize that this is not one of the 10 elementary steps discussed in Chapter 5. Unless you have encountered a particular elementary step, don't use it in a mechanism. In other words,

Don't make up your own elementary steps. When drawing mechanisms, do your best to use combinations of only those steps presented in Chapter 5.

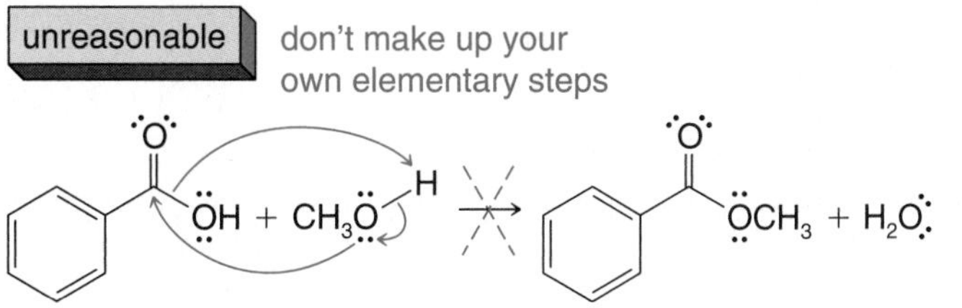

(7-30)

A second question to ask is, Why can't nucleophilic addition take place in the first step, as shown in Equation 7-31? The answer is that, if it did, the immediate product would have a localized negative charge on O, which represents a strong base (similar to HO^-). This is not reasonable under the acidic conditions that are given. Such a problem is solved by putting an H^+ ion on that O atom in the first step, as shown previously in Equation 7-29. When nucleophilic addition takes place in the second step, that O atom's charge goes from +1 to 0, so no strong base is ever produced.

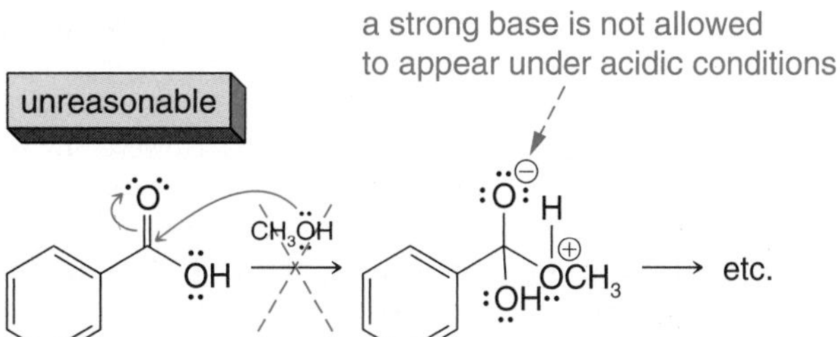

(7-31)

Yet another question to ask is, Why can't CH_3OH displace H_2O directly, as shown in Equation 7-32? In order to avoid the production of HO^- (a strong base), the OH group would have to be protonated first. The second step would be an S_N2 reaction.

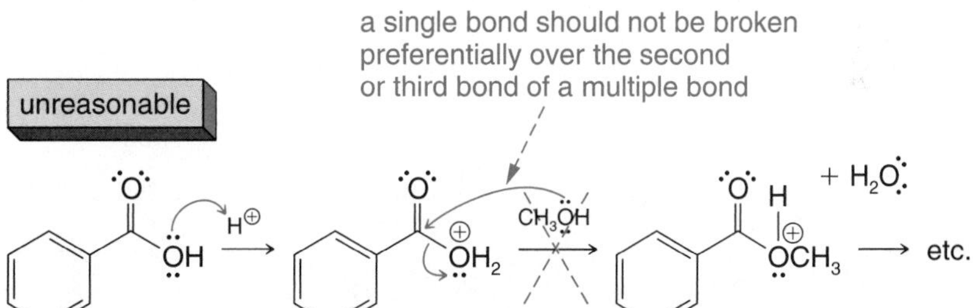

(7-32)

The reason this mechanism doesn't make sense has to do with a general rule we learned in Section 5.5: It is easier to break a second or third bond of a multiple bond than it is to break a single bond. When the nucleophile forms a bond to the C atom of the C=O group, another bond to the C atom must break in order to avoid exceeding the octet. The nucleophilic addition in step 2 of Equation 7-29 represents the breaking of one bond from a double bond. By contrast, for the S_N2 reaction to take place, as in Equation 7-32, the bond that is broken is a single bond. The easier bond-breaking is the former one.

A final question we'll ask is, Can steps 2 and 3 be swapped? Doing so would give us the partial mechanism shown in Equation 7-33.

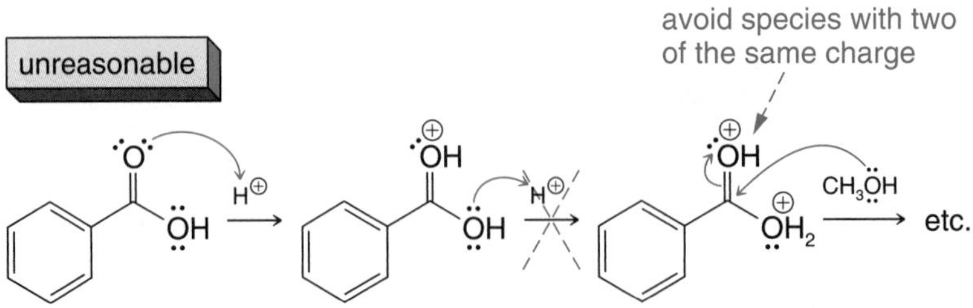

(7-33)

However, this is not reasonable because the second step produces a species that has two positive charges. As we saw in Chapter 6, a single formal charge in a species represents instability. Having two of the *same* charge is particularly unstable.

In general, mechanisms should not include species that have two of the same charge.

As we saw previously in Equation 7-29, the way around this problem is to first have a step that removes a formal charge on one atom before there is a step that adds a formal charge to another atom.

There are many other variations you can imagine for the mechanism of Fischer esterification, and you are encouraged to come up with others. However, for each variation you come up with, you should challenge yourself to figure out why it is unreasonable—that is, what rule is being violated? Furthermore, going the extra mile in this way with each new mechanism you encounter is what is going to separate you from the rest of the pack. This is what will allow you to actually *master* the material, as opposed to suffering through it.

QUICK CHECK

Draw the complete mechanism for the following reaction.

$$CH_3CO_2H + HOCH_3 \xrightarrow[H^+]{} CH_3CO_2CH_3 + H_2O$$

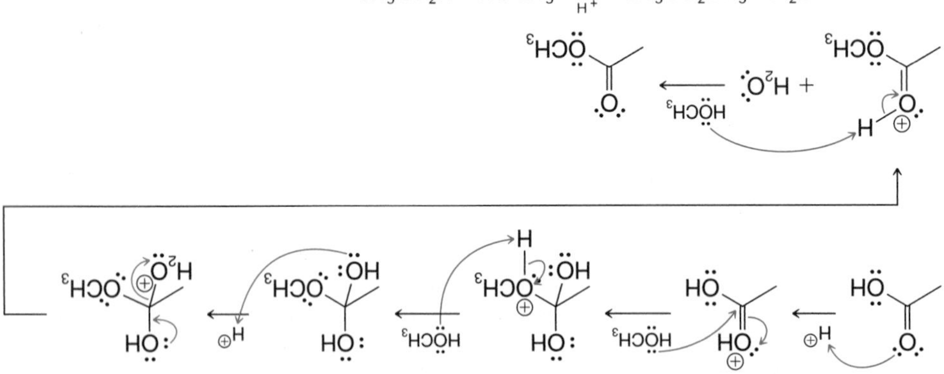

Let's now shift gears to imine formation, an example of which is shown in Equation 7-34. The imine, which is labeled, is characterized by the presence of a C=N bond, with only H or alkyl groups attached to those two atoms.

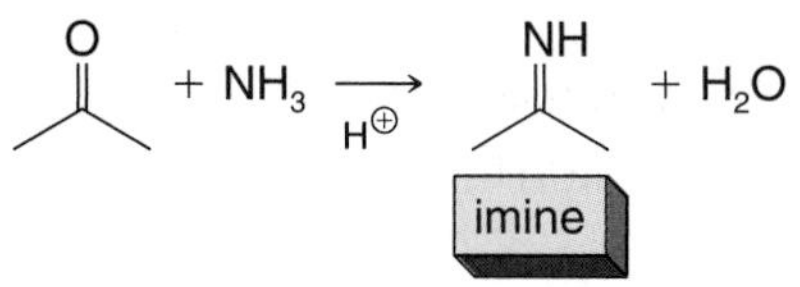

(7-34)

Once again, this is but one example of an imine formation reaction. In general,

In an **imine formation,** a ketone (having a $C_2C{=}O$ group) or an aldehyde (having a $C{-}CH{=}O$ group) reacts with NH_3 or an amine (having a $C{-}NH_2$ group), under acidic conditions, to produce a new $C{=}N$ bond.

The mechanism for the above imine formation is shown in Equation 7-35.

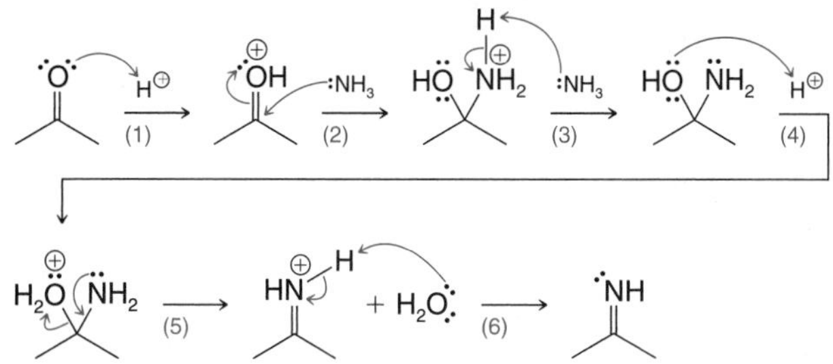

(7-35)

Just as with Fischer esterification, an imine formation reaction has six steps in its mechanism, which are numbered. In step 1, an H^+ ion is added to the O atom via a proton transfer. Step 2 is nucleophilic addition of NH_3. Steps 3 and 4 are proton transfers, and step 5 is nucleophilic elimination. Finally, step 6 is another proton transfer, in which an H^+ ion is removed from the N atom.

TIME TO TRY

In Equation 7-35, next to each of the numbers below the reaction arrows, specify the name of that elementary step. (Try to do so without looking at the previous paragraph.)

Similar to how we dissected the mechanism for Fischer esterification, we could ask several questions as to why certain variations are unreasonable. We will not do that here, but you are encouraged to do so on your own.

Rather, we will complete this section by comparing the mechanism for imine formation (Equation 7-35) with that for Fischer esterification (Equation 7-29). Notice, in particular, that the two reactions consist of precisely the same elementary steps, in the same order. The two mechanisms have a subtle difference, however, in step 5. In the imine formation mechanism (Equation 7-35), the double bond that is formed in step 5 involves the N atom from the original nucleophile, resulting in a positive charge on that N atom. This is okay because the newly generated positive charge on the N atom is subsequently removed—notice that in step 6 the newly generated +1 charge on that N atom goes to 0 upon the removal of the H^+ ion from the N atom. By contrast, in the Fischer esterification mechanism (Equation 7-29), the double bond that is formed in step 5 does *not* involve the O atom from the initial CH_3OH nucleophile. That's because, if we tried to do the same thing as in step 5 of the imine formation mechanism, the positive charge that would be generated on the O atom of OCH_3 could not be removed in a subsequent step—there is no H^+ ion to remove from that O atom. This is illustrated in Equation 7-36.

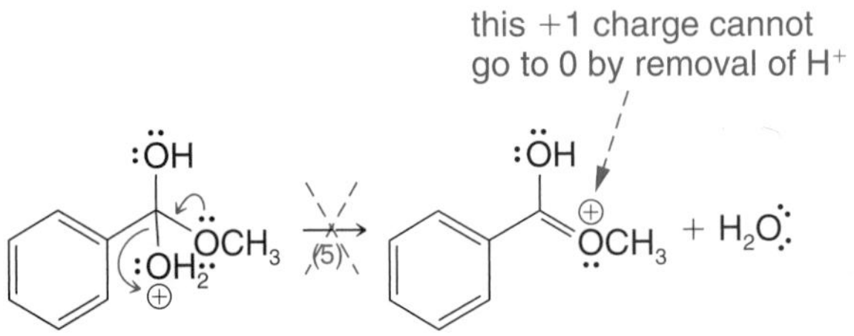

(7-36)

The reason for comparing the Fischer esterification and imine formation mechanisms is an important one. Examine once again the two *overall* reactions—Equations 7-28 and 7-34. Fischer esterification involves the transformation of a carboxylic acid group to an ester, whereas imine formation converts a ketone (or aldehyde) to an imine. The *overall* reactions appear to have very little in common. However, as we pointed out, their mechanisms are nearly identical! Thus, another way in which mechanisms can be used as powerful tools becomes apparent.

Many reactions that appear to differ greatly in their *overall* reactions have very similar (if not identical) mechanisms.

If you keep this in mind as you learn new reactions, you will be able to see that the hundreds of reactions you encounter throughout your organic chemistry course really boil down to just a handful of different mechanisms. As such, organic chemistry can be greatly simplified.

WHAT DID YOU LEARN?

7.1 Draw the complete, detailed mechanism for each of the following reactions undergoing (i) an S_N2 reaction and (ii) an S_N1 reaction. Pay attention to stereochemistry where appropriate.

(a) 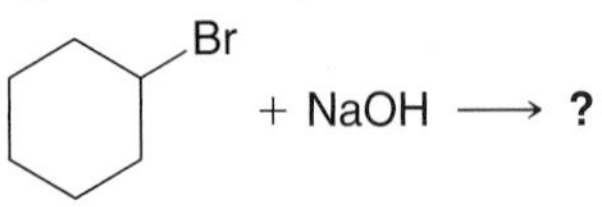

(b) 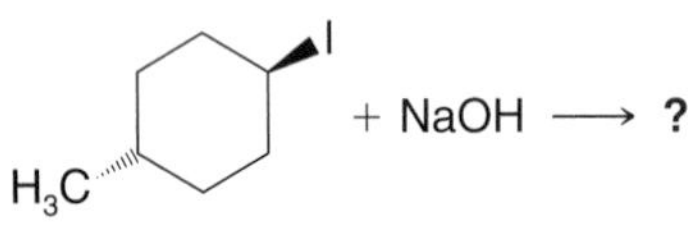

(c) 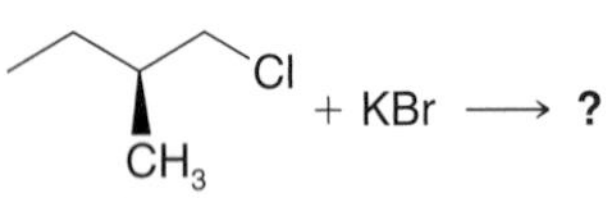

(d) 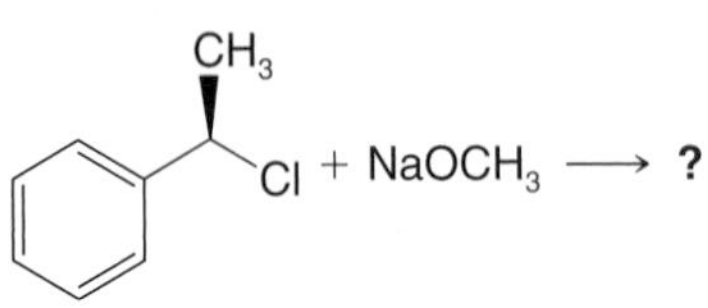

7.2 Draw the complete, detailed mechanism for each of the following reactions undergoing (i) an E2 reaction and (ii) an E1 reaction.

(a)

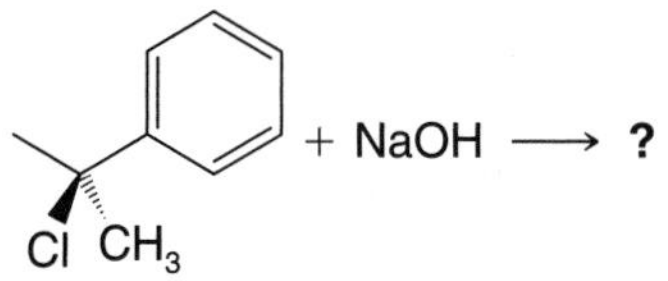

(b)

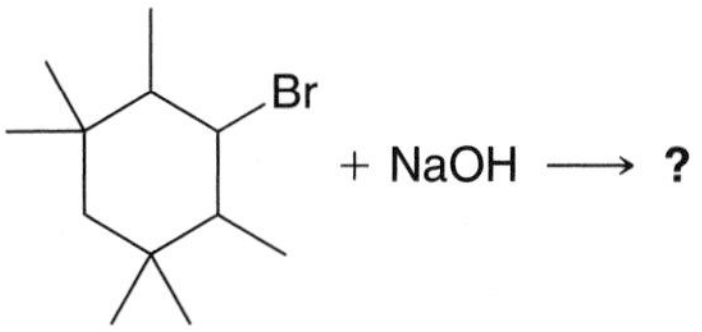

7.3 Which of the following substrates would you expect to undergo a carbocation rearrangement in an S_N1 or S_N2 reaction? Explain. Draw the curved arrow notation for the carbocation rearrangement that is likely to occur.

(a)

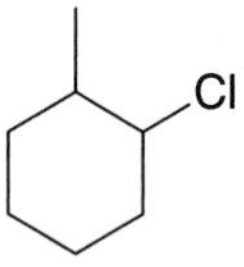

(b)

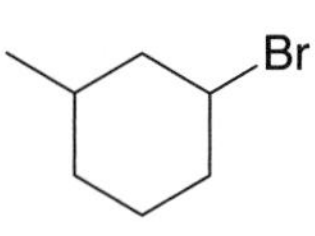

(c)

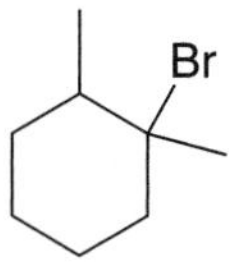

(d)

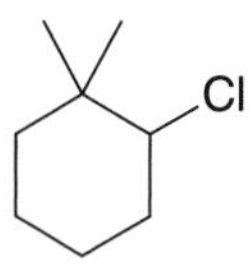

(e)

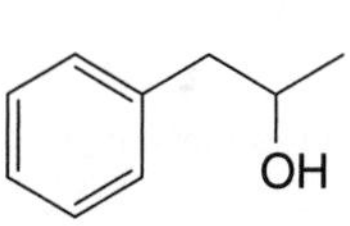

(f)

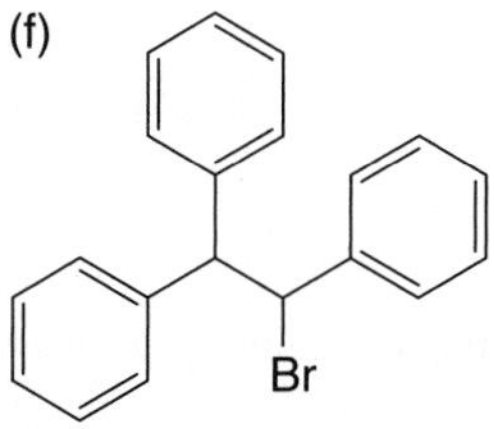

7.4 Consider the following *overall* reaction.

Overall:

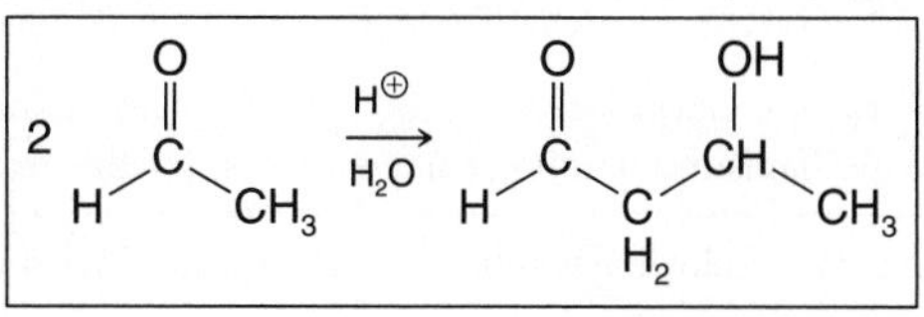

Referring to the proposed mechanisms below, determine whether each step is reasonable based on the rules presented in this chapter and Chapter 5.

(a)

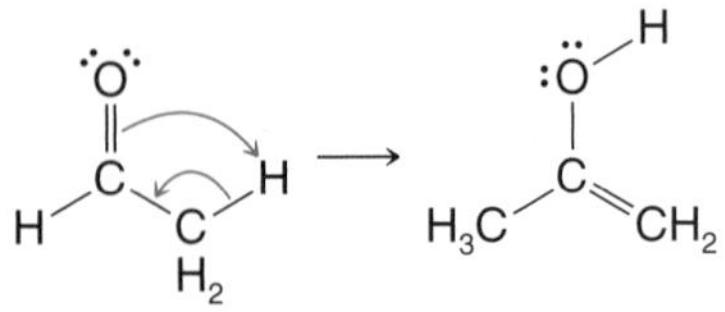

(b)

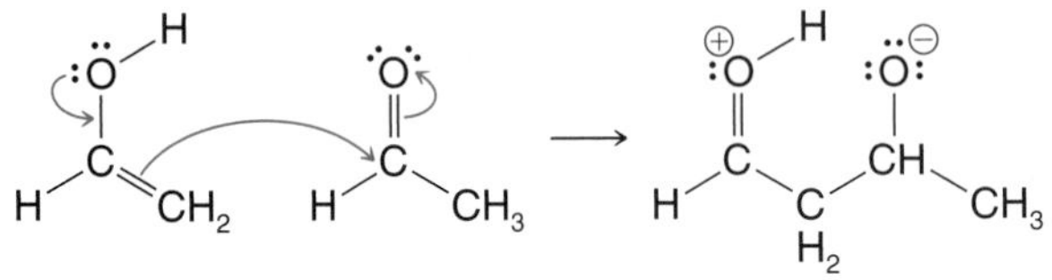

(c)

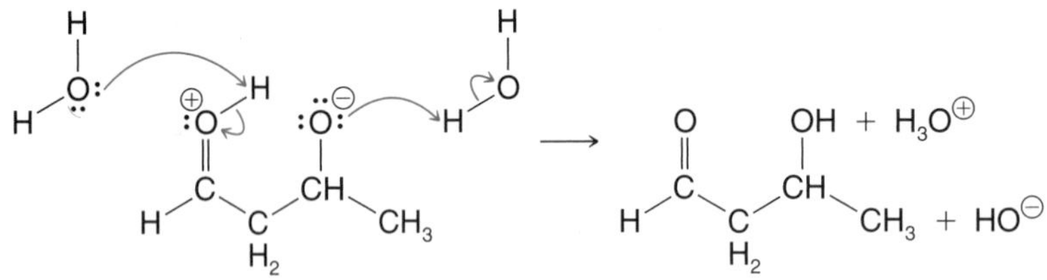

7.5 Consider the following E1 reaction.

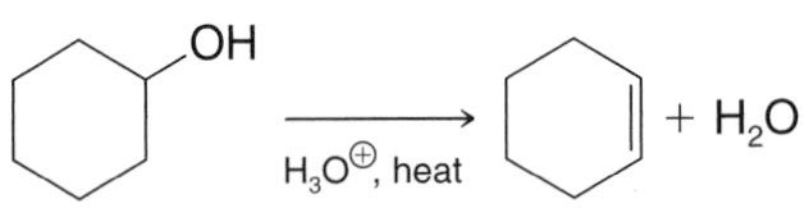

a Draw a complete, detailed mechanism for this reaction.

b Draw a reaction free-energy diagram that agrees with that mechanism, labeling overall reactants, overall products, all transition states, and all intermediates.

7.6 Consider the following nucleophilic substitution reaction.

$$(CH_3)_3CCHBrCH_2CH_2CH_3 + KI \rightarrow (CH_3)_2CICH(CH_3)CH_2CH_2CH_3$$

a Argue whether this reaction takes place via an S_N2 or an S_N1 mechanism.

b Draw the complete mechanism (including curved arrows) for this reaction.

7.7 Propose a mechanism for the following reaction, which produces 1,4-dioxane.

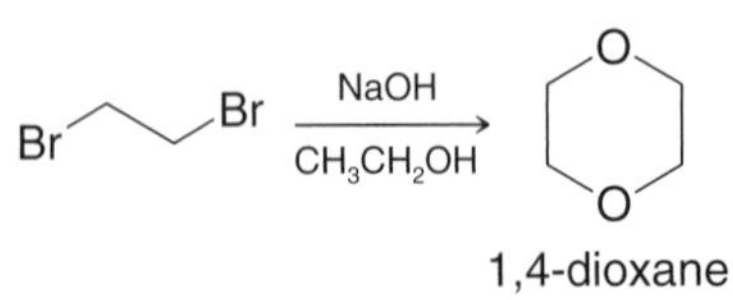

7.8 Propose a mechanism for the following reaction, which takes place under conditions that favor an S_N1 reaction.

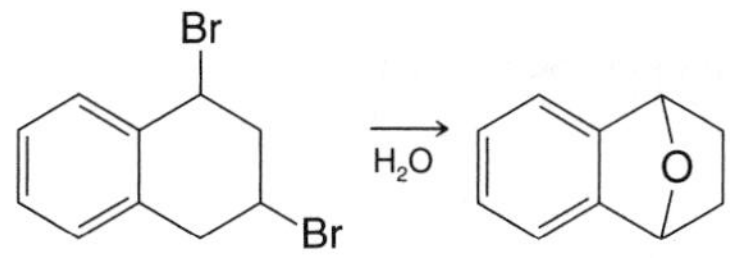

7.9 Draw a complete, detailed mechanism for the following reaction. (*Hint:* It may help to review the mechanism for keto-enol tautomerization.)

$$HC{\equiv}CCH_2CH_3 + NaOH \rightarrow CH_3C{\equiv}CCH_3$$

8 $S_N1/S_N2/E1/E2$ Reactions: The Whole Story

When you complete this chapter, you should be able to:

- Recognize when there is a competition among S_N1, S_N2, E1, and E2 reactions.
- Explain what a rate-determining step is and identify the rate-determining step of S_N1, S_N2, E1, and E2 reactions.
- Describe the role of the attacking species in the S_N1, S_N2, E1, and E2 rate-determining steps.
- Write the rate laws for S_N1, S_N2, E1, and E2 reactions and rationalize the form of each rate law.
- Explain how the strength of the attacking species governs the relative rates of S_N1, S_N2, E1, and E2 reactions.
- Classify an attacking species as a strong or weak nucleophile and as a strong or weak base and, based on this classification, determine which reactions are favored.
- Explain how the concentration of the attacking species governs the relative rates of S_N1, S_N2, E1, and E2 reactions.
- Recognize when an attacking species is in high or low concentration and subsequently predict which reactions are favored by such a concentration.
- Determine relative leaving group abilities and explain their roles in governing the rates of S_N1, S_N2, E1, and E2 reactions.
- Classify a leaving group as good, moderate, moderately poor, or terrible and, based on this classification, predict which reactions are favored by the leaving group.
- Explain how the type of carbon atom to which a leaving group is attached will govern the relative rates of S_N1, S_N2, E1, and E2 reactions.
- Determine which reactions are favored or disfavored by each specific type of carbon atom.

- Recognize a solvent as either protic or aprotic and explain how the type of solvent governs the relative rates of S_N1, S_N2, E1, and E2 reactions.
- Determine which reactions are favored by the specific type of solvent being used.
- Explain the role of heat in governing the outcome of substitution versus elimination reactions.
- Use a systematic method to predict the outcome of the competition among S_N1, S_N2, E1, and E2 reactions, given only the reactants and reaction conditions.

Your Starting Point

Answer the following questions to assess your knowledge about S_N1, S_N2, E1, and E2 reactions.

1. What enables a species to act as a base? As a nucleophile? ______

2. What is a rate-determining step, and how is it identified in a multistep mechanism? ______

3. If the rate law for a reaction $A + B \rightarrow C$ is Rate $= k[A]$, what happens to the rate if the concentration of A is doubled and the concentration of B is tripled? ______

4. For each pair, identify the stronger nucleophile: (a) H_2O or HO^-; (b) HO^- or F^-. ______

5. For each pair, identify the stronger base: (a) H_3C^- or HO^-; (b) H_2N^- or H_2O. ______

6. For each pair, identify the more stable species: (a) HO^- or H_2O; (b) HO^- or F^-; (c) F^- or Cl^-. ______

7. What happens to the stability of a carbocation when the number of alkyl groups attached to the C^+ atom increases? ______

8. Which carbocation is more stable—one in which the positive charge is delocalized by resonance or one in which the positive charge is localized? ______

9. Identify each solvent as either protic or aprotic: DMSO, CH_3OH, H_2O. ______

10. In what type of solvent is Cl^- more stable—protic or aprotic? ______________________

11. What is entropy? All else being equal, which is more favored—a system with greater entropy or one with less entropy? ______________________

Answers: 1. A species can act as a base if it has an atom with a lone pair of electrons and a partial or full negative charge, allowing that atom to form a bond to an electron-poor H atom. A species can act as a nucleophile if it has an atom with a lone pair of electrons and a partial or full negative charge, allowing that atom to form a bond to an electron-poor non-H atom. 2. A rate-determining step is an elementary step whose reaction rate is the same as that of the overall reaction, and is identified as the step that is significantly slower than all the other steps of the mechanism. 3. The reaction rate is directly proportional to the concentration of A, but it is independent of the concentration of B, so doubling the concentration of A will double the rate and tripling the concentration of B will have no effect. Increasing both concentrations will therefore only double the rate. 4. (a) HO^- is the stronger nucleophile because its negative charge makes it less stable. (b) HO^- is the stronger nucleophile because it is less stable, due to having the negative charge on a less electronegative atom. 5. (a) H_3C^- is the stronger base because it is less stable, as a result of the negative charge being on a less electronegative atom. (b) H_2N^- is the stronger base because it is less stable, as a result of having a negative charge. 6. (a) H_2O is more stable because it is uncharged. (b) F^- is more stable because the negative charge is on a more electonegative atom. (c) Cl^- is more stable because the negative charge is on a larger atom. 7. With additional alkyl groups attached, the stability of the carbocation increases. 8. Resonance delocalization of the positive charge gives rise to a more stable carbocation. 9. DMSO is aprotic, whereas CH_3OH and H_2O are both protic solvents. 10. Cl^- is more stabilized in a protic solvent because protic solvents solvate negative charges very strongly. 11. Entropy is a measure of disorder. All else being equal, a system with greater entropy is the one that will be favored.

8.1 Introduction

There are two major reasons for this chapter. First, S_N1, S_N2, E1, and E2 reactions are closely related to one another, and as such, they turn out to be the most difficult set of reactions for many organic chemistry students. These reactions are typically encountered in the middle of the first semester and are integral in whether or not a student does well in the course. Second, those four reactions bring together all of the major concepts we have learned thus far. As you read through this chapter, keep note of how we incorporate aspects of bonding, isomerism (in particular, stereoisomerism), charge stability, mechanisms, and intermolecular interactions (especially solvent effects).

These reactions tend to be challenging because, in general, if one reaction occurs, then *all four will occur simultaneously* (although at different reaction rates) and will thus compete with one another. Why is this so? We can see why by examining all four mechanisms together, which are shown in Figure 8-1.

Notice that all four reactions require a substrate with a suitable leaving group and they also require an **attacking species**—either a nucleophile for the substitution reactions (Figures 8.1a and 8.1b) or a base for the elimination reactions (Figures 8.1c and 8.1d). However, it is important to realize that an attacking species can in general have a dual character.

FIGURE 8-1 Complete mechanisms of the (a) S_N2, (b) S_N1, (c) E2, and (d) E1 reactions. All four reactions involve a substrate with a suitable leaving group. The nucleophilic substitution reactions involve a nucleophile whereas the elimination reactions involve a base. However, nucleophiles can act as bases, and vice versa.

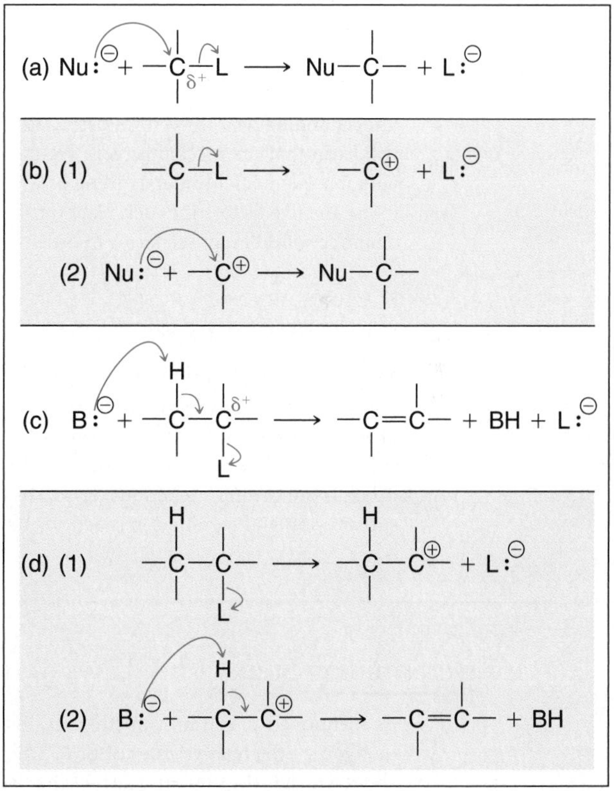

Species that are nucleophiles can also act as bases, and vice versa.

The reason for this dual nature of bases and nucleophiles is relatively straightforward. A base generally contains an atom with a partial or full negative charge and a lone pair, allowing it to form a bond to an electron-poor proton (H^+). A nucleophile generally contains an atom with a partial or full negative charge and a lone pair, allowing it to form a bond to an electron-poor non-H atom (usually a C atom), often displacing a leaving group in the process. Therefore, whether we call a species a base or a nucleophile depends entirely on the reaction it undergoes.

Despite the fact that all four substitution and elimination reactions occur simultaneously, there is often a **major product**, which is the one that is *formed in the greatest amount*. In your yearlong organic chemistry course, you will be expected to determine the major product of these reactions. To do so, you must first determine the *predominant mechanism* (i.e., the mechanism that proceeds the fastest) that occurs under the specific set of reaction conditions you are given. Keep in mind, though, that *the predominant mechanism can be different under different conditions*! Once you have determined the predominant mechanism, predicting the products is only a matter of working through that mechanism, as you did in Chapters 5 and 7.

Determining the predominant mechanism is quite often the single most difficult aspect of dealing with the competition among S_N1, S_N2, E1, and E2 reactions. This is because there are *five major factors* that dictate how favorable each reaction is. For a given set of reaction conditions, it is your job to figure out which reaction each of the five factors favors and then weigh all of the results to determine the predominant reaction.

By and large, students' frustration with these reactions stems from trying to memorize which set of conditions favors which reaction. Textbooks usually add to the temptation, providing a table that summarizes such things. *Don't attempt to memorize it!* Doing so without understanding *why* leads to disaster. Instead, throughout the remainder of this chapter, you will learn that the mechanisms in Figure 8-1 lead to different rate laws. The rate laws, along with a fundamental understanding of what drives each mechanism, provide insight into how to use the five factors for each reaction. Toward the end of this chapter, we will work through a number of different scenarios in order to gain experience in applying the five factors.

WHY SHOULD I CARE?

Nucleophilic substitution and elimination reactions will be introduced in your full-year course separately from other reactions. However, many of the concepts you will encounter in your treatment of these reactions will be applied to other reactions as well. Furthermore, you will find that nucleophilic substitution and elimination reactions comprise portions of other, more complex reactions. It is therefore important that you master these nucleophilic substitution and elimination reactions before encountering more complex reactions.

8.2 Rate-Determining Steps: Rate Laws and the Role of the Attacking Species

As mentioned in the introduction to this chapter, it is important to understand the factors that govern the relative rates of S_N1, S_N2, E1, and E2 reactions in order to make predictions about which is the predominant mechanism in the competition. In doing so, it is best to begin with the concept of a rate-determining step. As its name suggests, a **rate-determining step** of a mechanism is an elementary step that governs the rate of the overall reaction. For an S_N2 or E2 reaction, it is rather trivial to identify the rate-determining step because each of these reactions has just one step in its mechanism. For a multistep mechanism, such as that of an S_N1 or E1 reaction, it takes some additional consideration.

Probably the best way to grasp the concept of a rate-determining step is by analogy. Imagine that each elementary step of a mechanism is represented by a pipe and that the rate at which an elementary step takes place is represented by the rate at which water flows through that pipe. Quite simply, the rate at which water can flow through a pipe increases as the diameter of the pipe increases. Therefore, a fast elementary step is represented by a relatively fat pipe, and a slow elementary step is represented by a skinny pipe. Moreover, a multistep mechanism is represented by multiple pipes connected end to end.

Now consider what happens if two pipes are connected (representing a two-step mechanism) and one pipe is much skinnier than the other. Regardless of which pipe comes first, the diameter of the fatter pipe has essentially no effect on the rate at which water can flow through the entire system. Rather, as illustrated in Figure 8-2, the diameter of only the skinnier pipe dictates the flow of water. By analogy, the rate of an overall reaction is essentially the same as the rate of the slowest step of the mechanism. In other words,

The rate-determining step of a multistep mechanism is an elementary step that is much slower than any of the other steps.

With this in mind, let's examine, once again, the S_N1 and E1 mechanisms in Figure 8-1b and 8-1d. For each mechanism, the first step is the departure of the leaving group, producing two additional formal charges. By contrast, the second step of each reaction reduces the number of formal charges by two. In an S_N1 mechanism, this happens by the coordination of the nucleophile to the carbocation. In an E1 mechanism, it happens by the removal an H^+ ion from the carbocation.

As we saw in Section 7.6, an increase in the number of species with a formal charge is highly unfavorable. Therefore, the first step of an S_N1 or E1 mechanism is rather slow. The opposite is true for each reaction's second step, which proceeds rather quickly, due to the decrease in the number of formal charges. Consequently, in each case the first step is much slower than the second. In other words,

FIGURE 8-2 **Rate-determining steps.** Two pipes connected end to end represent the sequential steps of a reaction mechanism. Regardless of which comes first—the fatter pipe or the skinnier pipe—the rate at which water can flow does not change. It is governed only by the rate at which water can flow through the skinnier pipe. The skinnier pipe is therefore analogous to the rate-determining step of a multistep mechanism.

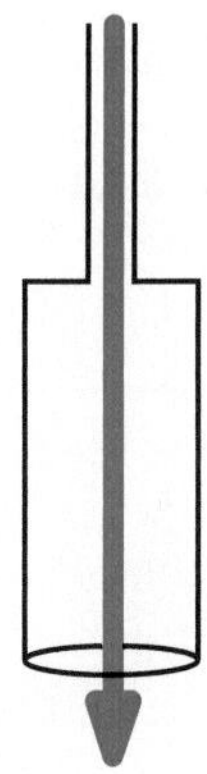

In an S_N1 or E1 reaction, the first step of the mechanism—that is, the departure of the leaving group—is the rate-determining step.

✔ QUICK CHECK

Identify the rate-determining step in each of the following mechanisms.

(1) 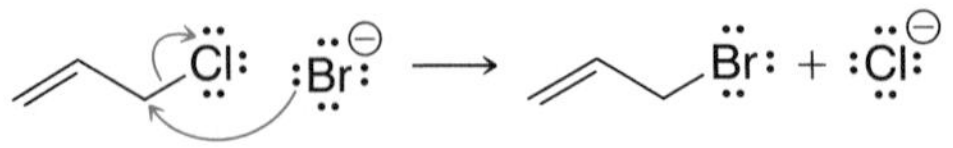

(2) 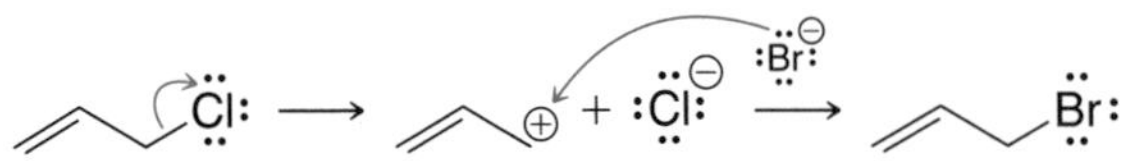

Answers: 1. This is an S_N2 mechanism, which takes place in a single step, so that step must be the rate-determining step. 2. This is an S_N1 mechanism. The first step is much slower than the second, so the first step is the rate-determining step.

Now that we know the rate-determining steps for S_N1, S_N2, E1, and E2 reactions, we can begin to develop two tools that will help us predict how various factors affect the relative reaction rates. One of these tools is a particular way in which to view the role of the attacking species (i.e., the nucleophile or base) in the rate-determining step. The second is the rate law for each reaction, which specifies the way in which each reaction's rate depends on the reactant concentrations.

Let's begin with the role of the attacking species in an S_N2 reaction (Figure 8-1a). In such a reaction, a nucleophile displaces the leaving group on the substrate in a single step. Therefore, the nucleophile can be thought of as *forcing the action.* More specifically,

In an S_N2 reaction, the nucleophile *forces off* the leaving group by direct displacement.

This is in contrast to what we see for an S_N1 mechanism (Figure 8-1b). In the rate-determining step of an S_N1 mechanism, the leaving group departs in a heterolysis step, producing a carbocation. Only then does the nucleophile enter the picture. In other words,

In an S_N1 reaction, the nucleophile *waits* for the leaving group to leave.

In an E2 mechanism (Figure 8-1c), the role of the attacking species (this time as a base) is similar to what we observe for an S_N2 mechanism. Namely, the base forces the action. More specifically,

In an E2 reaction, the base forces off the leaving group by pulling off an adjacent proton.

The role of the base in an E1 mechanism (Figure 8-1d) is similar to that of the nucleophile in an S_N1 mechanism. In the E1 rate-determining step, the only thing that happens is the departure of the leaving group, producing a carbocation. The base does not enter the picture until the second step. Therefore, the role of the attacking species can be viewed as follows.

In an E1 reaction, the base *waits* for the leaving group to leave.

By viewing the role of the attacking species in these ways, we can arrive at the rate law for each reaction. Beginning with the S_N2 reaction, notice that in the rate-determining step the nucleophile must collide with the substrate to form a bond. The greater the number of nucleophile species present, the more frequently these collisions should happen, and the faster the rate should be. Similarly, with additional substrate species present, such collisions should take place more frequently, leading to an increase in reaction rate. In other words, the rate of an S_N2 reaction should increase with an increase in the concentration of either the nucleophile or the substrate. This idea is captured in Equation 8-1, which states that the rate of the reaction is proportional to the product of the concentration of the nucleophile and the concentration of the substrate—k_{S_N2} is a constant, called the **rate constant** for the reaction, and concentrations are indicated by square brackets, [ ].

$$\text{Rate}(S_N2) = k_{S_N2}[\text{Nu}][\text{R–L}] \tag{8-1}$$

A very similar story can be told for the E2 reaction, due to the similarity of its mechanism to that of the S_N2 reaction. Namely, because the base and the substrate must collide during the course of the E2 rate-determining step (Figure 8-1c), the rate of the E2 reaction depends on the concentrations of both species. This is reflected by the rate law for the E2 reaction, shown in Equation 8-2—k_{E2} is the rate constant for the E2 reaction.

$$\text{Rate}(E2) = k_{E2}[\text{Base}][\text{R–L}] \tag{8-2}$$

The rate laws for the S_N1 and E1 reactions have a somewhat different form. For both reactions, the substrate is the only reactant that appears in the rate-determining step (the first step of Figures 8-1b and 8-1d). Therefore, the reaction rates should depend on the concentration of only the substrate, as indicated by their rate laws in Equations 8-3 and 8-4. The difference in the two equations lies in the rate constants—k_{S_N1} and k_{E1}.

$$\text{Rate}(S_N1) = k_{S_N1}[\text{R–L}] \tag{8-3}$$
$$\text{Rate}(E1) = k_{E1}[\text{R–L}] \tag{8-4}$$

With these tools—that is, specific views of the role of the attacking species in each reaction and the respective rate laws—we are now in a position to examine the factors that affect the various reactions' rates—and hence the outcome of the competiton among the four reactions.

Before moving on, however, it is worth noting that, with what we have discussed in this section, we can understand the significance of the "1" and the "2" in the names of the four reactions. Quite simply, these numbers refer to the molecularity of the rate-determining step. For example, the rate-determining step of an S_N2 reaction (i.e., the only elementary step of the mechanism) has two separate reactant species, so it is a bimolecular step. By contrast, there is just one reactant molecule in the rate-determining step of an S_N1 mechanism, making it a unimolecular step. The same can be said of E2 and E1 reactions as well.

8.3 Factor #1: Strength of Attacking Species

In the previous section, we saw that one of the key differences between the S_N1 and S_N2 reactions is the role of the nucleophile. The S_N2 reaction is primarily driven by the nucleophile displacing the leaving group—that is, the nucleophile forces the action. Therefore, it should be no surprise that the strength of the nucleophile has a major impact on the rate of an S_N2 reaction.

The stronger the nucleophile, the faster the rate of an S_N2 reaction.

Another way to see this is that the nucleophile appears in the rate law for the S_N2 reaction (Equation 8-1), suggesting that the nucleophile's identity is important in governing the reaction rate.

In the S_N1 mechanism, on the other hand, the nucleophile essentially sits back and waits for the leaving group to come off before it attacks the highly reactive carbocation. Therefore,

The strength of the nucleophile has essentially no effect on the rate of an S_N1 reaction.

Whether the nucleophile is strong or weak, the S_N1 mechanism has it sit back and wait. A second way of seeing this is that the nucleophile does not appear in the rate equation for the S_N1 reaction (Equation 8-3), suggesting that the nucleophile's identity is not important in governing the reaction rate.

Because an increased strength of the nucleophile speeds up an S_N2 reaction, but not an S_N1 reaction, it is possible to control the outcome of a competition between the two reactions simply by choosing the nucleophile wisely. If a nucleophile is strong enough, the rate of an S_N2 reaction could surpass that of a competing S_N1 reaction. Conversely, if the nucleophile is weak enough, the S_N2 reaction rate could be slowed substantially, leaving the S_N1 reaction as the faster one. This idea can be summarized as follows.

Strong nucleophiles tend to favor S_N2 reactions, and weak nucleophiles tend to favor S_N1 reactions.

With this understanding, it is important to be able to recognize strong nucleophiles and weak nucleophiles. Although there is no clear cutoff, the following generalities can be made.

Strong nucleophiles typically have a full negative charge.

Weak nucleophiles are typically uncharged.

There are certainly exceptions, but such a discussion is not appropriate here.

✔ QUICK CHECK

Identify each of the following species as either a strong nucleophile or a weak nucleophile.
H_3C^- H_2O HO^- Cl^- H_3N CH_3CH_2OH $CH_3CO_2^-$

Answer: The strong nucleophiles are the negatively charged ones: H_3C^-, HO^-, Cl^-, and $CH_3CO_2^-$. The uncharged nucleophiles—H_2O, H_3N, and CH_3CH_2OH—are weak.

The E1 and E2 mechanisms mirror the S_N1 and S_N2 mechanisms, respectively, in the way they are affected by the strength of the attacking species—the base. In the E2 mechanism, the base is viewed as forcing the action. Therefore,

The stronger the base, the faster the rate of an E2 reaction.

Notice, also, that this is supported by the fact that the base appears in the rate law of an E2 reaction (Equation 8-2).

In the E1 mechanism, however, the base does not come into play in the rate-determining step. Consequently,

The strength of the base does not affect the rate of an E1 reaction.

This is consistent with the fact that the concentration of the base does not show up in the E1 rate law (Equation 8-4).

Similar to substitution reactions, we can control the outcome of the competition between E2 and E1 reactions because the rates depend differently on the strength of the base. We do so by taking advantage of the fact that a stronger base speeds up an E2 reaction, but not an E1 reaction. If a base is strong enough, the E2 rate could surpass that of an E1, and if it is weak enough, the reverse could be true. The main lesson from this is as follows.

Strong bases tend to favor E2 reactions, whereas weak bases tend to favor E1 reactions.

In light of this, it is important to be able to distinguish strong bases from weak ones. Just as we saw in Section 7.4, the benchmark is the hydroxide ion, HO^-.

For elimination reactions, strong bases are ones that are roughly as strong as or stronger than HO^-.

8.4 Factor #2: Concentration of Nucleophile/Base

After having examined the rate laws for the S_N1, S_N2, E1, and E2 reactions, the effect of nucleophile/base concentration on the reaction rate of each reaction is straightforward. According to Equations 8-1 and 8-3, the rate of the S_N2 reaction depends on the concentration of the nucleophile, whereas the rate of the S_N1 reaction does not. Therefore, *increasing the concentration of the nucleophile will increase the rate of the S_N2 reaction, but will have essentially no effect on the rate of the S_N1 reaction.* Similarly, according to Equations 8-2 and 8-4, the E2 reaction rate depends on the concentration of the base, but the E1 reaction rate does not. Consequently, *increasing the concentration of the base will significantly increase the rate of the E2 reaction, but not the rate of the E1 reaction.*

Because of these different dependencies on the concentration of the attacking species, we can partly control the outcome of the competition among the four reactions by adjusting the concentration of the attacking species. The rate of an S_N2 reaction can surpass that of an S_N1 reaction if the concentration of a sufficiently strong nucleophile is high enough. Conversely, with a low enough concentration of the nucleophile, the S_N2 reaction can be made to be slower than that of the S_N1 reaction. In general, then, we can say the following.

A high concentration of a nucleophile tends to favor S_N2 reactions over S_N1 reactions.

A low concentration of a nucleophile tends to favor S_N1 reactions over S_N2 reactions.

We have a very similar story with E1 and E2 reactions. That is, with a high concentration of a sufficiently strong base, an E2 reaction can be made to be faster than an E1 reaction. And with a low enough concentration of the base, the reverse can be true. Thus,

A high concentration of a base tends to favor E2 reactions over E1 reactions.

A low concentration of a base tends to favor E1 reactions over E2 reactions.

We must be careful, however, about making such statements when the nucleophile or base is weak. A nucleophile is classified as weak because it is incapable of forcing off the leaving group in an S_N2 reaction. Instead, it waits for the leaving group to leave on its own before attacking the highly reactive carbocation that is produced. A high concentration of such a weak nucleophile does nothing to allow it to force off the leaving group. All that would be different is the number of molecules of the nucleophile waiting for the leaving group to come off on its own.

A similar picture exists with elimination reactions. A base is classified as weak because of its inability to pull off a proton in an E2 reaction. Rather, it waits for the leaving group to come off on its own before attacking the highly reactive carbocation intermediate. A high concentration of such a weak base would serve only to increase the number of molecules of the base that must wait.

The above ideas can be summarized as follows.

A weak nucleophile tends to favor the S_N1 mechanism, regardless of the concentration of that nucleophile.

A weak base tends to favor the E1 mechanism, regardless of the concentration of that base.

✔ **QUICK CHECK**

For each of the following conditions, determine whether the factor of concentration favors an S_N1, S_N2, E1, or E2 reaction.

1. High concentration of Br^-.
2. High concentration of CH_3OH.
3. Low concentration of Br^-.
4. Low concentration of CH_3OH.
5. High concentration of HO^-.
6. Low concentration of HO^-.

Answers: 1. Br^- is a strong nucleophile, but a weak base, so the high concentration favors S_N2, but not E2. 2. CH_3OH is a weak base and a weak nucleophile, so S_N1 and E1 are favored, despite the high concentration. 3. The low concentration favors S_N1 and E1, despite Br^- being a strong nucleophile. 4. The low concentration of the weak nucleophile and weak base favors S_N1 and E1. 5. HO^- is a strong base and a strong nucleophile, so the high concentration favors S_N2 and E2. 6. The low concentration favors S_N1 and E1, despite HO^- being a strong nucleophile and a strong base.

8.5 Factor #3: Stability of the Leaving Group

In each of the four reactions we are examining, a leaving group must leave from the substrate. Furthermore, the step in which the leaving group leaves is the rate-determining step of each reaction. Therefore, it should be no surprise that the identity of the leaving group has an impact on the rate of all four reactions. More specifically,

The better (more stable) the leaving group, the faster the reaction—in all four cases.

However, the stability of the leaving group affects the reaction rates differently. Compare, for example, the S_N2 and S_N1 reactions. In the S_N2 reaction, the leaving of the leaving group is *assisted* by the attack of the nucleophile—that is, the nucleophile is helping to force off the leaving group. By contrast, in the rate-determining step of an S_N1 mechanism, the only thing that happens is the leaving of the leaving group. Nothing is helping the leaving group to leave. With this in mind, it should make sense that the identity of the leaving group plays a greater role in S_N1 reactions than in S_N2 reactions. More specifically, with an increased stability of the leaving group, there is a more pronounced increase in the rate of an S_N1 reaction than in the rate of an S_N2 reaction. Thus, the identity of the leaving group can control the outcome of the competition between the two reactions.

Very good leaving groups tend to favor S_N1 reactions over S_N2 reactions.

Similar things can be said about the competition between E1 and E2. Namely, E1 reaction rates are more sensitive to the identity of the leaving group than are E2 reaction rates. This is because in the rate-determining step of an E1 reaction, the only thing that happens is the leaving of the leaving group. By contrast, in an E2 reaction, the attack by the base helps to force off the leaving group. Consequently,

Very good leaving groups tend to favor E1 reactions over E2 reactions.

The cutoff between what are considered good leaving groups and bad leaving groups is somewhat fuzzy, but we can provide some benchmarks. The benchmark for a very good leaving group is Br^-—the Br atom, because of its large size and relatively high electronegativity, accommodates the negative charge quite well. Therefore, any leaving group that is at least as stable as Br^- should also be considered a very good leaving group, favoring the S_N1 and E1 mechanisms over the S_N2 and E2 mechanisms.

Somewhat less stable is the Cl^- ion because it is significantly smaller in size than the Br^- ion. It is, however, considered a moderate leaving group—decent enough that substrates containing such a leaving group can undergo S_N1 and E1 reactions. On the other hand, it is not a good enough leaving group to say that it overwhelmingly tends to favor S_N1 and E1 reactions over S_N2 and E2 reactions. Instead, we can say that it tends to favor all four mechanisms roughly the same.

The F^- ion is less stable than the Cl^- ion, again because of its smaller size. Consequently, substrates that have the F^- leaving group do not undergo S_N1 or E1 reactions, meaning that F^- should be considered a moderately bad leaving group. Therefore, if a substrate containing F^- as the leaving group is to undergo any of the four mechanisms, it must be either S_N2 or E2—the F^- leaving group must be forced off by a strong nucleophile. However, F^- tends to be a bad enough leaving group that such S_N2 and E2 reactions are rare.

Finally, HO^- is less stable still. The O atom cannot accommodate the negative charge as well as F can because O has a smaller electronegativity. Because of that instability, we never see alcohols, R—OH, undergo S_N1 or E1 reactions without modification of the leaving group (discussed in Section 8.5A)—the leaving group would otherwise come off as HO^-. Moreover, HO^- is such a bad leaving group that we almost never see alcohols undergo S_N2 or E2 mechanisms.

The HO^- ion is therefore our benchmark for terrible leaving groups. Any leaving group that would come off as a species as stable as or less stable than HO^- is also considered a bad leaving group. Examples include CH_3O^-, H_2N^-, CH_3^-, and H^-. Any substrate containing only one of these (or similar) leaving groups tends to be unreactive toward any of the four mechanisms in this chapter.

QUICK CHECK

Determine whether each of the following substrates would favor S_N1, S_N2, E1, or E2 reactions.

1. $C(CH_3)_4$.
2. $(CH_3)_2CHCl$.
3. $(CH_3)_2CHOH$.
4. $(CH_3)_2CHNH_2$.
5. $(CH_3)_2CHI$.
6. $(CH_3)_2CHSH$.

Answers: The identity, quality, and favored reaction(s) of each leaving group are as follows: 1. H_3C^-, terrible leaving group, no reactions. 2. Cl^-, moderate leaving group, all four reactions. 3. HO^-, terrible leaving group, no reactions. 4. H_2N^-, terrible leaving group, no reactions. 5. I^-, excellent leaving group, S_N1 and E1. 6. HS^-, moderately poor leaving group, S_N2 and E2.

8.5A MAKING A GOOD LEAVING GROUP OUT OF A BAD ONE

As just discussed, HO^- is a terrible leaving group because the O atom does not accommodate the negative charge very well. However, it is rather straightforward to convert the HO^- leaving group into a good one. One method is to protonate the OH group at the O atom, using a somewhat strong acid like H_2SO_4 or H_3PO_4. This results in $R{-}OH_2^+$, which enables the leaving group to come off in the form of OH_2 (i.e., H_2O), or water. Since water is quite a stable molecule, bearing no formal charge, it is an *excellent* leaving group! With such a good leaving group, both the S_N1 and the E1 mechanisms become quite favorable.

Notice that in protonating the O atom before it comes off, the leaving group is stabilized by having the formal negative charge removed (HO^- versus H_2O). There are other ways of converting a poor HO^- leaving group into a very good one without removing the negative charge. Although we will not discuss it here, one way is through extensive resonance of the −1 formal charge that develops on the leaving group—recall from Chapter 6 that resonance can stabilize a charged species significantly. The leaving groups that work in this way will be left for your full-year course.

Amines, $R{-}NH_2$, present a situation similar to that of alcohols. As is, amines are essentially unreactive toward substitution and elimination reactions because the leaving group that would come off is H_2N^-, which is even more unstable than HO^-. However, because $R{-}NH_2$ is mildly basic at the N atom, acidic conditions will convert the substrate to $R{-}NH_3^+$, so that the leaving group that can come off is NH_3—a neutral molecule.

QUICK CHECK

For each reaction, identify the leaving group, and determine whether or not it would be a good one.

(1)

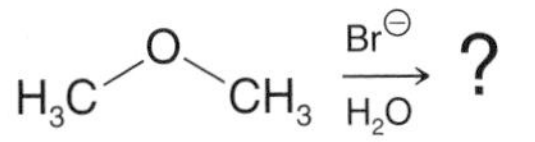

(2)

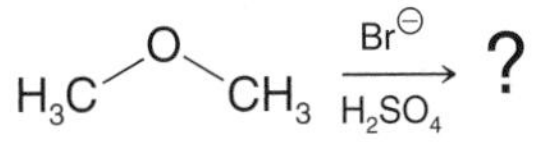

Answers: 1. The conditions are neutral, so the leaving group has to come off as CH_3O^-, a terrible leaving group.
2. Under strongly acidic conditions, the O is protonated, so the leaving group is CH_3OH, a very good leaving group, due to its lack of a negative charge.

8.6 Factor #4: Type of Carbon Atom Bonded to the Leaving Group

The number of alkyl (R) groups attached to the C atom bearing the leaving group (L) can range from zero to three. If there are no R groups and three Hs, then the C is part of a *methyl* (CH_3) group. If there is one R group, it is a **primary** (1°) **C atom.** Two and three R groups give rise to **secondary** (2°) and **tertiary** (3°) **C atoms,** respectively.

The rate of the S_N2 reaction depends heavily on whether the leaving group is on a methyl, primary, secondary, or tertiary C atom. The reason is that the S_N2 reaction rate is dictated by the nucleophile's ability to force the action, and in order to do so, it must form a bond with that C. As it turns out,

The more alkyl groups attached to the C atom that has the leaving group, the slower the S_N2 rate.

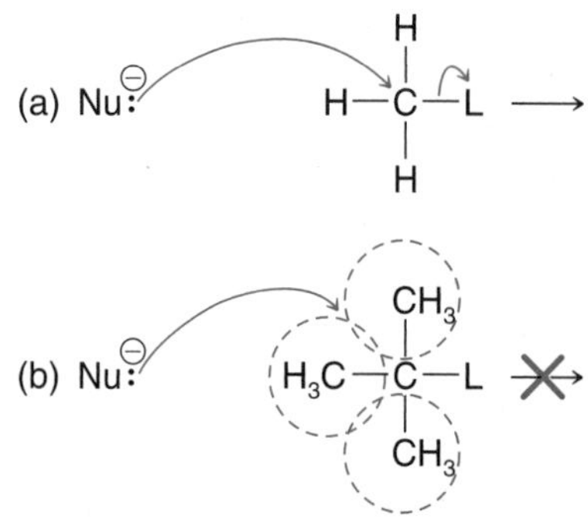

FIGURE 8-3 **Steric hindrance in S_N2 reactions.** The S_N2 reaction with different numbers of alkyl groups on the C atom bearing the leaving group. (a) The H atoms are small, allowing the nucleophile to attack the C atom. (b) The R groups are bulky, providing steric hindrance that prevents the nucleophile from being able to attack the C atom. As the number of alkyl groups on the C atom increases, the S_N2 rate decreases.

The reason is that those additional R groups are bulky, making it more difficult for the nucleophile to approach the C atom and to actually form a bond with it (Figure 8-3).

This phenomenon is an example of *steric hindrance* by the R groups, which we first encountered in Section 5.11. Therefore, a substrate that has the leaving group on a tertiary C atom slows down the S_N2 reaction significantly. By contrast, if the C atom is either a methyl or a primary carbon atom, steric hindrance is not a problem, and the S_N2 reaction is allowed to proceed quickly. In other words,

Methyl and primary C atoms favor the S_N2 mechanism, whereas tertiary C atoms disfavor it.

Secondary C atoms are somewhat neutral in this regard.

Unlike in S_N2 reactions, in E2 reactions, the number of R groups on the C atom attached to the leaving group has little influence on the rate of the overall reaction. This is because the base does not attack that C atom—in fact, it does not attack any C atom. Rather, it attacks the H atom on the C atom next door, and that H atom is exposed, ready to be picked off by a base (Figure 8-4).

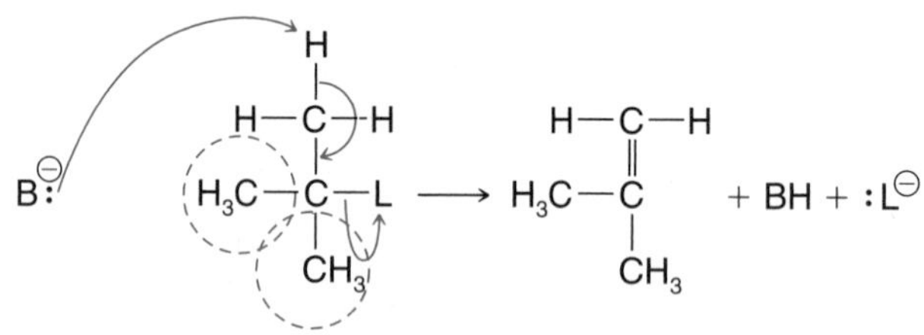

FIGURE 8-4 **Accessibility of the proton in E2 reactions.** Unlike in S_N2 reactions, an increase in the number of alkyl groups to the C atom with the leaving group does not cause much steric hindrance in an E2 reaction. Even with tertiary C atoms, the protons on an adjacent C atom remain well exposed to the base.

Therefore, the number of R groups on the C atom that is attached to the leaving group has little impact on the exposure of that H atom. As such,

E2 reactions can proceed readily with all three types of C atoms—primary, secondary, and tertiary.

The S_N1 and E1 mechanisms have exactly the same rate-determining step, so we can examine these reactions together. In the rate-determining step for each reaction, the leaving group departs on its own, leaving behind a carbocation, C^+. As we saw in Section 7.2, this step is a substantially uphill elementary step, due to the loss of an octet from C and also due to the increase in the number of formal charges. Thus, this step has quite a large energy barrier, which is why it is rather slow. However, recall the Hammond postulate from Section 6.8, which allows us to say that, when comparing two of the same type of elementary step (in this case, heterolysis steps), the one that is more downhill (less uphill) tends to be faster. Applying this to the current situation, we can say the following:

The rate of the rate-determining step of an S_N1 or E1 reaction tends to increase as the carbocation that is produced becomes more stable.

One way the carbocation is stabilized is with additional R groups attached to the C^+ atom. Consequently, S_N1 and E1 rates increase with an increasing number of alkyl groups attached to the C atom that has the leaving group. Therefore,

Tertiary substrates ($R_3C—L$) tend to favor S_N1 and E1 reactions.

Methyl and primary substrates ($CH_3—L$ and $RCH_2—L$) tend to disfavor S_N1 and E1 reactions.

Secondary substrates ($R_2CH—L$) are somewhat neutral in this regard.

We must be careful with the above generalizations, however, because a more important contribution to the stability of a carbocation could come from *resonance delocalization* of the positive charge. Two important types of substrates that give rise to such resonance stabilization are **benzylic substrates** and **allylic substrates**, which are shown in Equations 8-5 and 8-6.

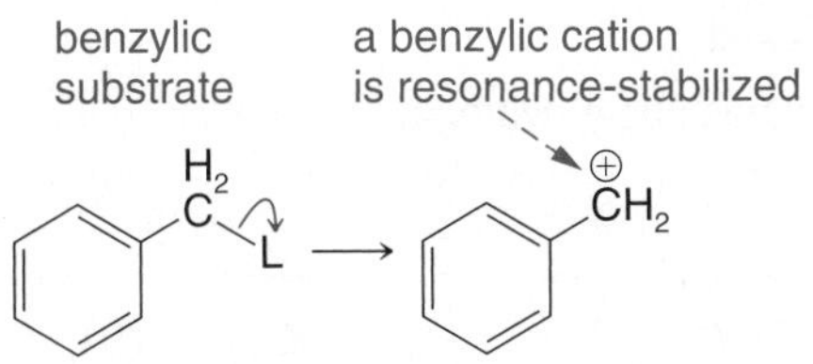

(8-5)

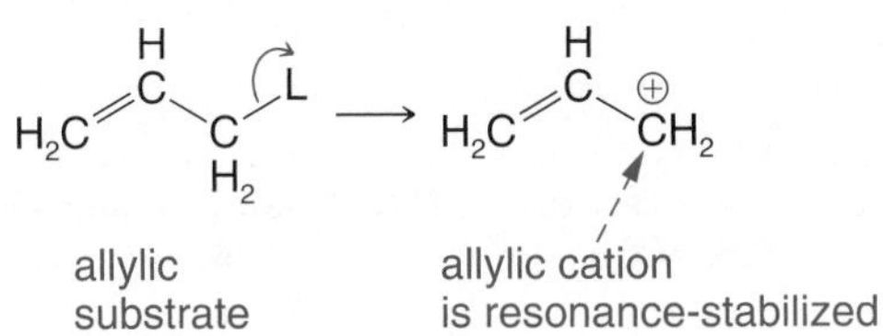

(8-6)

The substantial stabilization that comes from this resonance goes a long way toward facilitating the leaving of the leaving group. Consequently,

Benzylic and allylic substrates tend to favor S_N1 and E1 reactions, even if the leaving group is attached directly to a primary C atom.

QUICK CHECK

Based on the type of C atom to which the leaving group is attached, determine whether each substrate favors an S_N1, S_N2, E1, or E2 reaction.

(1) (2) (3) (4) (5)

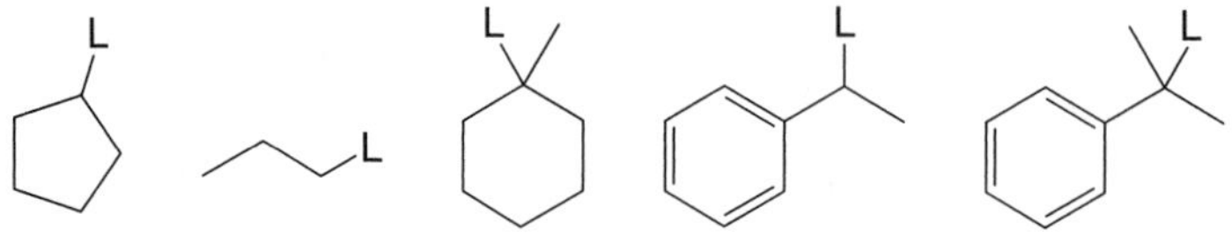

Answers: 1. The C atom that has the leaving group is secondary, so it favors all four reactions. 2. The C atom is primary, so it favors S_N2 and E2 only. 3. The C atom is tertiary, so it favors S_N1, E1, and E2 only. 4. The C atom is secondary, so all four reactions are favored. Because the leaving group is on a benzylic C atom, the S_N1 and E1 reactions are particularly favored. 5. The C atom is tertiary, so only S_N1, E1, and E2 reactions are favored. Because the leaving group is on a benzylic C atom, the S_N1 and E1 reactions are particularly favored.

8.7 Factor #5: Solvent Effects

The choice of solvent can have a dramatic effect on the outcome of S_N1, S_N2, E1, and E2 reactions. The two types of solvent important in these reactions are protic and aprotic solvents, which we introduced in Section 6.9. Recall that a *protic solvent* contains a hydrogen-bond donor, whereas an *aprotic solvent* does not. Examples of common protic solvents include water and alcohols (ROH). Common aprotic solvents include dimethyl sulfoxide (DMSO), dimethyl formamide (DMF), and acetone, shown again below.

common aprotic solvents

none of these compounds has a hydrogen-bond donor (NH, OH, or FH)

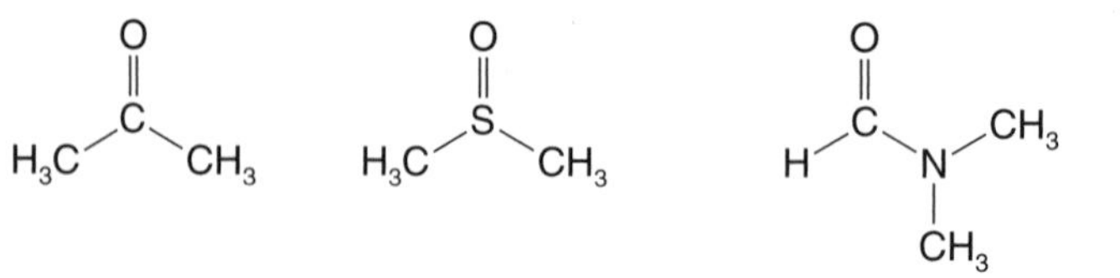

acetone — dimethyl sulfoxide (DMSO) — *N,N*-dimethylformamide (DMF)

A solvent can affect the outcome of nucleophilic substitution and elimination reactions by altering the strength of the nucleophile or base differently, leading to different effects on the relative rates of the reactions. An attacking species that is a strong nucleophile or base has a significant concentration of negative charge centered on a single atom. The large concentration of positive charge on the H atom of protic solvents enables those solvent molecules to stick quite strongly to strong nucleophiles and bases. As discussed previously in Section 6.9, this dramatically stabilizes and thus weakens the attacking species as a nucleophile and a base because the concentration of negative charge is tied up with the numerous partial positive charges on the H atoms of the solvent molecules (see again Figure 6-17).

Recall from Section 8.3 that weak nucleophiles and weak bases tend to favor S_N1 and E1 mechanisms, respectively. As a result, *conditions that normally favor an S_N2 or E2 reaction—strong nucleophile/base—can be altered to favor an S_N1 or E1 reaction if we choose the solvent to be protic, which effectively weakens the nucleophile/base* (we will see examples of this later).

By contrast, we saw in Section 6.9 that aprotic solvent molecules do not stick to anions very well because the solvent molecules do not have a large partial positive charge on an H atom. Rather, the positive charge is buried inside the solvent molecules (as shown previously in Figure 6-21), making the positive charge less accessible. Therefore, aprotic solvent molecules do not substantially weaken strong nucleophiles and bases; strong nucleophiles and bases are allowed to remain strong.

The lesson from the above can be stated as follows:

Aprotic solvents do not weaken the attacking species substantially—and therefore tend to favor S_N2 and E2 reactions.

Protic solvents weaken the attacking species substantially—and therefore tend to favor S_N1 and E1 reactions.

QUICK CHECK

Determine whether or not each of these solvents will favor an S_N1, S_N2, E1, or E2 reaction.

1. CH_3CH_2OH.
2. $[(CH_3)_2N]_3P{=}O$.

Answers: 1. This solvent is protic, as the OH group is a hydrogen-bond donor, so it favors S_N1 and E1 reactions.
2. This solvent is aprotic because there are no hydrogen-bond donors, so it favors S_N2 and E2 reactions.

8.8 Substitution Versus Elimination

Having looked at the five major factors thus far, you should be able to see the following trend:

Conditions that favor S_N2 reactions often favor E2 reactions as well.

The reason is that the same concentration of negative charge that makes a nucleophile a nucleophile also makes a base a base (as discussed in the introduction to this chapter). In other words, strong nucleophiles are usually strong bases, examples of which include HO^-, CH_3O^-, H_2N^-, and H_3C^-. Therefore, if HO^- reacts with a primary alkyl halide, such as CH_3CH_2Br, in an aprotic solvent, we expect the S_N2 mechanism to proceed at a substantial rate. This is because the nucleophile is rather strong, forcing the leaving group to leave. However, under the same conditions, we should also expect the E2 mechanism to proceed at a substantial rate. This is because the base is quite strong, which forces the removal of a proton and the leaving of the leaving group. The result is therefore a mixture of predominantly the S_N2 and E2 products.

If the nucleophile/base is chosen wisely, however, it is possible to promote the S_N2 reaction over the E2 reaction, and vice versa. This can be done because *the rate-determining step of the S_N2 reaction is not exactly the same as that of the E2 reaction.* A strong nucleophilic character of the species that attacks the substrate enhances the rate of the S_N2 reaction, while a strong basic character enhances the rate of the E2 mechanism. Therefore, in order to promote the S_N2 mechanism over the E2 mechanism, we must simply choose an attacking species that is a strong nucleophile, but a weak base. Examples include Cl^-, Br^-, HS^-, and HCO_3^-. Notice that they all have −1 formal charges, characterizing them as strong nucleophiles, but they are weaker bases than HO^-.

Conversely, in order to promote the E2 mechanism over the S_N2 mechanism, we must find an attacking species that is a strong base, but a weak nucleophile. A good example is the ***t*-butoxide anion,** $(CH_3)_3CO^-$. This species is a strong base because the negative charge is localized on an O atom, and it should therefore have a base strength similar to that of HO^-. (In fact, because of the inductive effects of the alkyl groups, this base is even stronger than HO^-—convince yourself of this!) However, it is a relatively poor nucleophile, despite the negative charge. The reason is that the entire species is bulky, with the three methyl groups surrounding the nucleophilic O atom. Those methyl groups introduce steric hindrance that makes it difficult for the nucleophilic O atom to get close to the C atom of the substrate (Figure 8-5)—something that is otherwise necessary to form a C—O bond and to force the leaving group to leave.

✔ QUICK CHECK

The diisopropylamide anion, $(CH_3)_2CH—N^-—CH(CH_3)_2$, is another example of a strong base that is a relatively weak nucleophile.

1. How can you tell it is a strong base?
2. Why is it a relatively weak nucleophile?

(*Hint:* To see what's going on, you should construct this species using a model kit!)

Answers: 1. It is a strong base because it is a stronger base than HO^-, given that a negative charge on N is less stable than a negative charge on O. 2. It is relatively weak nucleophile because of the substantial steric hindrance introduced by the rather bulky isopropyl groups, $—CH(CH_3)_2$, attached to the N atom.

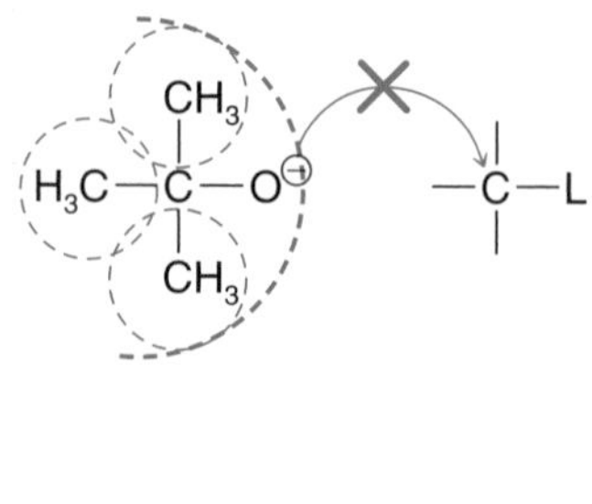

FIGURE 8-5 Steric hindrance with the *t*-butoxide anion. The steric hindrance introduced by the CH_3 groups diminishes the nucleophilicity of $(CH_3)_3CO^-$. Therefore, whereas the *t*-butoxide anion is a very strong base, it is a relatively poor nucleophile, allowing it to promote E2 reactions over S_N2 reactions.

Just as S_N2 and E2 reactions are favored under very similar reaction conditions, a parallel observation is made with S_N1 and E1 reactions.

Conditions that favor S_N1 reactions will favor E1 reactions as well.

Both the S_N1 and the E1 reactions are favored when the substrate has the leaving group on a tertiary C atom, the leaving group is a good leaving group, the solvent is protic, and the attacking species (the base or nucleophile) is weak. As it turns out, weak nucleophiles are often weak bases as well—both typically have low concentrations of negative charge on a single atom (recall arguments from Chapter 6).

Unlike S_N2 versus E2 reactions, it is very difficult to manipulate the percentage of S_N1 product compared to that of E1. The reason is that *both the S_N1 and the E1 reactions share exactly the same rate-determining step* (this is not the case for S_N2 versus E2). As a result, just about anything that can speed up the S_N1 mechanism will also speed up the E1 mechanism. Therefore, with S_N1 and E1 reactions, it is very difficult to choose reaction conditions that favor one over the other.

In general, S_N1 products will always be contaminated with a significant percentage of E1 products, and vice versa.

However, as will be discussed in the next section, the temperature at which the reaction is carried out can influence the outcome of the competition.

8.8A HEAT

Although S_N1 and E1 reactions are both favored under similar conditions, we can sometimes control the outcome by taking advantage of *entropy*, which we discussed briefly in Section 3.9, and which you should have encountered in general chemistry in the discussion of Gibbs free energy. Recall that entropy can be considered *a measure of disorder*. This is important because, all else being equal, a more disordered (i.e., higher entropy) system is more highly favored.

When we compare substitution and elimination, we can examine the amount of disorder in the overall products to determine the role that entropy plays. Compare the general substitution and elimination reactions (Equations 8-7a and 8-7b), where a negatively charged species (acting either as a nucleophile or as a base) reacts with a substrate containing a leaving group. For the substitution reaction, there are two product species formed overall (Equation 8-7a), but for the elimination reaction, there are three product species formed (Equation 8-7b). Consequently, *there is more disorder—that is, more entropy—in the products of an elimination reaction than in the products of a substitution reaction.*

Substitution: $\text{Nu:}^- + \text{R–L} \rightarrow \text{Nu–R} + \text{:L}^-$ (8-7a)

Elimination: $\text{B:}^- + \text{H–C–C–L} \rightarrow \text{B–H} + \text{C=C} + \text{:L}^-$ (8-7b)

Because competing substitution and elimination reactions start with exactly the same reactant molecules, we can say that there is a greater increase in entropy in an elimination reaction than there is in a substitution reaction. In other words,

Entropy favors elimination over substitution.

We have yet to discuss how we can take advantage of the additional entropy in the products of an elimination reaction over that in the products of a substitution reaction. The answer has to do with heat. From experience, we know that *heat favors disorder.* We see this when heat is added to ice, thereby melting the ice and forming liquid water. Additional heat converts the water to steam. Heat favors the liquid over the solid because the liquid has greater entropy—in the solid, the molecules are well ordered, but in the liquid, they are tumbling and moving around in space. Furthermore, heat favors the gas over the liquid because that gas has still more entropy.

Relating this to chemical reactions, all else being equal, *increasing the temperature tends to favor the reaction that produces products with higher entropy.* In this case,

Increasing the temperature of competing substitution/elimination reactions tends to favor elimination because its products have higher entropy—there are more of them.

There are several ways to indicate that a reaction is taking place at high temperatures. One way is simply to provide the temperature at which the reaction is occurring. For example, a reaction arrow might be written $\xrightarrow[150°C]{}$ to indicate that the reaction is being heated to a temperature significantly above room temperature. Another way the arrow could be written is $\xrightarrow[\text{heat}]{}$ or $\xrightarrow[\Delta]{}$ to explicitly show that heat is being added to the reaction. Although these reaction arrows may not automatically mean that elimination is occurring more prevalently than substitution, the fact that heat is being added to the reaction should suggest that elimination is a strong likelihood.

✔ **QUICK CHECK**

Which of the following reactions will produce more elimination product?

$(CH_3)_2CHCl + NaSH \longrightarrow ?$ $\quad\quad$ $(CH_3)_2CHCl + NaSH \xrightarrow[150\ ^\circ C]{} ?$

Answer: The reaction on the right will produce more elimination product because the higher temperature provides more heat.

8.9 Sample Problems—Putting It All Together

As was mentioned at the beginning of this chapter, for many students, the most difficult thing to do is predict the products of a reaction that can proceed by S_N1, S_N2, E1, or E2. Even though we just went through all of the individual factors that dictate the outcome, predicting the products can still be challenging. This is because there are several pieces of information that must be considered at the same time in order to make the correct prediction. However, there is a rather straightforward method to make sense of all the information in this chapter. Getting comfortable with it requires practice, which is why this section walks you through applying what you have learned in a variety of different situations.

This method has you begin by considering the two factors that have the ability to completely cut off certain reactions. The first factor is the identity of the leaving group. As we saw in Section 8.5, F^- is the benchmark for a moderately poor leaving group, whereas HO^- is the benchmark for a terrible leaving group. Thus,

In order for nucleophilic substitution or elimination reactions to take place, the leaving group must be at least as good as F^-.

If this condition is not satisfied, there will be no S_N2, S_N1, E2, or E1 reactions—end of story!

If this condition is satisfied, we then construct a table with one column for each of the four types of reactions (S_N1, S_N2, E1, and E2) and one row for each of the five main factors that govern reaction rates of nucleophilic substitution and elimination reactions. An example of a blank table is shown in Table 8-1.

The second factor that can cut off certain reactions is the type of C atom to which the leaving group is attached. For example, recall from Section 8.6 that S_N2 reactions are not feasible if the leaving group is attached to a tertiary C atom, due to excessive steric hindrance. Also, S_N1 and E1 reactions are not feasible if the leaving group is attached to a primary C atom unless the carbocation that is produced is resonance-stabilized (such as a benzylic or allylic carbocation). This allows us to do the following.

If the substrate has the leaving group attached to a tertiary C atom, cross out the S_N2 column.

If the substrate has the leaving group attached to a primary C atom, cross out the S_N1 and E1 columns unless the substrate is benzylic or allylic.

Once this has been done, it's time to fill in the table. To do so, we examine one factor at a time and determine which of the four reactions each factor favors.

For each reaction favored by a particular factor, place a check mark in the corresponding column.

Once the factors have all been considered, we tally the score.

The reaction with the most check marks in its column is the winner.

TABLE 8-1 Blank table for the analysis of substitution and elimination reactions.

Factor	S_N1	S_N2	E1	E2
Strength of attacking species				
Concentration of attacking species				
Leaving group				
Type of C				
Solvent				
Total				

Sometimes the result is a tie between substitution and elimination. If this is the case, then we expect the reaction to produce a mixture of products from the two types of reactions, but to predict the major product, we can consider heat as a tie-breaking factor.

If tallying the five factors results in a tie between substitution and elimination, the addition of heat tips the balance toward elimination, whereas the absence of heat generally tips the balance toward substitution.

Finally, once the predominant mechanism is determined, it's time to obtain the major products.

To determine the major products of the reaction, feed the specific reactants into the predominant mechanism.

Not only will this give you the appropriate connectivity of the products, but also it will give you the appropriate stereochemistry. Let's now apply this method to a variety of specific reactions.

8.9A SAMPLE PROBLEM 1

In the first problem, you are asked to predict the products of the reaction in Equation 8-8, including stereochemistry if appropriate.

$$CH_3CH_2CH_2Cl + NaBr \xrightarrow[DMSO]{} ? \qquad (8\text{-}8)$$

According to the method outlined above, we begin our analysis by making sure that the leaving group is suitable—at least as good as F^-. In this case, the leaving group will depart as Cl^-, which is indeed more stable than F^-, so we can proceed with constructing the reaction table—Table 8-2. Next, we examine the type of C atom to which the leaving group is attached in order to see if any of the reaction columns need to be crossed out. In this case, the leaving group is attached to a primary C atom that is not benzylic or allylic, so we cross out the S_N1 and E1 columns—this is indicated by shading the two columns gray.

TABLE 8-2 Table for the analysis of substitution and elimination reactions for equation 8-8.

Factor	S_N1	S_N2	E1	E2
Strength of attacking species		√	√	
Concentration of attacking species		√		–
Leaving group	√	√	√	√
Type of C		√		√
Solvent		√		√
Total	1	5	2	3

Now we are ready to fill in the table by determining which reactions each factor favors. We begin with the strength of the attacking species, which we identify as Br^- (recall from Section 5.9 that Na^+ is a spectator ion). As a nucleophile, Br^- is strong, given that it bears a full negative charge, so it favors an S_N2 reaction over S_N1—a check mark is therefore put in the S_N2 column of Table 8-2. As a base, Br^- is weak, because it is substantially more stable than HO^-, so it favors an E1 reaction over E2. We write a check in the E1 column of Table 8-2, but realize that the E1 colum has been crossed out, so this will ultimately be ignored.

The next factor to look at is the concentration of the nucleophile/base. With the reaction written the way it is, we can assume that both the Br^- and the substrate are present in significant amounts. If this were *not* the case, it would have been made explicit to us that one of the reactants was dilute. For example, "dil. NaBr" may be written above the reaction arrow. Assuming that Br^- is fairly concentrated, S_N2 will be favored over S_N1—being a strong nucleophile, Br^- is viewed as forcing off the leaving group, and the greater the concentration of this nucleophile, the more it will do so. We indicate this in Table 8-2 by writing a check mark in the S_N2 column.

Using the same arguments, we might also expect the high concentration of Br^- to favor E2 over E1. However, we identified Br^- as a weak base, meaning that it does not have the ability to force the action in an E2 step. The large concentration of a weak base does not change the situation. Therefore, we ignore this factor in considering elimination reactions and remind ourselves by placing a dash in the E2 column.

The third factor to examine is the stability of the leaving group. In this case, the leaving group comes off in the form of Cl^-. In Section 8.5, we identified Cl^- as a moderate leaving group, which favors all four reactions roughly the same. This is indicated in Table 8-2 by placing a check mark under each of the four mechanisms (alternatively, we could have left all of the entries blank).

The next factor to take into account is the type of C atom bonded to the leaving group. In this case, it is a primary C atom, and the substrate is neither benzylic nor allylic. As we saw in Section 8.6, this type of C atom tends to favor S_N2 over S_N1 for two reasons. First, the absence of steric hindrance allows the nucleophile to approach the C atom from the opposite side of the leaving group and to force the action. Second, the rate-determining step of the S_N1 mechanism is heavily disfavored because the product of that step is an unstable primary carbocation. For the same reason, the rate of the E1 mechanism is quite slow, suggesting that the E2 mechanism is favored over the E1 mechanism. These conclusions are noted in Table 8-2 by writing check marks in the columns for S_N2 and E2.

The final factor that we examine is solvent effects. From the information that we are given, we can see that the solvent is DMSO, which is an aprotic solvent. As we learned in Section 8.7, aprotic solvents tend to favor S_N2 and E2 reactions because the strength of the nucleophile/base is not weakened. This is indicated in Table 8-2 by placing check marks in the columns for S_N2 and E2.

Now that we have considered all five factors, let's tally them. This is done at the bottom of Table 8-2. The S_N2 mechanism appears to be the most heavily favored—it is favored by all of the five factors. We therefore predict that the reaction we were presented with in Equation 8-8 will proceed via the S_N2 mechanism faster than any of the others and that the S_N2 products will be the major products.

Recall that we said, in general, if reaction conditions favor S_N2, they also favor E2. However, that does not appear to be the case with this example. Clearly, from the table, it appears that the S_N2 reaction is favored over the E2. Why is this? It primarily has to do with the strength of Br^- as a nucleophile compared to its strength as a base. We concluded that it is a strong nucleophile, but a weak base. It is therefore able to force the action in an S_N2 mechanism, but not in an E2 mechanism, allowing it to discern between the two mechanisms.

Finally, let's predict the products. We do so by feeding the reactants into the S_N2 mechanism, as shown in Equation 8-9. The Br^- nucleophile attacks the C nucleophile from the opposite side of the leaving group, and the leaving group leaves at the same time. Notice that the C atom involved in the reaction is not a stereocenter, meaning that there is no stereochemistry that needs to be shown explicitly.

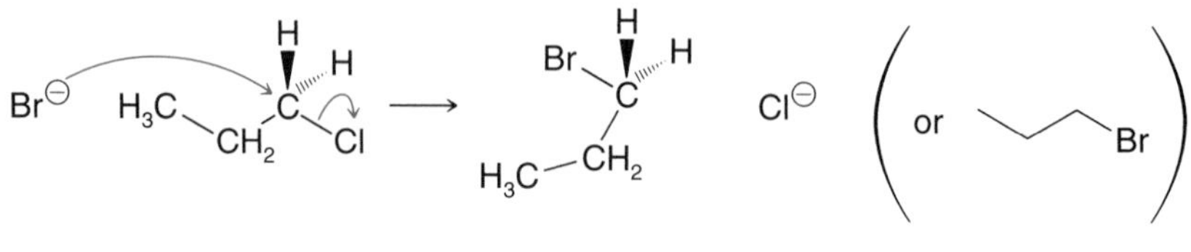

(8-9)

8.9B SAMPLE PROBLEM 2

Let's change the problem slightly, so that the substrate is the one given in the reaction in Equation 8-10.

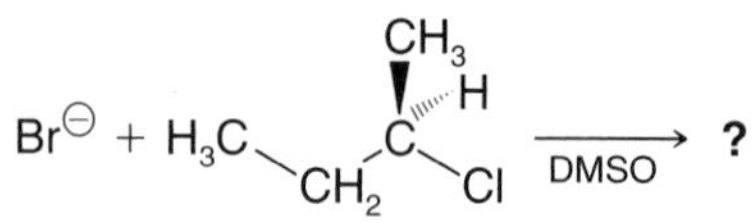

(8-10)

Once again, knowing that the leaving group is at least as good as F^-, we construct our table for analysis—Table 8-3. This time, however, the leaving group is attached to a secondary C atom, so no columns are crossed out.

When filling in the table, realize that all the arguments pertaining to all the factors remain the same as we saw for the reaction in Equation 8-8 (Table 8-2) except for the type of C atom to which the leaving group is bonded. Here the C with the leaving group is bonded to two alkyl (R) groups, making it a secondary C atom. As we saw in Section 8.6, a secondary C atom favors all four reactions roughly equally, so we can place a check mark in each of the columns for this factor (or, alternatively, leave them all blank). This makes the table somewhat different from Table 8-2. However, when the columns are totaled, the S_N2 reaction still wins.

TABLE 8-3 Table for the analysis of substitution and elimination reactions in equation 8-10.

Factor	S_N1	S_N2	E1	E2
Strength of attacking species		√	√	
Concentration of attacking species		√		–
Leaving group	√	√	√	√
Type of C	√	√	√	√
Solvent		√		√
Total	2	5	3	3

Once again, finishing the problem requires us to feed the reactants into the S_N2 mechanism, as shown in Equation 8-11. Notice that this time the C atom bonded to the leaving group is a stereocenter, so stereochemistry will be important. The Br^- nucleophile again attacks the C atom from the side opposite the leaving group, the leaving group comes off, and the remaining three groups on that C atom—two R groups and an H atom—flip over to the other side. As we can see, the correct three-dimensional representation, and therefore the correct stereochemistry of the product, follows directly from the mechanism and is as shown. Notably, the product is *not* a mixture of stereoisomers.

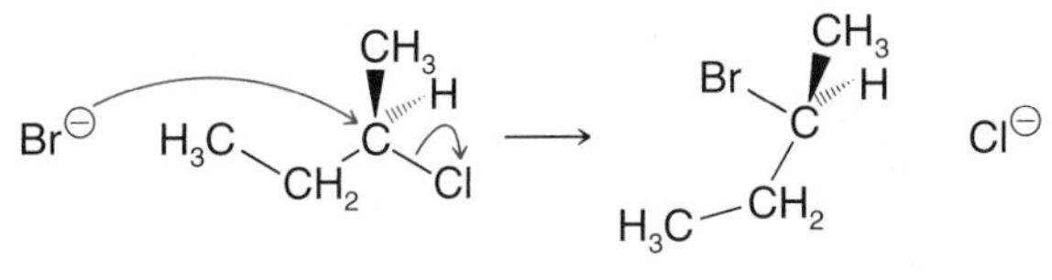

(8-11)

8.9C SAMPLE PROBLEM 3

For the next problem, let's predict the products of the reaction in Equation 8-12, which denotes that the substrate is dissolved in water.

$$(CH_3)_3CBr \xrightarrow[H_2O]{} ? \qquad (8\text{-}12)$$

Before constructing the table, we must make sure that the leaving group, Br^-, is at least as good as F^-. It is, so we construct the table (Table 8-4) and then determine which columns to cross out by examining the type of C atom bonded to the leaving group. In this case, the C atom is tertiary, which means that we must cross out the S_N2 column. This is indicated by shading that column gray.

Now we can fill in the table according to which reactions each factor favors. The first factor is the strength of the nucleophile/base. But what is the nucleophile/base? Clearly there is a substrate—the species being dissolved that contains the Br^- leaving group. That must mean that the H_2O is the nucleophile/base, such that the solvent is behaving as a reactant. (This type of reaction is called **solvolysis**, and because water is the particular solvent, it is more specifically

TABLE 8-4 Table for the analysis of substitution and elimination reactions in equation 8-12.

Factor	S_N1	S_N2	E1	E2
Strength of attacking species	√		√	
Concentration of attacking species		–		–
Leaving group	√		√	
Type of C	√		√	√
Solvent	√		√	
Total	4	0	4	1

called **hydrolysis**). Since water is a weak nucleophile (uncharged) and a weak base (weaker than HO^-), this factor favors S_N1 and E1 reactions.

The second factor, concentration of the nucleophile/base, is irrelevant in this reaction. There is a large concentration of water, which would normally favor S_N2 and E2, but as we saw previously, a large concentration of a weak nucleophile/base does not enable that species to force the action. To remind ourselves of this, we put dashes in those two columns.

The third factor, stability of the leaving group, favors S_N1 and E1. The leaving group comes off as Br^-, which, as was stated in Section 8.5, is a benchmark for a very good leaving group.

The fourth factor, the type of C atom bonded to the leaving group, favors S_N1, E1, and E2. It favors S_N1 and E1 because the C atom is tertiary and the leaving of the leaving group would produce a relatively stable tertiary carbocation. The E2 reaction is also favored—remember that, even though there is substantial steric hindrance sourrounding the C atom that has the leaving group, the protons on the *adjacent* C atoms remain accessible.

The final factor is the solvent. Water is a protic solvent and therefore favors S_N1 and E1 over S_N2 and E2.

Summing the factors in the table, we see that there is a tie between S_N1 and E1. Therefore, we expect this reaction to produce a significant amount of product from each of these reactions.

To predict which reaction should produce more product, however, we turn to the factor of heat to break the tie. In this case, no heat appears to be added, so we predict the balance to be tipped toward substitution, making S_N1 the predominant mechanism.

To complete the problem, it is now a matter of feeding the reactants into the S_N1 mechanism, as shown in Equation 8-13. The first step of the S_N1 mechanism is heterolysis of the bond between the C atom and leaving group, producing a carbocation intermediate. Next, water acts as a nucleophile to form a bond to C^+, which results in a positively charged O atom. To produce the final uncharged product molecule, that O^+ atom is deprotonated by another molecule of water.

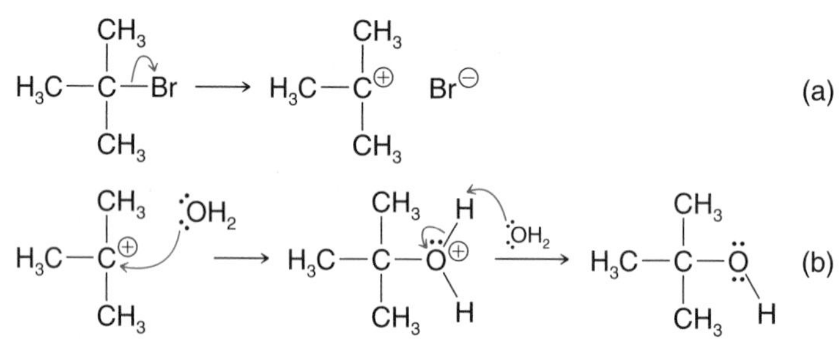

(8-13)

LOOK OUT

Solvolysis reactions are a favorite among many professors—and also among those who write standardized exams, such as the MCAT! Therefore, it would be to your benefit to learn how to spot them, how to fill out the table containing the five factors, and how to feed the reactants into the appropriate mechanism that wins out.

8.9D SAMPLE PROBLEM 4

For yet another problem to solve, let's predict the product of the reaction in Equation 8-14.

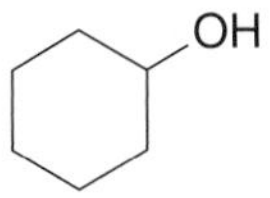

$$+ \ 85\% \ H_3PO_4 \xrightarrow[\Delta]{} ? \qquad (8\text{-}14)$$

As usual, our analysis begins by determining if there is a suitable leaving group for nucleophilic substitution and elimination reactions to take place. In this case, the leaving group, as written, is HO^-, which is our benchmark for a *terrible* leaving group. So unless things change, no reaction whatsoever will take place. However, H_3PO_4 is a strong acid, which will rapidly protonate the OH group of the substrate. This will produce an OH_2^+ group on the substrate, which will come off as H_2O, a very good leaving group. Therefore, we can proceed with constructing the table, Table 8-5.

Before filling in the table, we once again examine the type of C atom that has the leaving group in order to determine if there are any columns to cross out. In this case, it is a secondary C atom, so all columns remain.

Now we can begin to fill in the table, considering the strength of the nucleophile/base first. Looking at Equation 8-14, it is not immediately obvious what that species is. Realize, however, that the acid solution is only 85% H_3PO_4, meaning that the remaining 15% of the solution is water. Consequently, water, which we recognize as a weak nucleophile and a weak base, can act as the attacking species. This factor therefore favors S_N1 and E1.

As we saw previously in Section 8.9C, a substantial concentration of a nucleophile or base normally favors S_N2 and E2, but because water is a weak nucleophile and a weak base, concentration is irrelevant—we put dashes in those columns to remind us. The leaving group (H_2O) is very stable, so this factor favors S_N1 and E1. The C atom bonded to the leaving group is a secondary C atom and therefore favors all mechanisms roughly equally. Lastly, we consider solvent. In this case, the solvent can be water or the alcohol (R—OH) reactant itself (each is present in a significant amount), both of which are protic and favor S_N1 and E1. At this point, tallying the scores in Table 8-5 suggests that S_N1 and E1 are both favored, at four factors apiece.

Because of this tie, we consider heat as the tie-breaker. In this case, heat is clearly shown to be added to the reaction, which tips the balance in favor of elimination—thus, we predict the E1 reaction gives us the major products.

TABLE 8-5 Table for the analysis of substitution and elimination reactions in equation 8-14.

Factor	S_N1	S_N2	E1	E2
Strength of attacking species	√		√	
Concentration of attacking species		–		–
Leaving group	√		√	
Type of C	√	√	√	√
Solvent	√		√	
Total	4	1	4	1

To complete the problem, we now feed the reactants into the E1 mechanism, as shown in Equation 8-15.

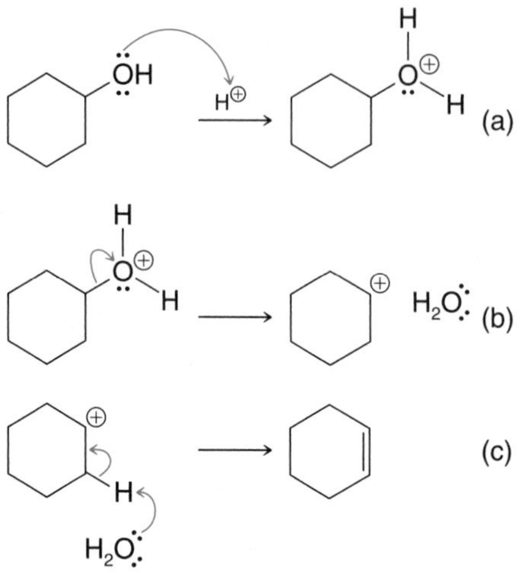

(8-15)

8.9E SAMPLE PROBLEM 5

Let's examine one more reaction, shown in Equation 8-16.

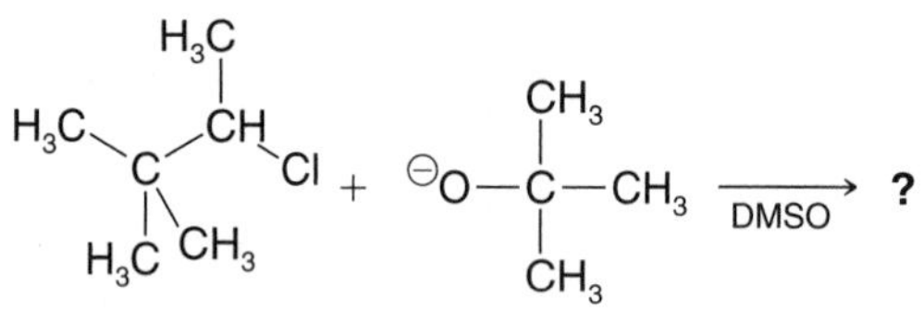

(8-16)

We begin by recognizing that the Cl^- leaving group is suitable for nucleophilic substitution and elimination reactions, so we proceed with constructing the table for analysis, Table 8-6.

TABLE 8-6 Table for the analysis of substitution and elimination reactions in equation 8-16.

Factor	S_N1	S_N2	E1	E2
Strength of attacking species	√			√
Concentration of attacking species		–		√
Leaving group	√	√	√	√
Type of C	√	√	√	√
Solvent		√		√
Total	3	3	2	5

The leaving group is attached to a secondary C atom, so we can begin filling in the table without crossing out any of the columns. The attacking species, $(CH_3)_3CO^-$, is our example of a strong base, but a weak nucleophile. This is because the negative charge on the base/nucleophile is on an O atom, similar to HO^- (a strong base), but the bulkiness gets in the way of the O atom forming a bond to the C atom that has the leaving group. The fact that it is a strong base favors E2, and the fact that it is a weak nucleophile favors S_N1. Next, the concentration of the attacking species is high—this favors E2 because the base is strong, but does not favor S_N2 because the nucleophile is weak. The leaving group (Cl^-) is moderate, so it favors all four mechanisms roughly equally. The C atom with the leaving group is secondary, which also favors all four mechanisms roughly equally. Finally, the solvent is aprotic, which favors S_N2 and E2. Tallying all the results, it appears that E2 is the favored reaction. The resulting mechanism is shown in Equation 8-17.

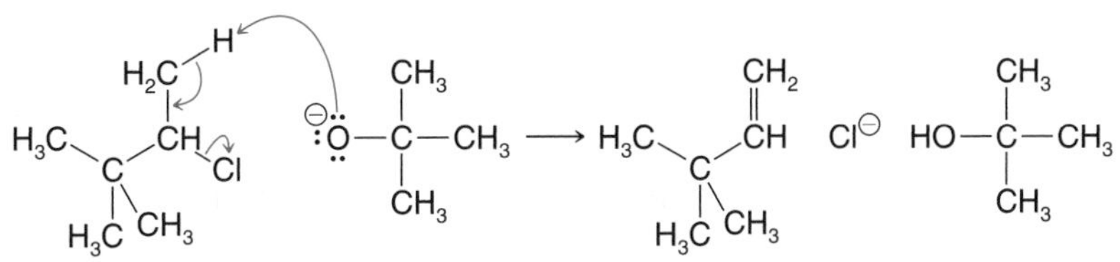

(8-17)

WHAT DID YOU LEARN?

8.1 Rank the following attacking species in order of increasing S_N2 reaction rate in DMSO.

a NaOH
b CH_3NH_2
c H_2O
d LiF
e NaI
f HCO_2K
g NH_3

8.2 Rank the following attacking species in order of increasing E2 reaction rate in DMSO.

a NaOH
b $(CH_3)_3COK$
c $(CH_3)_2NLi$
d NaF
e KBr
f C_6H_5ONa
g NH_3

8.3 Rank the following substrates in order of increasing S_N2 reaction rate in DMSO.

a $(CH_3)_3CCH_2CH_2Cl$
b $(CH_3)_2CClCH_2CH_3$
c $(CH_3)_3CCH_2CH_2Br$
d $(CH_3)_3CCH_2CH_2I$
e $(CH_3)_3CCHClCH_3$
f $(CH_3)_3CCH_2CH_2OH$

8.4 Rank the following substrates in order of increasing E1 reaction rate in water.

a $(CH_3)_3CCH_2CH_2Br$
b $(CH_3)_2CBrCH_2CH_3$
c $(CH_3)_3CCH_2CH_2Cl$
d $C_6H_5CHBrCH_3$
e $(CH_3)_3CCHBrCH_3$
f $(CH_3)_3CCH_2CH_2F$
g $C_6H_5CHICH_3$

8.5 For each reaction here, predict the predominant mechanism(s); draw out the detailed mechanism(s), including all curved arrows; and using the mechanism, predict the major product(s), including stereochemistry where appropriate.

(a) 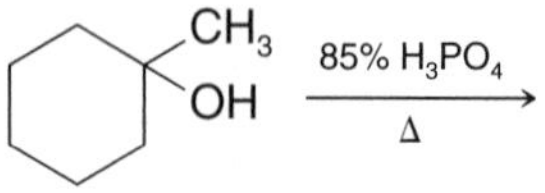

(b) 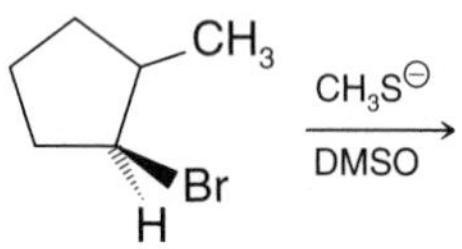

(c) 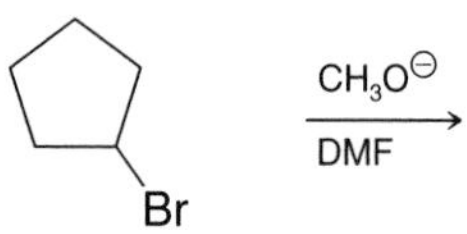

(d) 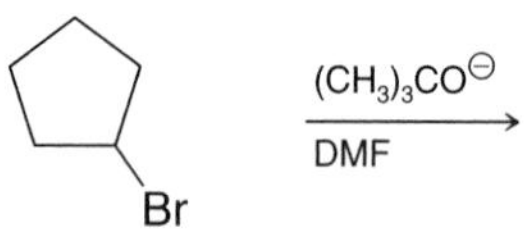

(e) 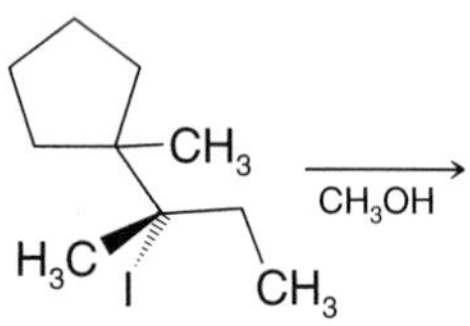

8.6 Provide the missing reagents and/or reaction conditions (e.g., solvent, heat) necessary to produce each of the following products from S_N1, S_N2, E1, or E2 reactions. Reaction conditions are placed in the box below the arrow. Pay attention to stereochemistry where appropriate.

(a) 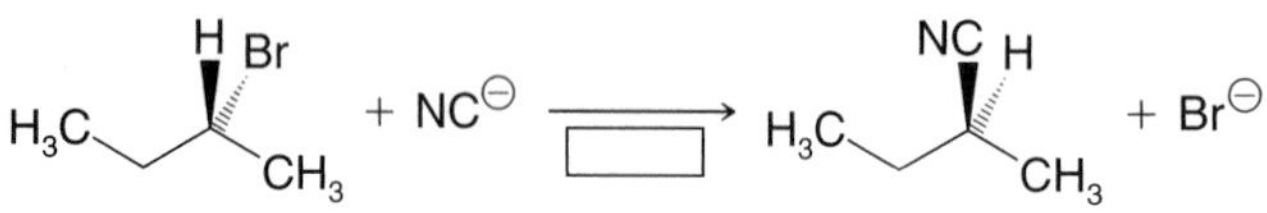

(b) 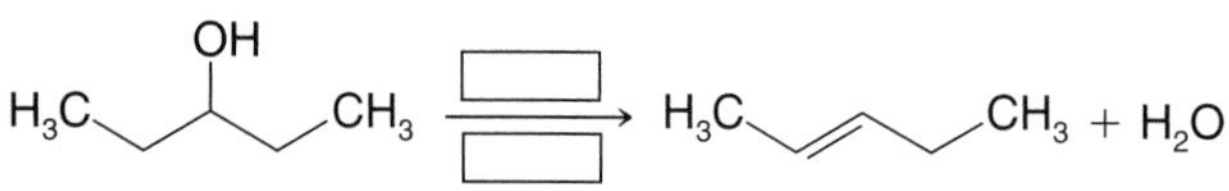

(c) 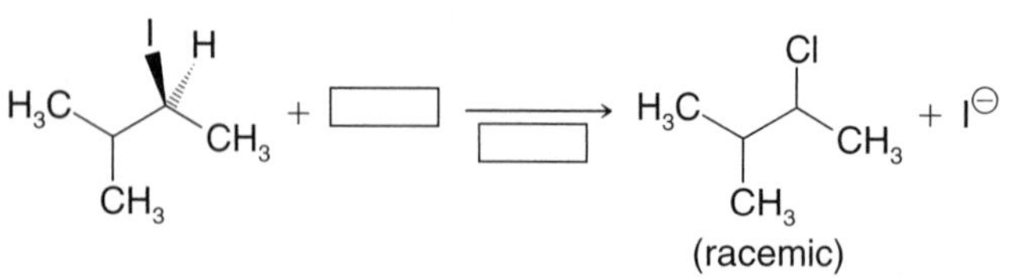

(d) 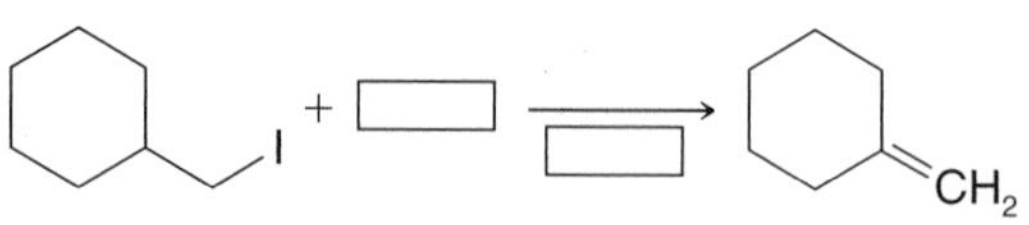

8.7 The molecule here cannot be used as a substrate for an E2 reaction, but can be for an E1 reaction. Explain why.

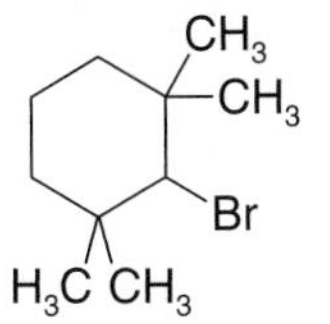

8.8 A student attempted to carry out a substitution reaction using molecule (a) as the substrate. The resulting product was molecule (b). Which nucleophilic substitution mechanism had occurred—S_N1 or S_N2? How can you tell?

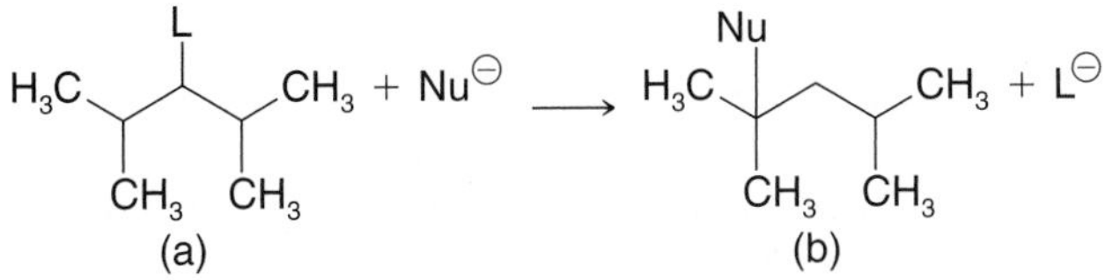

8.9 Predict the major product(s) of the following reaction.

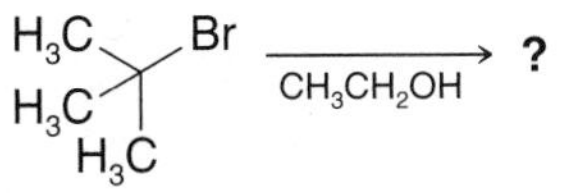

8.10 Predict the product of the following reaction in which oxygen-18-labeled water is used as the solvent.

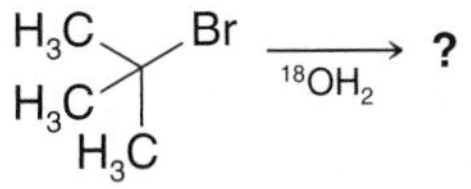

9 Concluding Remarks—What Now?

When students from my summer organic prep course complete this book, they overwhelmingly feel that they are prepared to face their year long organic chemistry course. However, they unanimously express one concern: What do I do if my professor doesn't teach like you?

Quite frankly, you should expect your course to be different, simply because of the way that most organic chemistry textbooks (and courses) are organized. Whereas this book is organized primarily around concepts, most organic chemistry textbooks are organized around classes of compounds. It may be that your professor *does* focus heavily on fundamental concepts and stress understanding mechanisms (thank them if they do), but it may feel different simply because multiple fundamental concepts are introduced at the same time in order to explain the variety of reactions in which a given class of compounds can participate. In other instances, your course may be very different in that your professor places less emphasis on reaction mechanisms.

In either case, this does *not* mean that you should abandon what you have learned here. Instead, it means that you must assimilate your understanding of fundamental concepts we have learned with the way material is presented in your class. This book provides you the tools, and the rest is up to you. If a reaction is presented to you without a mechanism, get ahold of the mechanism, either by looking it up or by asking your professor or TA. Convince yourself that each step of the mechanism is reasonable, according to the concepts you learned from this book. If even one step of a mechanism does not make sense to you, ask questions. If you don't take it upon yourself to do so, expect a lot of memorization!

A second concern from my students typically arises after they have returned to their respective institutions and begun their yearlong courses. Up through the middle of the first semester, they have great success without resorting to flash cards and memorization, but the increasing number of reactions often makes flash cards more appealing. They essentially ask me: "What do I do now?" The following summarizes the advice I give them, which, they tell me, helps considerably.

If you get overwhelmed by the sheer number of reactions, then you are doing something wrong. You are probably straying from focusing on the mechanism. I say this because even though you face dozens upon dozens of reactions in the first semester, and will face many more in the second semester, the number of *mechanisms* that dictate those reactions is quite small. In the first semester, the total number of fundamentally different mechanisms is somewhere around 10. In the second semester, there may be an additional 10 or so, and several of them are nothing more than a simple twist on, or a combination of, the mechanisms you learn in the first semester. Therefore, if you focus on the mechanisms, all of the reactions you see in your year of organic chemistry—which may be *several hundred*—are reduced to only a couple dozen mechanisms at most.

Once you understand that all of the organic chemistry reactions you encounter are governed by a handful of mechanisms, things will get a lot more straightforward. But you still have to work. As mentioned earlier, you have to work at using the concepts you learned in this book to *understand* each mechanism. You also have to work at *using* the mechanism. Students have perhaps the most difficulty seeing where a mechanism is useful in problems that simply ask you to predict the products of a given reaction (problems that don't explicitly ask for a mechanism). But those are the types of problems where knowing and understanding mechanisms is the *most* useful! When you encounter a problem like this, write out the *complete* mechanism that those reactant molecules will undergo—even if the problem does not ask you to do so! That includes *every* step of the mechanism and *every* curved arrow, including formal charge. All along, remind yourself why *that* particular step occurs, and not some other one. Think about *electron-rich-to-electron-poor*. Think about steric hindrance and charge stability, and think about stereochemistry. Think about the rules of thumb for proton transfer and carbocation rearrangement steps. When the mechanism is finished, the answer will be sitting right there on the paper. If the problem asks for the stereochemistry of the product, the mechanism you just wrote out contains that information.

When you work on studying reactions in this way, it is vital that you do so on your own as much as possible—at least initially. The reason is so that you can continuously assess your own understanding of and ability to work with the material. When you get stuck on something, seek help—from your textbook, this book, a classmate, your TA, or your professor—but take note of where you're running into difficulty. Those are the areas that most likely will trip you up on exams. However, if you work through those problem areas and can get them "fixed" before the exam, your performance on the exam will be much better.

At first, writing out the mechanism when it is not asked for might require more time than not doing so. In the long run, however, the opposite is true—you begin to *save* time. This is especially true as you become more practiced with those mechanisms because you are eventually able to do many of the steps in your head. (That is different from skipping them entirely!) You will get faster, while maintaining your accuracy in predicting the products.

So where does this end up saving you time? First, by becoming well practiced with the mechanism, you will be well prepared to answer any question about that mechanism on an exam, especially one that presents an overall reaction and asks you for its detailed mechanism. You will therefore not need to study the reactions separately from the mechanisms, as those students who primarily memorize would be forced to do. Second, instead of spending your time memorizing the hundreds of seemingly different reactions, you will have to spend time working with only a handful of mechanisms. Many reactions share exactly the same mechanism, as we illustrated with some examples in Section 5.9. Other reactions might not have exactly the same mechanism, but might share many of the same steps, as we showed with an example in Section 7.7. For those mechanisms that are different by a simple "twist," all you must learn and understand is that twist.

Another great advantage to working with the mechanisms is what happens when you draw a blank on an exam. If you memorized, then all you can do is sit back and wait for divine inspiration. On the other hand, if you primarily focused on the concepts that go into driving the various elementary steps of a mechanism, such as *electron-rich-to-electron-poor*, you can begin to rederive the mechanism. It may be that all you need is the first step in front of you to get you back on the right track.

Most importantly, by working primarily with the mechanisms instead of memorization, you will be much less frustrated. Things will actually make sense! You will become much more accurate in your answers, while spending less time studying than your friends who memorize. With greater accuracy and less wasted time, you will wonder why everyone makes organic chemistry out to be such a beast.

Good luck!

SOLUTIONS TO SELECTED END-OF-CHAPTER PROBLEMS

Chapter 2

2.1 C^- has five valence electrons and two core electrons; O^+ has five valence electrons and two core electrons; F has seven valence electrons and two core electrons.

2.2

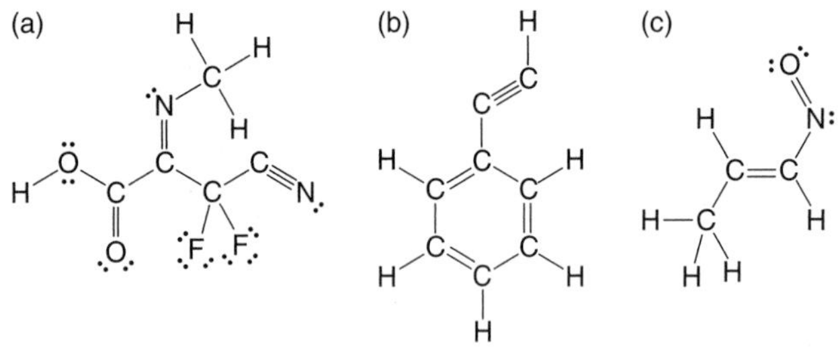

2.3

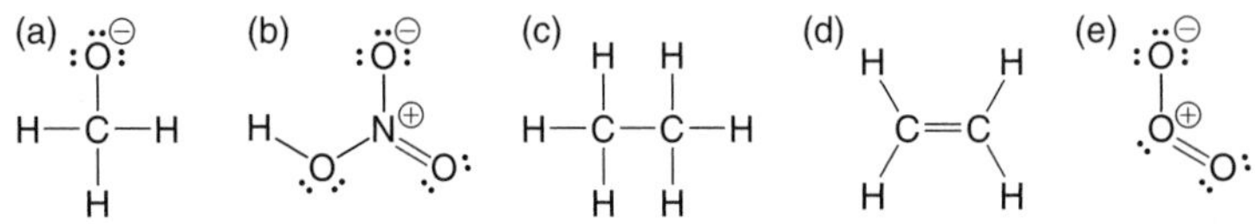

2.4 Each of the indicated N atoms has a formal charge of +1, and the indicated O atom has a formal charge of −1.

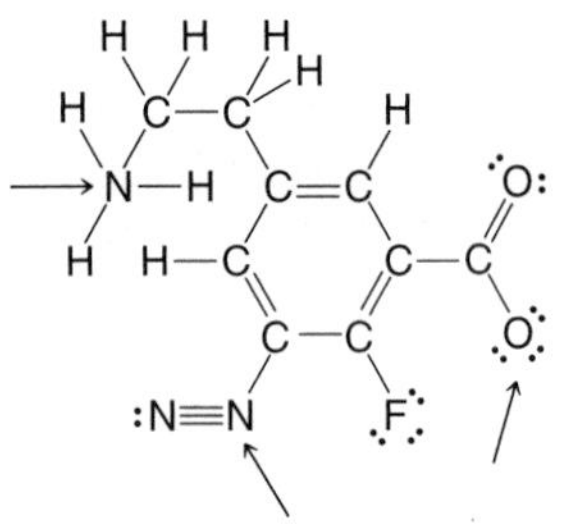

2.5 The condensed formula is shown below at the left, and the line structure is shown at the right.

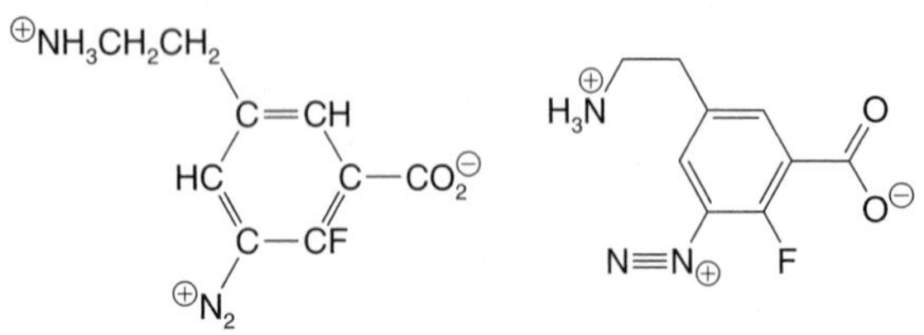

2.6

	Lewis Structure	Condensed Formula		Lewis Structure	Condensed Formula
(a)	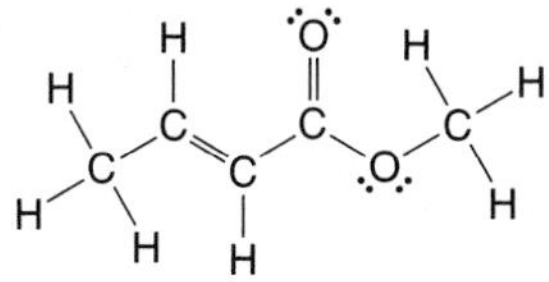	$CH_3CHCHCO_2CH_3$	(b)	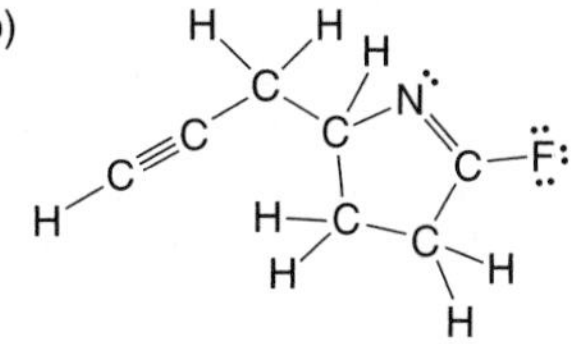	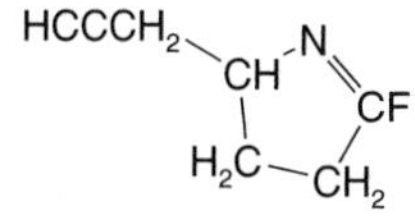

2.7

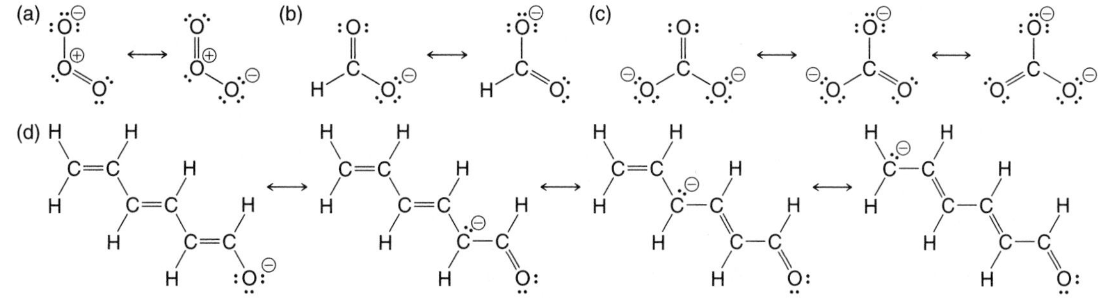

2.8

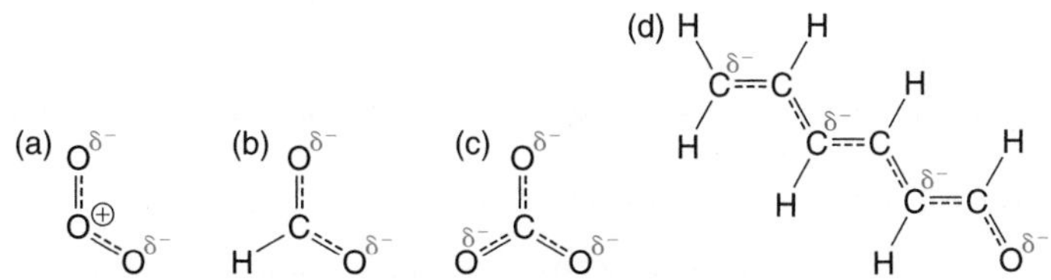

2.9 Pairs (a), (b), and (g) do not represent resonance structures of the same overall species because atoms must move in order to convert one structure into the other. Pair (f) represents two structures that are technically resonance structures, but the second structure is not a valid one because the number of charges has increased.

2.10 The resonance structures of the two species are shown below. The species are very similar, but because the first one has more resonance structures than the second, the first is more stable.

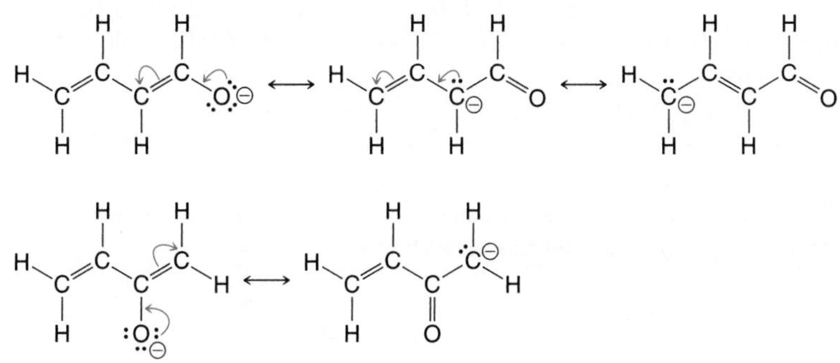

Chapter 3

3.1 (a) All atoms have a tetrahedral electronic geometry with an approximate bond angle of 109.5°. The molecular geometry for each C atom is tetrahedral, whereas that for the O atom is bent. (b) All atoms have a tetrahedral electronic geometry with an approximate bond angle of 109.5°. The molecular geometry for each C atom is tetrahedral, whereas that for the O atom is bent. (c) The electronic geometry of N is linear and has no molecular geometry or bond angle. The triply bonded C has a linear electronic geometry, a linear molecular geometry, and a bond angle of 180°. The positively charged C atom has a trigonal planar electronic geometry, a trigonal planar molecular geometry, and an approximate bond angle of 120°. The remaining C atoms have a tetrahedral electronic geometry, a tetrahedral molecular geometry, and an approximate bond angle of 109.5°.

3.2 (a) All atoms have a tetrahedral electronic geometry with an approximate bond angle of 109.5°. All C atoms have a tetrahedral molecular geometry, whereas the N atom has a trigonal pyramidal molecular geometry. (b) All C atoms have a trigonal planar electronic geometry, a trigonal planar molecular geometry, and an approximate bond angle of 120°. (c) The three C atoms that are part of the C═C bond and the C═O bond have a trigonal planar electronic geometry, a trigonal planar molecular geometry, and an approximate bond angle of 120°. The other C atoms have a tetrahedral electronic geometry, a tetrahedral molecular geometry, and an approximate bond angle of 109.5°. The O atom that is part of the C═O bond has a trigonal planar electronic geometry with no molecular geometry or bond angle. The other O atom has a tetrahedral electronic geometry, a bent molecular geometry, and an approximate bond angle of 109.5°. The N atom has a tetrahedral electronic geometry, a trigonal pyramidal molecular geometry, and an approximate bond angle of 109.5°. (d) All atoms have a tetrahedral electronic geometry with an approximate bond angle of 109.5°. The negatively charged C atom has a trigonal pyramidal molecular geometry, the O atom has a bent molecular geometry, and the remaining C atoms have a tetrahedral molecular geometry.

3.3 (a), (c), (d), (e), and (g) are polar. The net dipole in (a) points along the C—F bond toward F; the one in (c) bisects the two C—F bonds and points from the ring toward the center of the F atoms; the one in (d) points from Br toward F; the one in (e) points along the C—Cl bond from Cl toward the center of the F atoms; and the one in (g) bisects the S—F bonds and points from S toward the center of the two F atoms.

3.4 For each pair of molecules, the set of intermolecular interactions that exists is as follows (the most important type is *italicized*): (a) *Dipole–dipole*; dipole–induced dipole; induced dipole–induced dipole. (b) *Induced dipole–induced dipole.* (c) *Dipole–dipole*; dipole–induced dipole; induced dipole–induced dipole. (d) *Hydrogen-bonding*; dipole–dipole; dipole–induced dipole; induced dipole–induced dipole. (e) *Dipole–dipole*; dipole–induced dipole; induced dipole–induced dipole. (f) *Hydrogen bonding*; dipole–dipole; dipole–induced dipole; induced dipole–induced dipole.

3.5 For each pair of species the set of intermolecular interactions that exists is as follows (the most important type is *italicized*): (a) *Ion–ion*; induced dipole–induced dipole. (b) *Ion–dipole*; induced dipole–induced dipole. (c) *Dipole–induced dipole*; induced dipole–induced dipole. (d) *Hydrogen-bonding*; dipole–dipole; dipole–induced dipole; induced dipole–induced dipole. (e) *Induced dipole–induced dipole.* (f) *Dipole–dipole*; dipole–induced dipole; induced dipole–induced dipole.

3.6 Only (b) and (c) have *cis/trans* isomers that exist.

3.7 e < d < c < g < a < b < f.

3.8 This enol is stable because of the internal hydrogen bond that results. The OH is the hydrogen-bond donor and the doubly bonded O atom is the hydrogen-bond acceptor, as shown below.

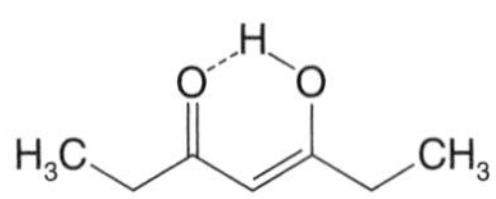

3.9 For O_2, CO_2, and CS_2, there is no permanent dipole. Therefore, the strongest interaction they can have with acetone is in the form of dipole–induced dipole interactions. This strength of interaction will increase with the number of electrons in the molecule—the greater the number of electrons, the stronger the induced dipole. Therefore, the strength of interaction, and consequently solubility in acetone, increases in the order $O_2 < CO_2 < CS_2$, or d < b < f.

Choices (a), (e), and (g) are polar, and the strongest interactions they can have with acetone are in the form of dipole–dipole interactions. Such interactions increase with increasing dipole moment: e < g < a.

Choices (h) and (c) can undergo hydrogen bonding with acetone, and the strength increases with increasing numbers of donors and acceptors: h < c.

The final order of increasing solubility is d < b < f < e < g < a < h < c.

Chapter 4

4.1 (a) Constitutional isomers. (b) Unrelated. (c) Diastereomers. (d) Enantiomers. (e) Same molecules. (f) Enantiomers. (g) Diastereomers.

4.2 Only identical molecules and enantiomers will have identical properties, which includes (d), (e), and (f).

4.3 (b), (d), (f), and (g) are the only chiral species.

4.4 (a) 4. (b) 5. (c) 2. (d) 0. (e) 4. (f) 2.

4.5

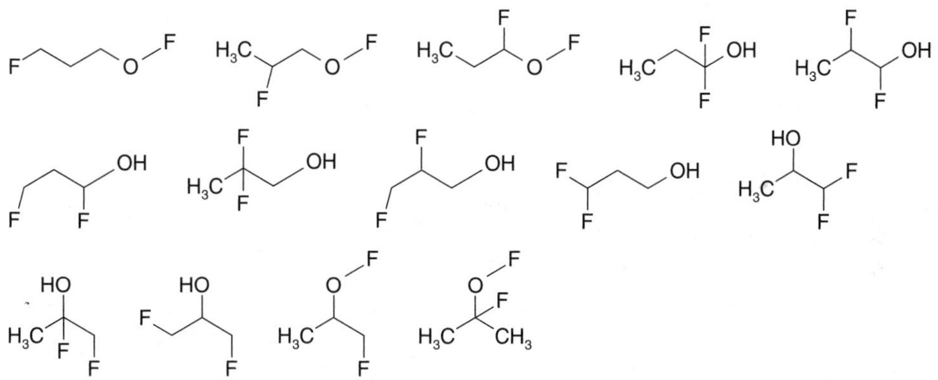

4.6

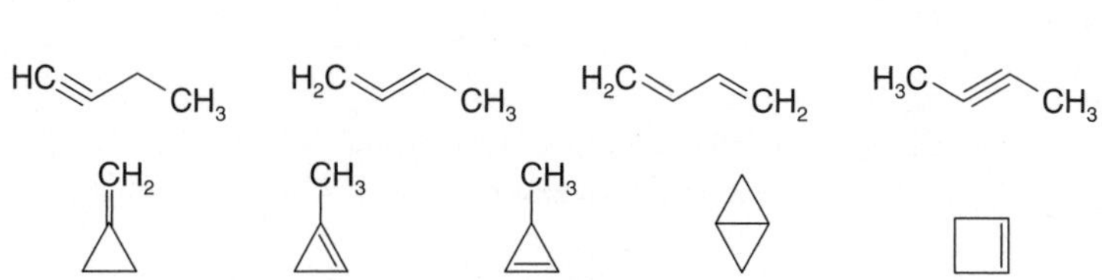

4.7 There are three stereoisomers for (a) and eight for (b).

(a)

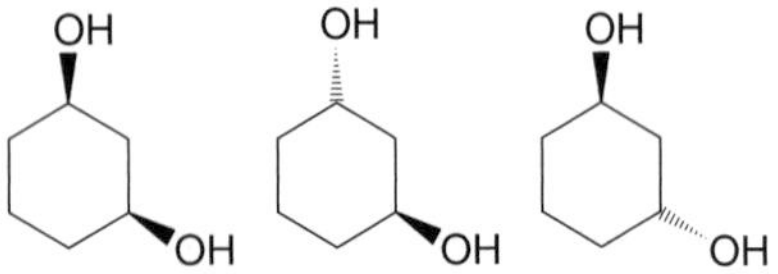

(b)

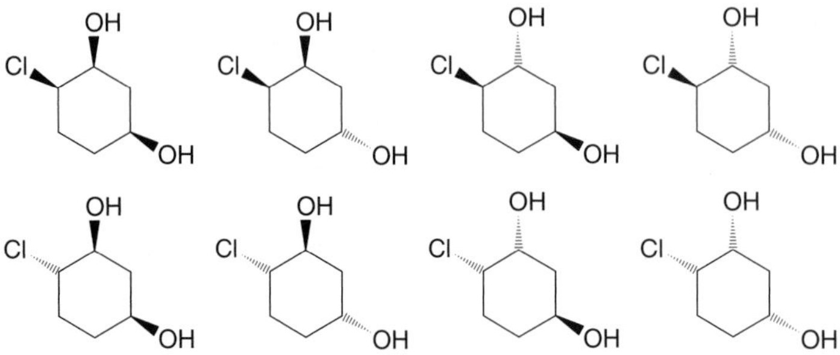

Chapter 5

5.1

(a)

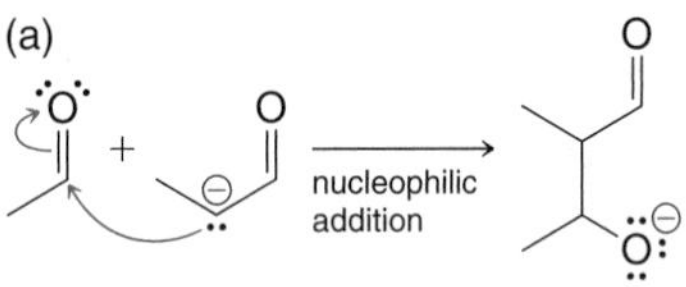

(b)

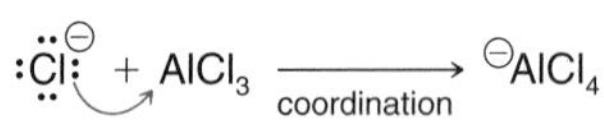

(c)

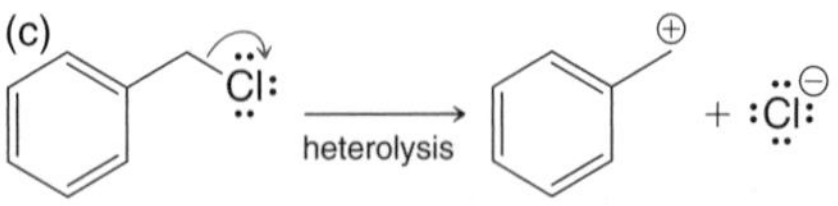

(d)

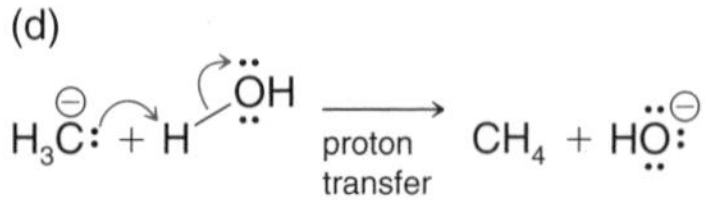

(e)

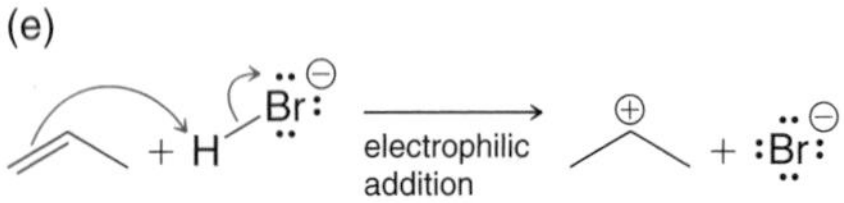

(f)

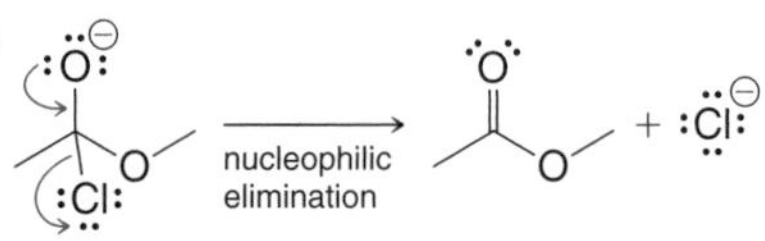

(g)

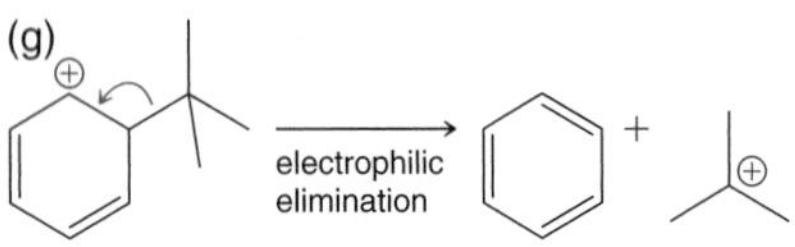

5.2 The products, electron-rich sites, and electron-poor sites are included below. The curved arrows that are colored blue are the ones involved in the flow of electrons from an electron-rich site, to an electron-poor site, or both.

(a)

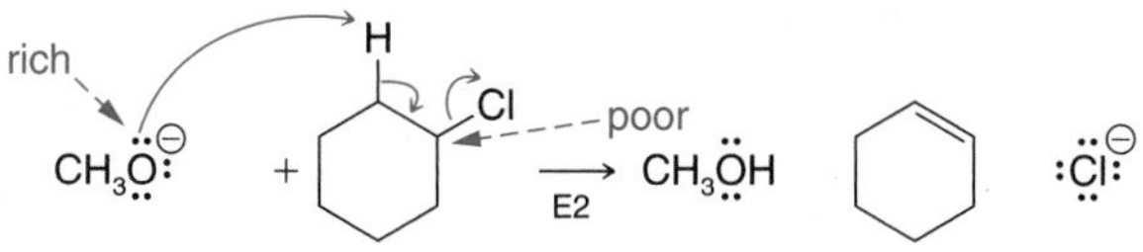

(b)

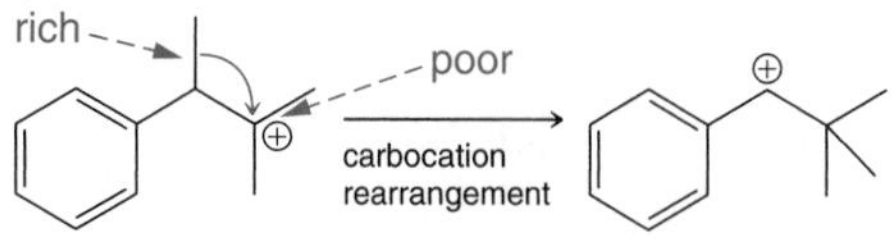

(c)

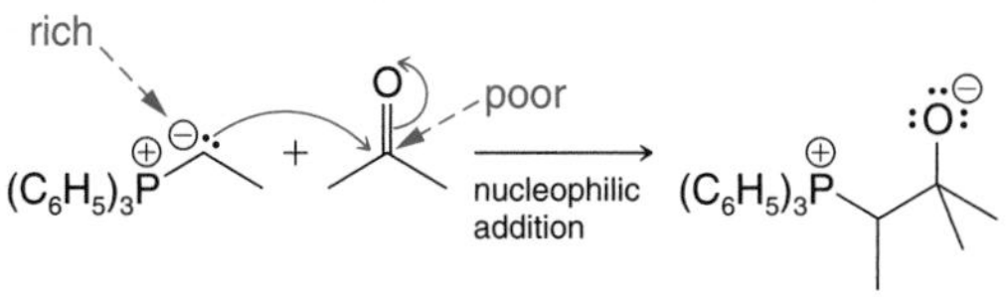

(d)

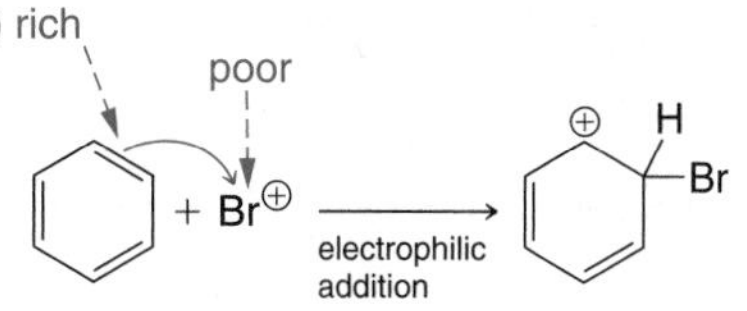

(e)

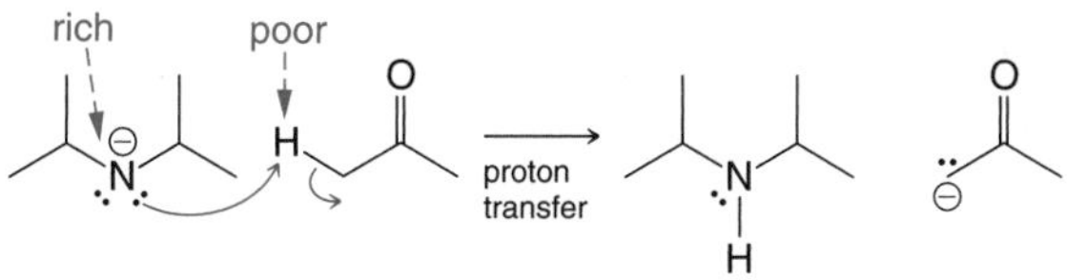

(f)

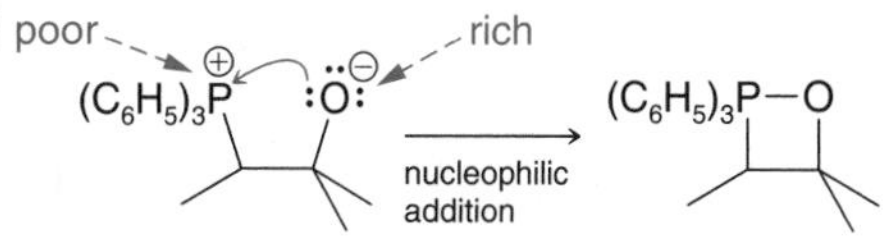

5.3

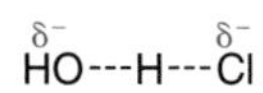

(Equation 5-7)

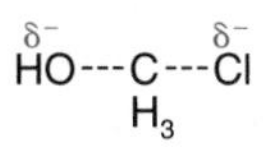

(Equation 5-8)

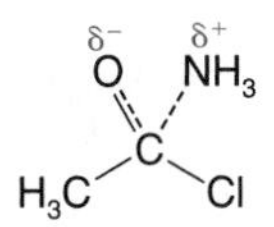

(Equation 5-10)

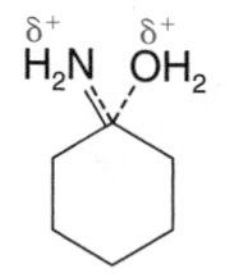

(Equation 5-11)

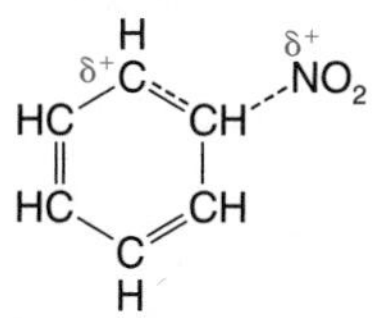

(Equation 5-13)

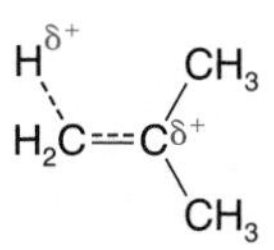

(Equation 5-16)

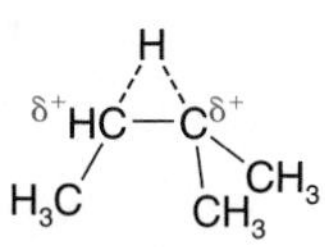

(Equation 5-17)

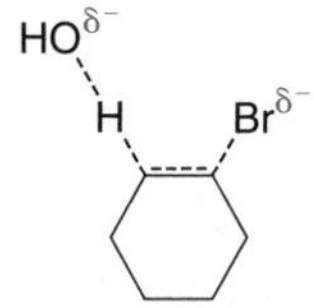

(Equation 5-19)

5.4

(a)

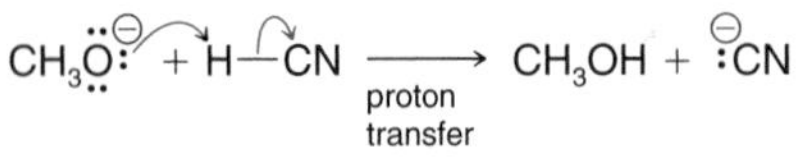

(b)

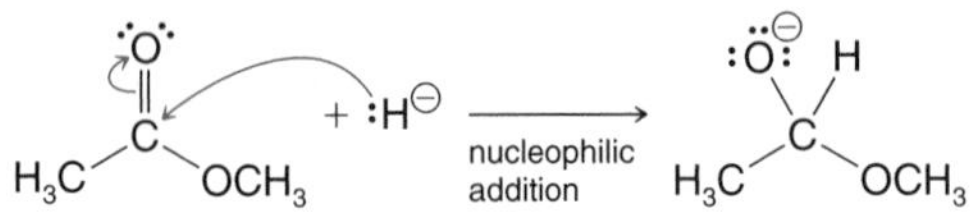

(c)

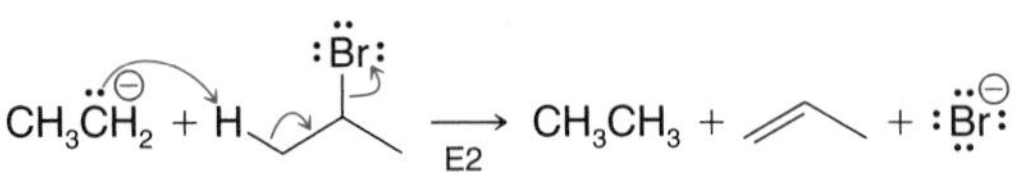

(d)

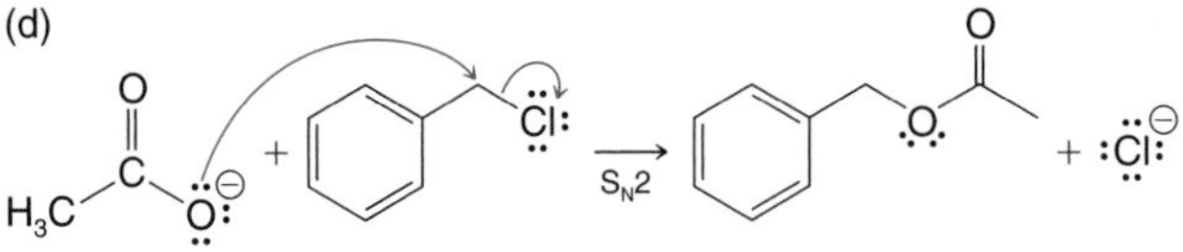

5.5

(a)

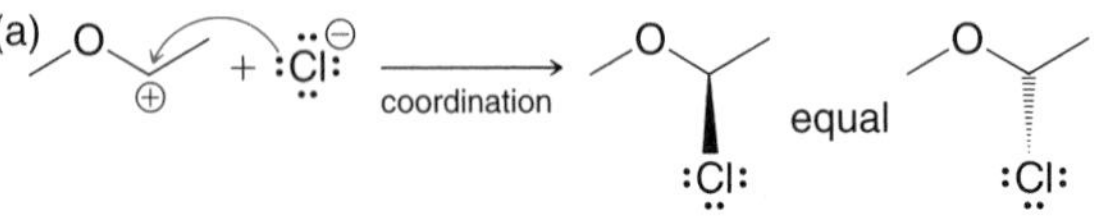

(b)

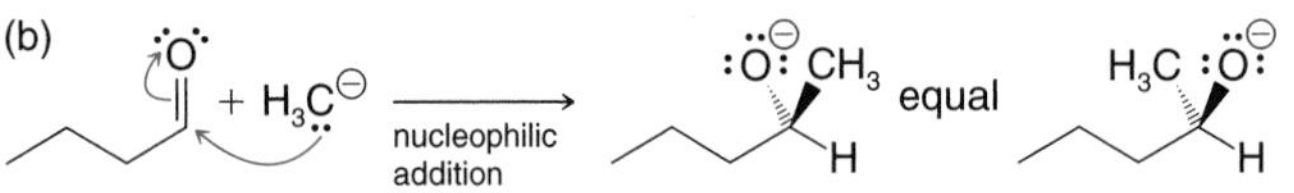

(c)

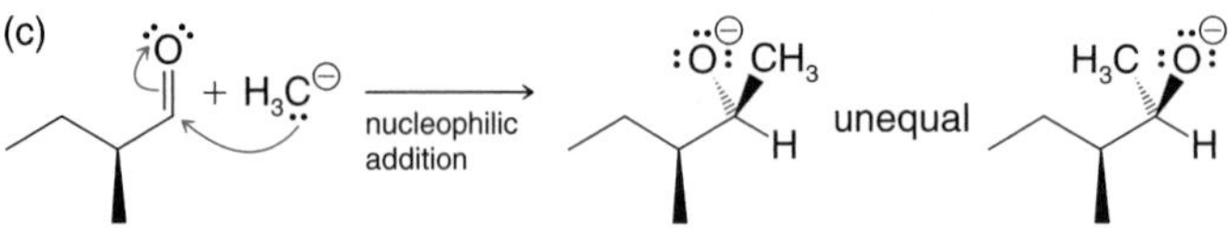

5.6

(a)

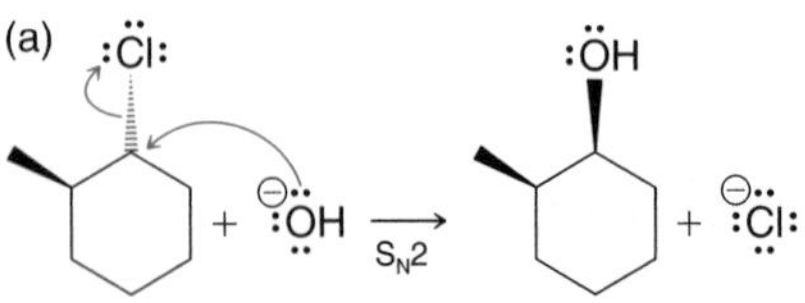

(b)

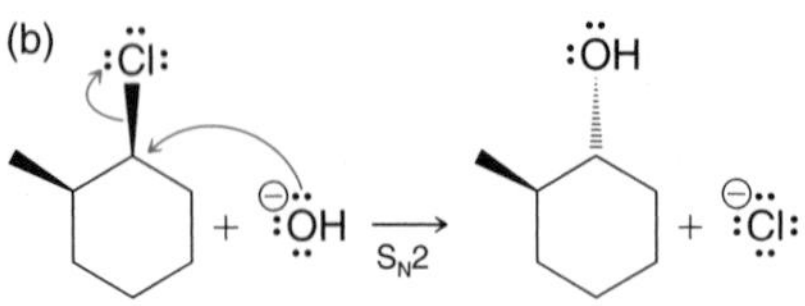

(c)

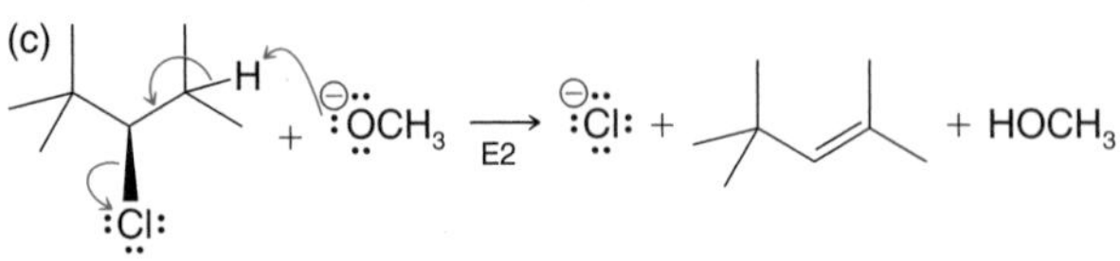

5.7

(a)

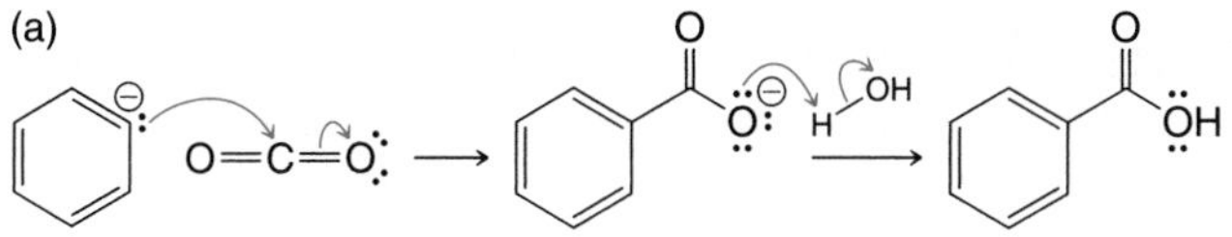

(b)

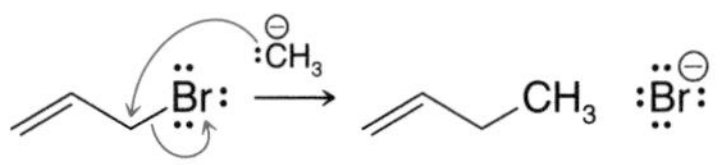

(c)

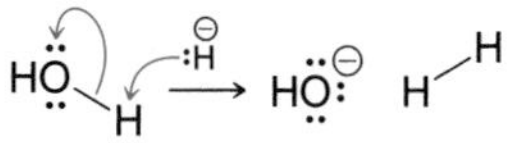

(d)

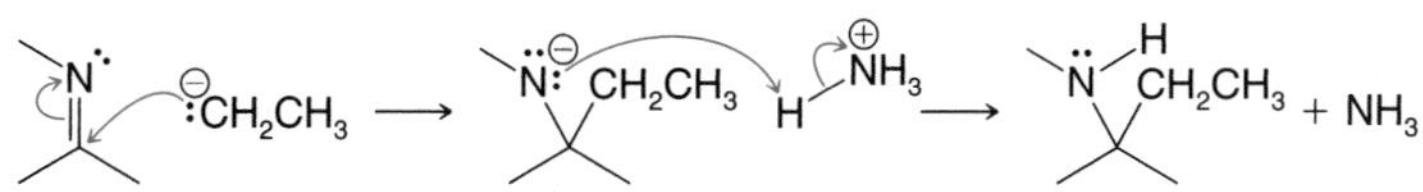

5.8

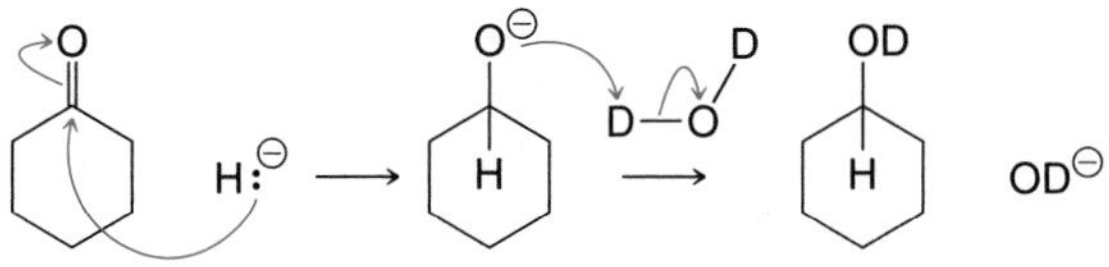

5.9

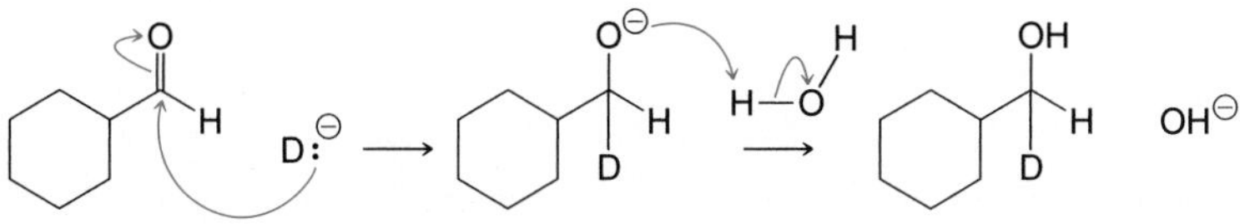

5.10

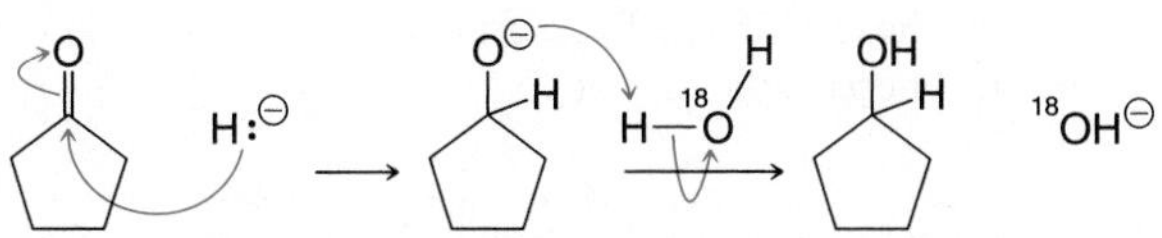

5.11

(a)

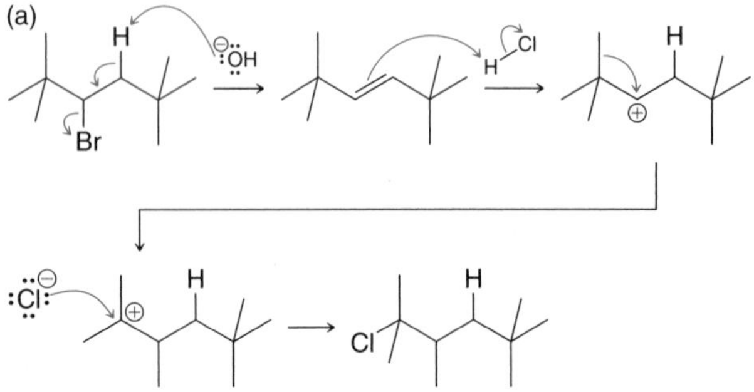

(b)

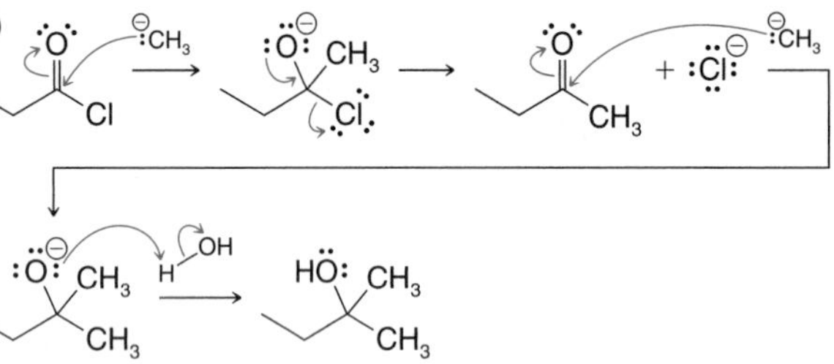

(c)

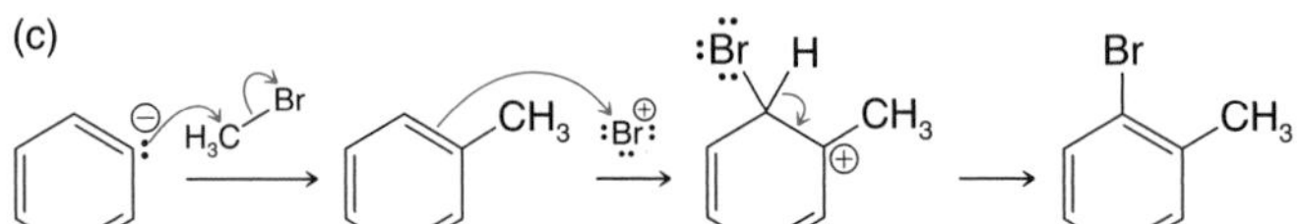

Chapter 6

6.1 It is tempting to say that the resonance contributor on the left is better because the positive charge would prefer to be on a C atom than on an O atom. However, with the positive charge on the O atom, all atoms in the contributor have their octets fulfilled, making it a better contributor.

6.2 According to arguments from the chapter, base strength increases in the order of $H_2O < NH_3 < Br^- < F^- < HO^- < NH_2^- < CH_3NH^- < CH_3CH_2^-$. If you look up the experimentally determined base strengths in water, you will find a slightly different order. This has to do with effects of the solvent.

6.3 f < c < a < d < g < b < e < h.
Notice that (f) is the only one for which there is no resonance contributor that would allow the negative charge in the deprotonated product to be shared between two O atoms. The rest of the order is determined by inductive effects—electron withdrawing by the Cl and F atoms and electron donating by the CH_3 group.

6.4 Using resonance arguments, HNO_3 is a stronger acid. This is because the resulting anion, NO_3^-, has a total of three resonance contributors, whereas the resulting anion from the other acid, HCO_2^-, has only two. Therefore, the negative charge is shared more, and the negative charge is more stabilized.

6.5 There are a total of four resonance contributors—the one provided in the problem and the following three:

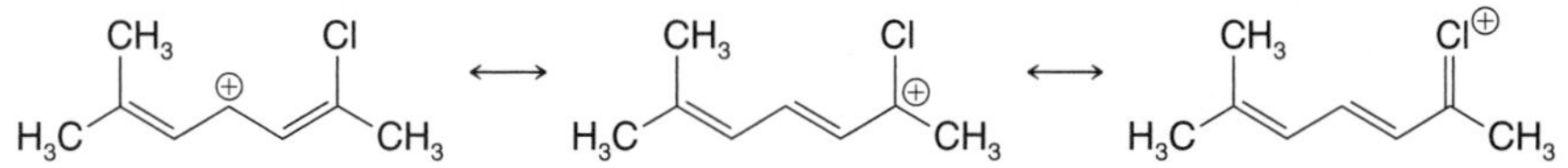

The best one is the one on the right because it is the only one of the four that has all atoms with octets. The worst one is the one in the middle. This is because the Cl atom is electron withdrawing compared to an H atom and therefore pulls more electron density away from the C^+ atom. This serves to increase the positive charge already on it, making it more unhappy.

6.6 The compound with the double bonds is a stronger acid because the anion that results from removing a proton is one that has resonance contributors—five in all. By contrast, the anion that results from removing the proton from the compound with all single bonds is one that has no other resonance contributors.

6.7 The products of protonating each of the two oxygen atoms are shown below. In the first of the two products, there is a resonance contributor, which serves to allow the positive charge to be shared on the second O atom. In the second of the two products, there is no other resonance contributor, suggesting that the positive charge is stuck on just that one O atom. The first of the two products is therefore more stable and is the one that is formed.

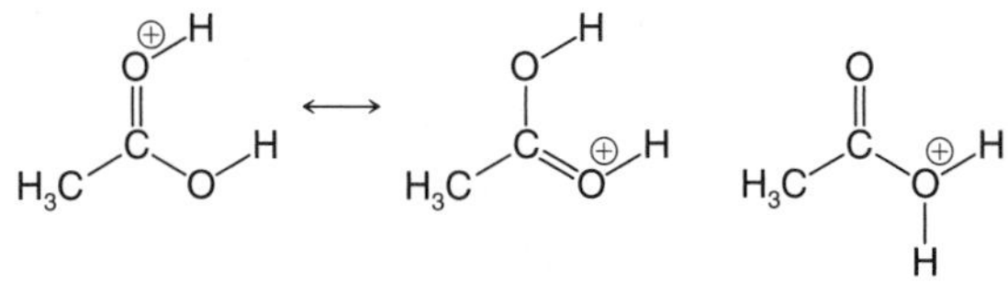

6.8 We can ignore X^- in each reaction because it is a reactant in each. We can also ignore the neutral species, taking them to be about the same stabilities. That leaves Cl^- and CH_3O^- to look at. Cl^- is a more stable anion than CH_3O^- because Cl is below F in the periodic table (making Cl a larger atom) and F is to the right of O (giving F a higher electronegativity). Because the more stable product will be formed faster and Cl^- is more stable, the first reaction will proceed faster.

6.9 Equations 6-6 and 6-7 are examples of reactions involving negatively charged nucleophiles and uncharged nucleophiles, respectively. In general, the reaction with the negatively charged nucleophile will be more downhill, making negatively charged species stronger nucleophiles than uncharged ones. Therefore, (b) and (e) will be weaker than the others. The relative nucleophile strengths of (b) and (e) are treated in Figure 6-16. Because the positive charge that is produced is more stable on S than on O, (e) is the stronger nucleophile. Among the remaining negatively charged nucleophiles, Figure 6-15 illustrates that the less stable the initial negative charge is on the nucleophile, the more downhill the reaction is, and the stronger the nucleophile is. The stability of the negative charges increases in the order $CH_3^- < CH_3CH_2O^- < CH_3CO_2^- < HS^- < Cl^- < I^-$, so the nucleophile strength increases in the reverse order. Overall, the order of nucleophile strength is $b < e < g < a < f < d < h < c$.

6.10 DMSO is an aprotic solvent, so the order of nucleophile strength does not change from the scenario without solvent. The order remains $b < e < g < a < f < d < h < c$.

6.11 Only CH_3^- will be protonated by CH_3CH_2OH because the proton transfer reaction trades a C^- atom for an O^- atom, which signifies a downhill reaction.

6.12 The solvent is protic, which will cause reversals of nucleophile strengths only to negatively charged species where the negative charges are on atoms in different rows of the periodic table. Nucleophile (g) has the negative charge on an atom in the fifth row of the periodic table, (a) and (f) have it on an atom in the third row, and (d) and

(h) have it on an atom in the second row. Therefore, (d) and (h) are shifted to be weaker than (a), (f), and (g). Also, (a) and (f) are shifted to be weaker than (g). Therefore, the order of increasing nucleophile strength is as follows: b < e < d < h < a < f < g. (We are ignoring nucleophile (c) because, as explained in the solution to the previous problem, it will react with the solvent.)

6.13 In all three reactions, a neutral reactant breaks apart into a cation and an anion. Comparing reactions (a) and (c), we can ignore the reactant molecules because they are neutral and therefore have about the same stabilities. We can also ignore the cations because they are exactly the same in both reactions. That leaves Br^- and Cl^- to compare as products. Br^- is more stable because Br is below Cl in the periodic table and is therefore larger. As a result, reaction (a) is more downhill and should be faster than reaction (c). Now comparing reactions (b) and (c), we can ignore everything except the two cations. In reaction (c), the cation is more stable because the C^+ atom is bonded to three alkyl groups, whereas in reaction (b) it is bonded to only two. Therefore, reaction (c) is more downhill and should be faster than reaction (b). Therefore, the order of the rates of reaction is b < c < a.

Chapter 7

7.1

(a)

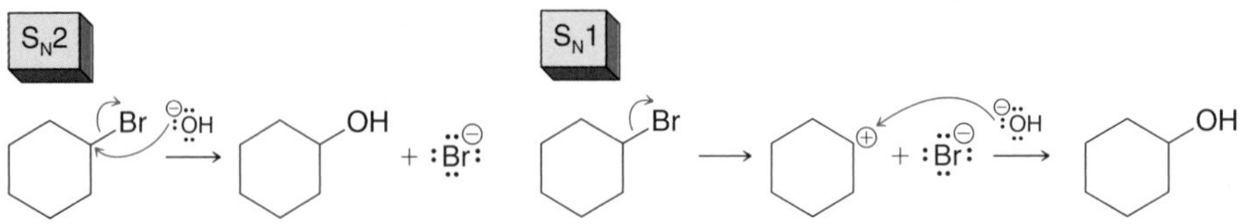

(b)

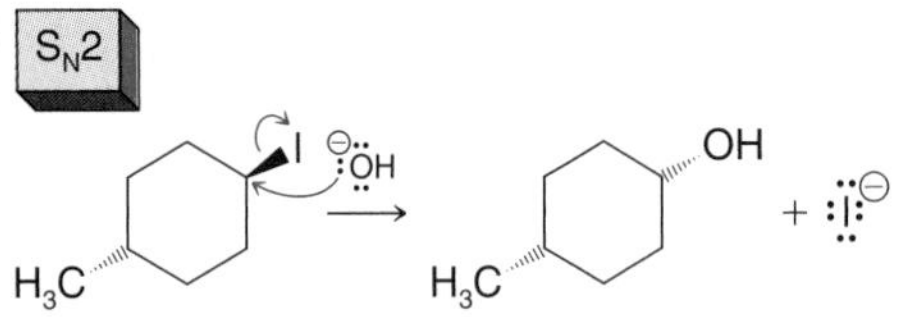

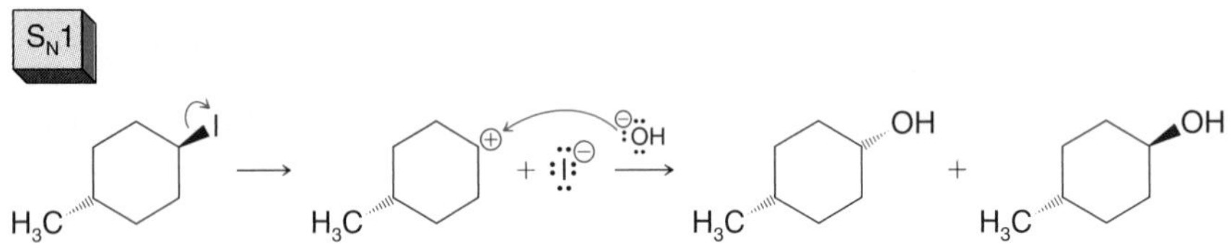

(c)

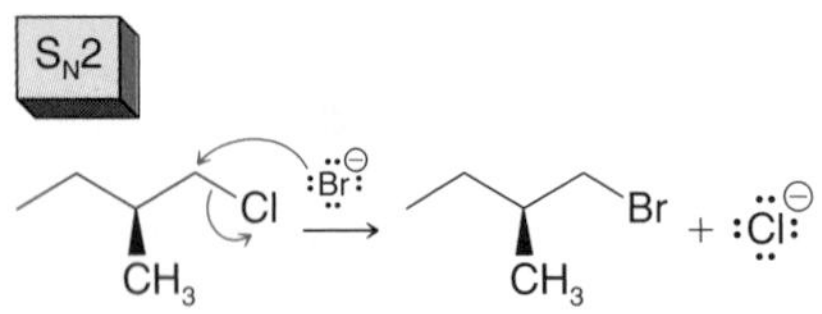

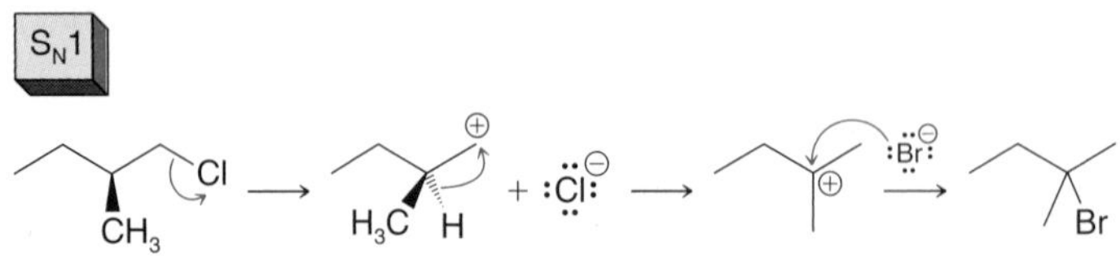

(d) S_N2

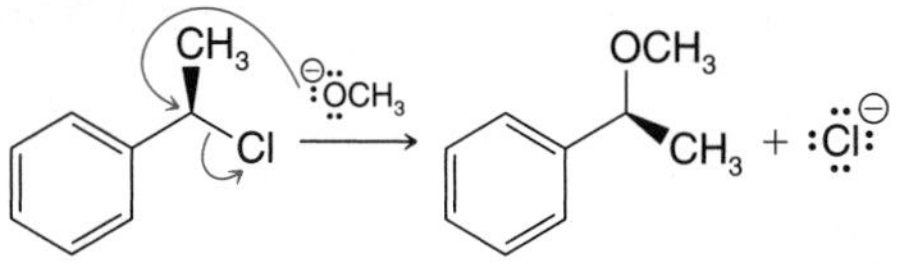

S_N1

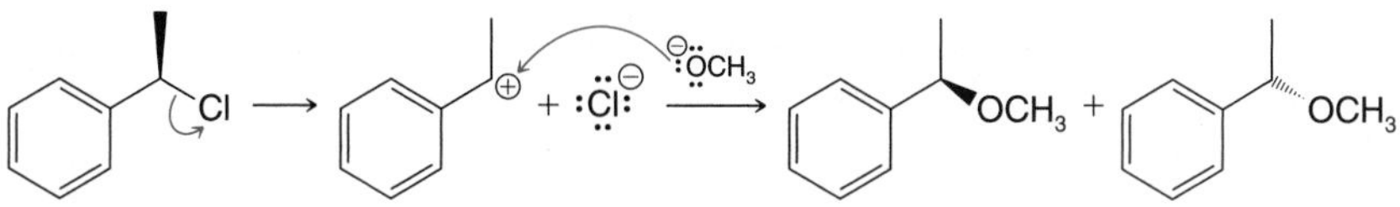

7.2

(a) E2

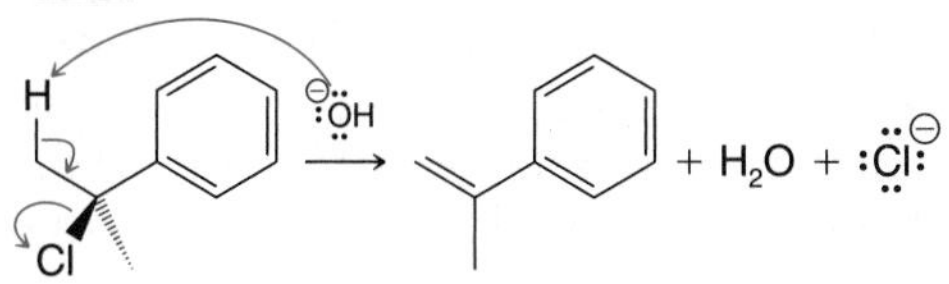

E1

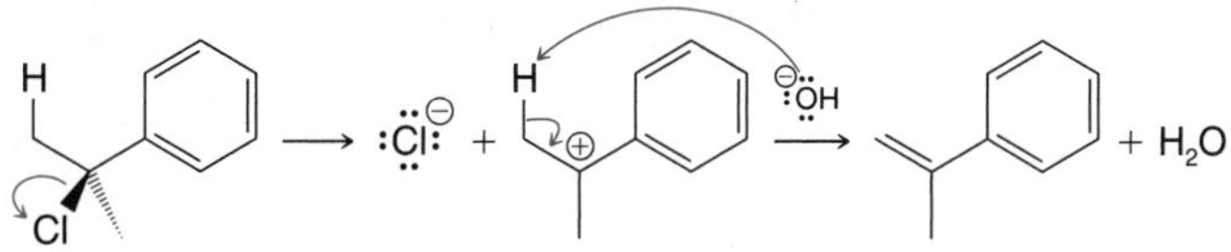

(b) E2

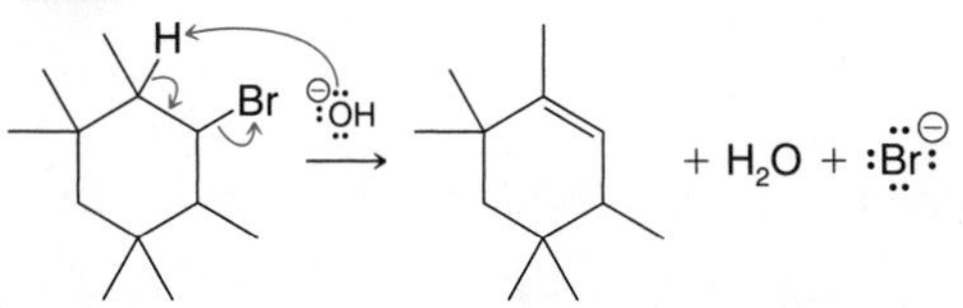

E1

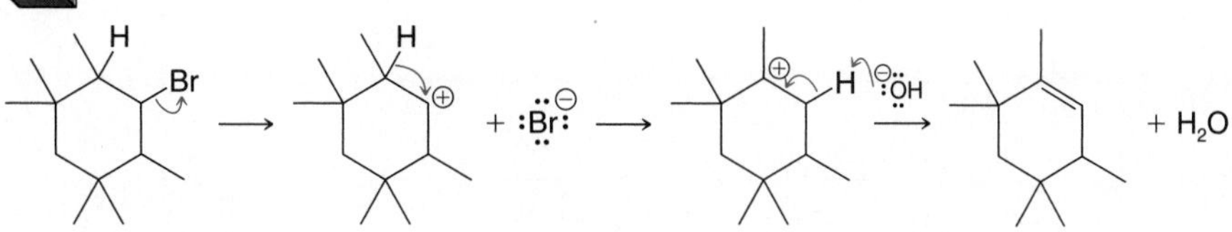

7.3 Carbocation rearrangements are not possible in S_N2 reactions because no carbocations are formed. When participating in S_N1 reactions, substrates (a), (d), (e), and (f) will undergo carbocation rearrangements, as shown below, in order to gain stability in the carbocation. By contrast, carbocations produced from substrates (b) and (c) cannot undergo rearrangements to gain stability.

(a)

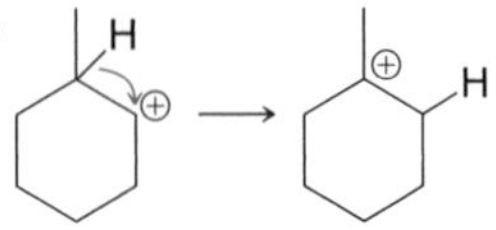

(d)

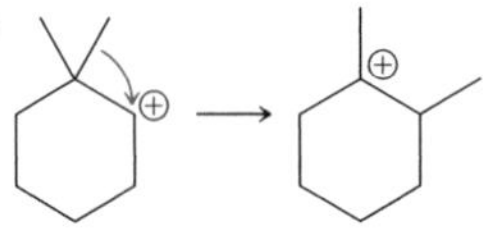

(e)

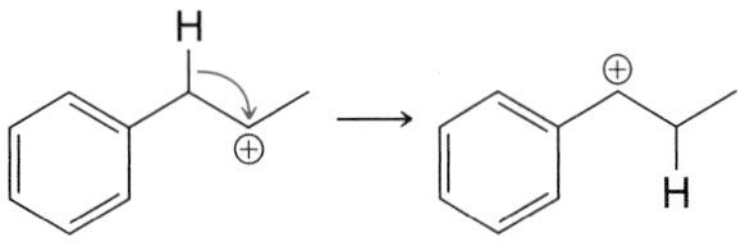

(f)

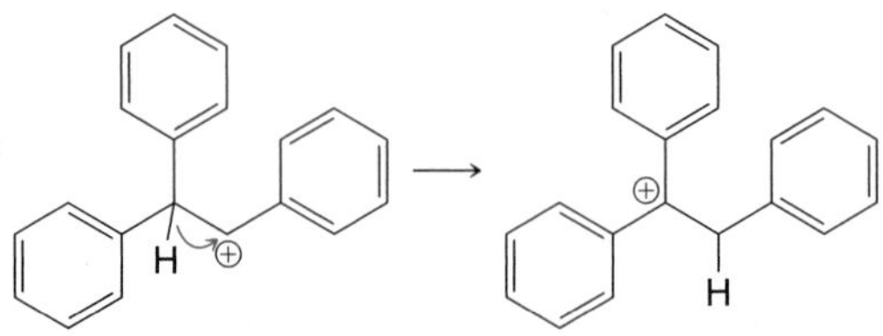

7.4 (a) Not reasonable because it is an internal proton transfer. (b) Reasonable. (c) Unreasonable because it is a termolecular step.

7.5 The mechanism and corresponding free-energy diagram are shown below. Each horizontal line in the energy diagram represents the energy of the set of species appearing directly above it. Notice that the reactants and the products of the first step are roughly the same in energy because a positive charge is localized on O in both cases. The energy of the products of the second step is significantly higher because of the loss of an octet. The energy of the products of the third step is roughly the same as that of the reactants and the products of the first step because the octet is regained and the positive charge is again on O.

(a)

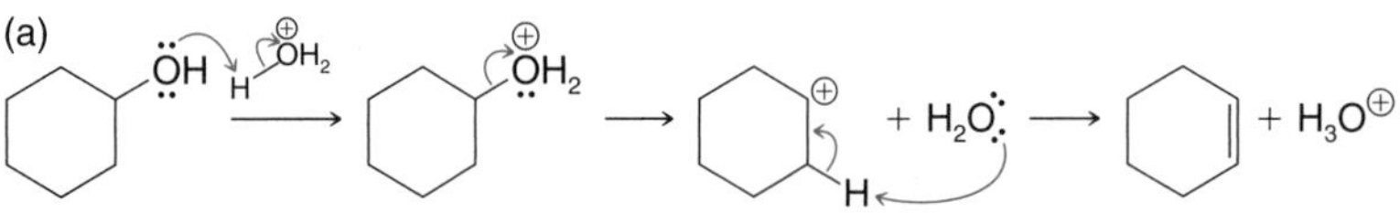

(b)

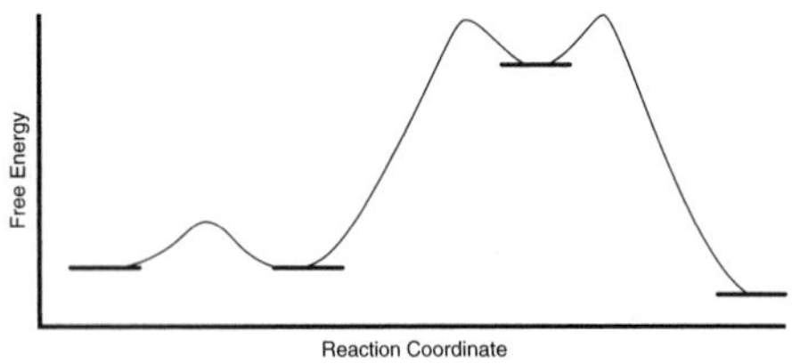

7.6 (a) The reaction must have been an S_N1 reaction in order for a 1,2-methyl shift to occur, as shown below.

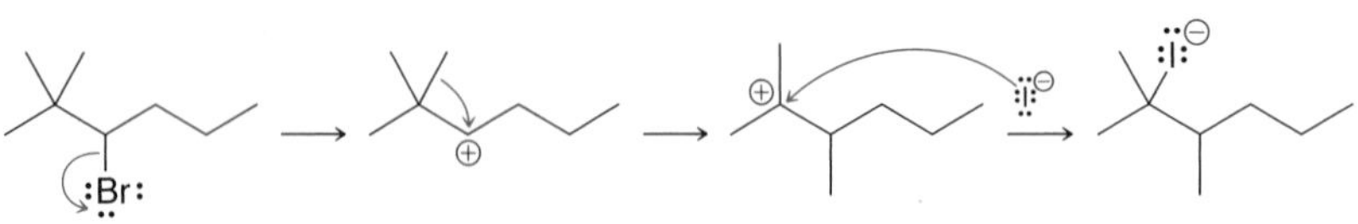

7.7

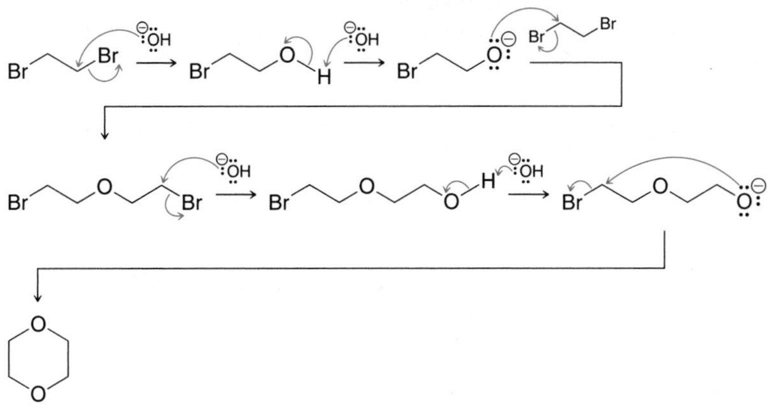

7.8

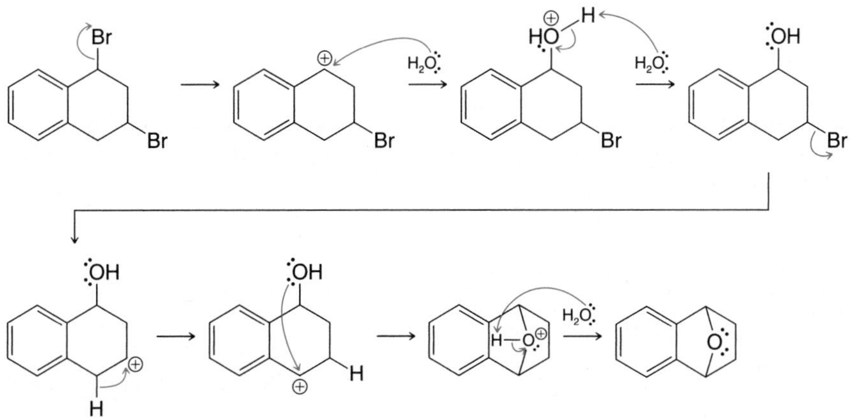

7.9

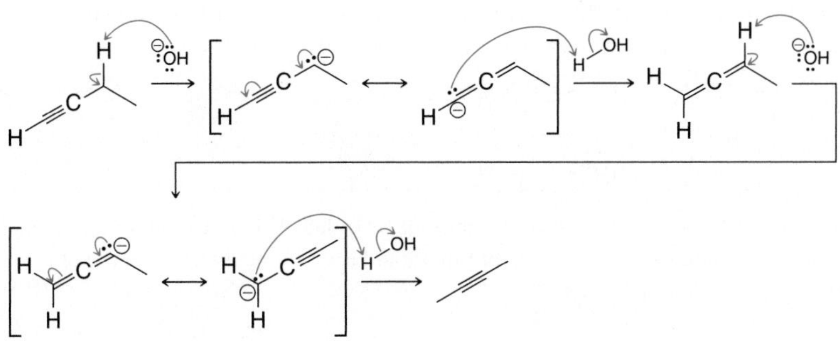

Chapter 8

8.1 Because the solvent is aprotic, the order is the same as the order of increasing nucleophile strength absent any solvent: c < g < b < e < d < f < a.

8.2 Because the solvent is aprotic, the order is the same as the order of increasing base strength absent any solvent: g < e < d < f < a < b < c.

8.3 The S_N2 reaction rate increases with increasing leaving group ability and with decreasing steric hindrance surrounding the C atom with the leaving group: f < b < e < a < c < d.

8.4 The E1 reaction rate increases with increasing leaving group ability and with increasing resonance and inductive stabilization of the carbocation that is produced as an intermediate: f < c < a < e < b < d < g.

8.5

(a) **E1 mechanism**. The leaving group (H_2O) is sufficient, and the three alkyl groups on the C atom with the leaving group rule out S_N2. Excellent leaving group (H_2O) favors E1 and S_N1. Weak nucleophile (H_2O) favors S_N1 and weak base (H_2O) favors E1. Tertiary C atom bonded to leaving group favors S_N1 and E1. Protic solvent (H_2O or alcohol reactant) favors S_N1 and E1. Heat favors an elimination reaction.

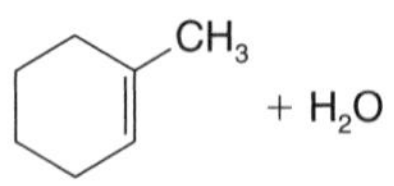

(b) **S_N2 mechanism.** The leaving group (Br^-) is sufficient, and because it is on a secondary C atom, we must consider all four reactions. Strong nucleophile favors S_N2. Weak base favors E1. High concentration of nucleophile favors S_N2. Aprotic solvent favors S_N2 and E2. Good leaving group favors S_N1 and E1. Secondary C atom, so favors all four reactions roughly the same.

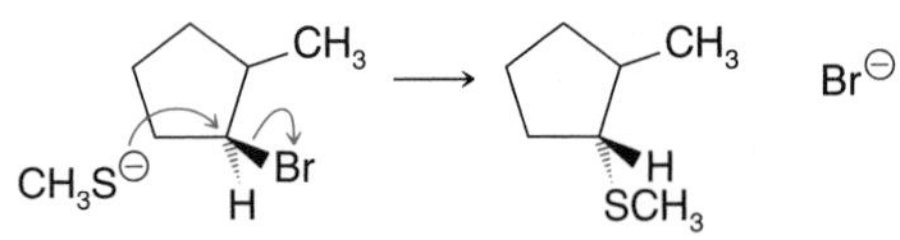

Notice in the mechanism that the nucleophile attacks from the end opposite the Br leaving group and therefore ends up on the opposite side.

(c) **Both S_N2 and E2 mechanisms.** The leaving group (Br^-) is sufficient, and because it is on a secondary C atom, we must consider all four reactions. Strong nucleophile favors S_N2. Strong base favors E2. High concentration of nucleophile/base favors S_N2 and E2. Secondary C atom, so favors all four reactions roughly the same. Aprotic solvent favors S_N2 and E2. Good leaving group favors S_N1 and E1.

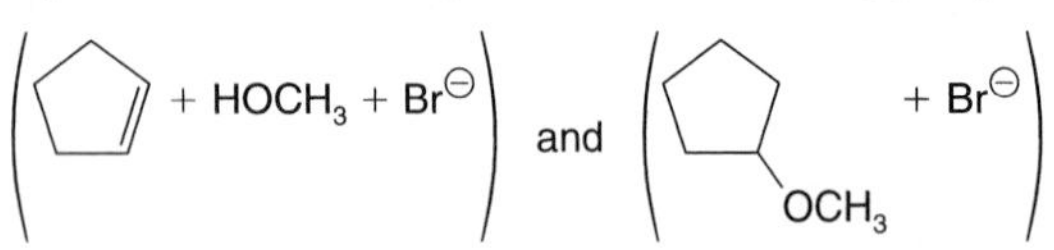

Note that the products are achiral, so no stereochemistry must be provided. Also, note that the lack of added heat will probably favor substitution over elimination, but we should expect a significant mixture of products from both reactions.

(d) E2 mechanism. Everything is exactly the same as in (c) except that the nucleophile is bulky—and therefore weak.

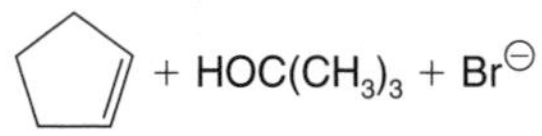

(e) Both S_N1 and E1 mechanisms. The leaving group (I^-) is sufficient, and because it is on a tertiary C atom, it rules out S_N2. Good leaving group favors S_N1 and E1. Weak nucleophile ($HOCH_3$) favors S_N1. Weak base ($HOCH_3$) favors E1. Protic solvent ($HOCH_3$) favors S_N1 and E1. Tertiary C atom favors S_N1 and E1. The S_N1 and E1 mechanisms are shown below. Note that this is a solvolysis reaction, and because there is lack of added heat, the substitution product will probably be favored over elimination, but we should expect a significant amount of products from both reactions.

S_N1 Mechanism

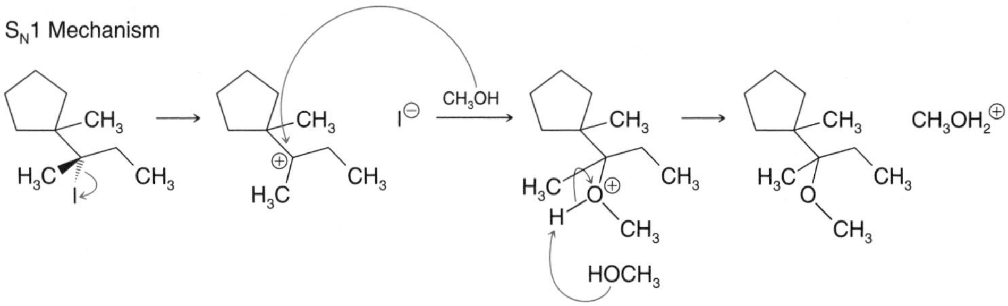

E1 Mechanism

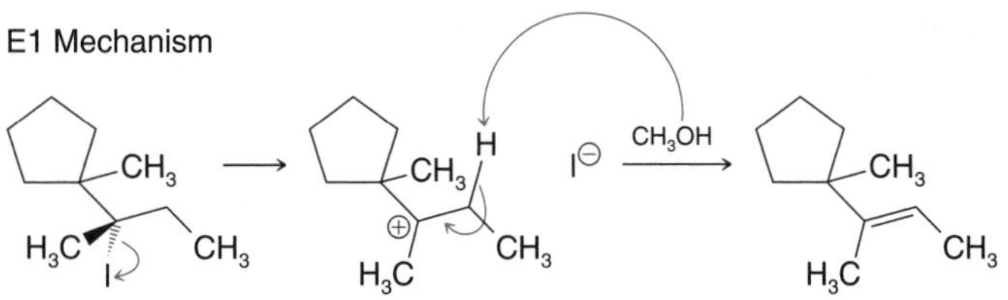

8.6

(a) This looks like it needs to be an S_N2 reaction. To ensure that, we choose a solvent that is aprotic, like DMSO, DMF, or acetone.

(b) This needs to be an elimination reaction, but the leaving group (HO^-) is not sufficient for a reaction to occur. A strong acid such as H_3PO_4 must be added to convert this bad HO^- leaving group into a good H_2O leaving group. Before it leaves, the H_2O leaving group would be an $-OH_2^+$ group, which is strongly acidic. Therefore, the elimination reaction cannot involve a strong base. The fact that the elimination reaction must involve an excellent leaving group (H_2O) and a weak base means the reaction will be E1 instead of E2. In order to help promote E1 over S_N1, we add heat.

(c) This is a substitution reaction, where a Cl atom substitutes for an I atom. Therefore, the nucleophile must be Cl^-. Because the product is a mixture of the R and S configurations at the stereocenter, it must be S_N1. To help promote S_N1, we should choose a protic solvent, like ethanol, CH_3CH_2OH.

(d) This is an elimination reaction. Although the leaving group is good (I^-), it is on a primary C atom, that is not allylic or benzylic, meaning that E1 is ruled out. Therefore, let's help ensure that it goes by E2 by choosing a strong base that is a weak nucleophile, such as $(CH_3)_3CO^-$.

8.7 E2 requires that an H atom be bonded to a C atom adjacent to the C atom with the leaving group. There are no such H atoms in this substrate. On the other hand, an E1 reaction can proceed if a carbocation rearrangement is included. First the leaving group would leave, resulting in a carbocation intermediate. Then, a 1,2-methyl shift would convert the secondary carbocation into a tertiary one. Finally, elimination of an H^+ ion would produce a $C{=}C$ double bond.

8.8 It must have been S_N1. The reason is that an S_N2 mechanism happens all in one step, so the nucleophile *must* substitute at the C atom bonded to the leaving group. Here it looks like the nucleophile attached at the adjacent C atom, which is possible after a carbocation rearrangement.

8.9 A mixture of S_N1 and E1 products. The leaving group (Br^-) is sufficient, and S_N2 is ruled out because the leaving group is on a tertiary C atom. The nucleophile and base ($HOCH_2CH_3$) are both weak, favoring S_N1 and E1. The leaving group (Br^-) is good, which favors S_N1 and E1. The solvent is protic, favoring S_N1 and E1. The leaving group is on a tertiary C atom, which favors S_N1 and E1. The S_N1 organic product is $(CH_3)_3COCH_2CH_3$, and the E1 organic product is $(CH_3)_2C{=}CH_2$. Note that, because this is a solvolysis reaction and because there is no added heat, substitution will probably be favored over elimination.

8.10 Same mechanisms as in Problem 8.9. The S_N1 product is $(CH_3)_3C{-}^{18}OH$ and the E1 product is the same as before.

INDEX